D1165806

CARTOGRAPHIES OF TRAVEL AND NAVIGATION

THE KENNETH NEBENZAHL, JR., LECTURES IN THE HISTORY OF CARTOGRAPHY

Published in Association with the Hermon Dunlap Smith Center for the History of Cartography, The Newberry Library

Series Editor, James R. Akerman

CARTOGRAPHIES
OF TRAVEL AND NAVIGATION

Edited by James R. Akerman

THE UNIVERSITY OF CHICAGO PRESS *Chicago and London*

JAMES R. AKERMAN is director of the Hermon Dunlap Smith Center for the History of Cartography at the Newberry Library, Chicago.

The University of Chicago Press, Chicago 60637
The University of Chicago Press, Ltd., London

Printed in the United States of America

15 14 13 12 11 10 09 08 07 06 1 2 3 4 5

ISBN-13: 978-0-226-01074-8 (cloth)
ISBN-10: 0-226-01074-0 (cloth)

Library of Congress Cataloging-in-Publication Data

Cartographies of travel and navigation/edited by James R. Akerman.
p. cm.
Includes bibliographical references and index.
ISBN 0-226-01074-0 (cloth : alk. paper)
1. Maps. 2. Navigation. 3. Map reading. 4. Maps—United States. 5. Navigation—United States. 6. Map reading—United States. I. Akerman, James R.
GA108.C35 2006
629.04′5—dc22 2005031379

⊗ The paper used in this publication meets the minimum requirements of the American National Standard for Information Sciences—Permanence of Paper for Printed Library Materials, ANSI Z39.48-1992.

Contents

Preface and Acknowledgments

In an advertisement for the venerable American map publisher Rand McNally that appeared in *Time* in 1938, a grinning gasoline service station attendant hands a road map to a young girl through an open passenger door of an automobile. "Does it show us the best way to get to Grandma's?" she asks. The advertisement, of course, draws upon a common perception that, in a most fundamental sense, maps are supposed to tell us how to get from one place to another. Navigational cartography and traveler's maps, moreover, are among the most ubiquitous and familiar genres of modern cartography. And yet, astonishingly little scholarship has examined the historical relationship between travelers, navigation, and maps. There is too much ground to cover here to pretend that the present volume can offer a comprehensive view of this field. Consequently, when we organized the Twelfth Kenneth Nebenzahl, Jr., Lectures in the History of Cartography in 1996, we invited papers that would offer a glimpse of the cartographic accommodation of the travel and navigational needs and motivations associated with the most common modes of transportation in the modern world. The studies published here of mapping for movement on land, on the sea, on rails of steel, in the air, and on modern highways sketch out many of the more important issues while offering a basis for comparisons across modes and technologies.

Most of the articles of this volume were presented on the occasion of the 1996 Nebenzahl Lectures, Maps on the Move. Chapters 2–6 were the core of the original series, and have been substantially updated and enlarged from the original papers presented by their authors. At the 1996 lectures, historical geographer James Vance's contribution offered an overview of the historical geography of transportation in the United States. Regrettably, Professor Vance died not long after the lectures, and we were unable to publish his talk. Another lecturer elected not to publish his paper, which offered an overview of recent developments in the digital navigation systems. We are very pleased to offer in its stead a chapter by Robert French, himself a pioneer in automotive navigation systems. Versions of his contribution were originally presented at the Institute of Navigation Fiftieth Anniversary Annual Meeting and at a 1997 conference titled Mapping the Earth and Seas at the University of Texas at Arlington. I thank the organizers of those conferences for allowing us to publish a revision and of those papers here. We also wish to acknowledge that Mr. French's contribution substantially enlarges and updates a previously published paper summarizing the history of automobile navigation systems that was originally published in the journal *Navigation,* which kindly agreed to the republication of its copyrighted material.

No work of this nature comes to pass without the help and good will of many people. My predecessor as director of the Smith Center, David Buisseret, first conceived the idea of a Nebenzahl series on transportation and communications mapping and helped shape its original outline. Susan Hanf, first as administrative assistant and then as program assistant in the Smith Center, and before her, Kristen Block, provided additional editorial and logistical support, without which this work would not have been completed. Sarah Fenton provided invaluable editorial assistance and advice on the final draft manuscript. Three anonymous readers provided extremely thoughtful and helpful comments on the entire work and its several parts. The staff of the Newberry Library's Map Room, particularly Bob Karrow and Patrick Morris, have helped along the way with the selection and acquisition of references and illustrations, and have been a constant source of good advice. John Powell and Catherine Gass, of the Library's photoduplication department have helped with much of the photography contained herein. I wish to acknowledge a particular debt of gratitude to the authors of the several essays, who have patiently seen this project through a prolonged transformation from lecture series to printed work. *Cartographies of Travel and Navigation* is the eleventh book published in what has now grown to be a very substantial Nebenzahl series all published by the University of Chicago Press. The press has been unwavering in its support for the volume, and our editor Christie Henry, has been generous throughout with her advice and support.

Most of all, we gratefully acknowledge the continuing support of the Nebenzahl Lectures by Ken and Jossy Nebenzahl. Their generous decision to launch and support a lecture series in the history of cartography at the Newberry Library honoring their late son has widened the horizons of four decades of lecture attendees, readers, students, and scholars. Through these lectures they have contributed enormously to the explosive growth in scholarly and public interest in the study of old maps, and we are truly in their debt.

1

Introduction

JAMES R. AKERMAN

Wayfinding is popularly thought to be among the primary uses of a map. But the historical record shows that, with perhaps the major exception of navigational sea charts, the use of maps by travelers for wayfinding only became common in the modern era. Catherine Delano-Smith writes in her contribution to this volume that "the paradox is that what today constitutes by far the commonest use of the commonest types of maps . . . is the one purpose for which maps were not used in pre-modern times." To be sure, we may point to a few scattered ancient examples of what have traditionally been considered road maps—the so-called Peutinger Table, for instance, which shows the major routes of the later Roman world, or Matthew Paris's thirteenth-century itinerary-style map of the route from London to The Holy Land. But as late as the dawn of the nineteenth century, Delano-Smith observes, maps were used mostly for planning journeys in advance, not for guiding travelers on the road. Sea charts emerged in the Mediterranean somewhat earlier, by the thirteenth century, but scholars are still debating the extent to which these early portolan charts, represent a developed tradition of on-board ship use.[1] Andrew Cook shows in his contribution to this book that late eighteenth-century sailors largely relied on simple instruments and their own knowledge of currents and winds to navigate the open ocean. By way of contrast, written travel directions—or itineraries—for use on land or sea date at least from Roman times, and were the preferred form for communicating navigational information on land well into the modern era.[2]

It is an open question even today whether automobile travelers need, much less prefer, to consult road maps on most journeys. The signage and insularity of modern superhighways have made it possible for American motorists to travel capably across the continent without the benefit of a map. The present push to develop the digital in-car navigation systems described here by Robert French—some of which rely on verbal navigational commands—suggests that a substantial number of motorists would prefer to do without maps altogether, or at least without a concentrated reading of them. Passive train travelers have little navigational need for maps; modern air passengers still less.

Yet, the simple fact remains that travelers' maps were among the most widely distributed of all map forms during the nineteenth and twentieth centuries, because as in the early modern past, their function extends far beyond simple navigation. Gerald Musich documents a broad array of railroad maps developed during the nineteenth century, in the first instance to help railroads and their customers conceptualize and operate in the railroad environment, and, in the second instance, to sell railroads to investors, shippers, and passengers. In my contribution I argue that American automobile road maps of the twentieth century similarly transcended their simple navigational requirements to promote and to provide meaning to the motoring experience. Yet, Ralph Ehrenberg's chronicle of the search by early air pilots for map formats suitable to the new experience of navigating in heavier-than-air craft reminds us that each modern form of transport presented new navigational challenges, but also new opportunities, that required new cartographies. We might summarize this volume as a contemplation of the cartographic response to the widening geographic range and the new modalities of modern travel brought about by the technological and institutional transformations of the age.

A number of considerations help us to characterize the level and nature of map use by travelers, including the role of the traveler in navigation, and the nature, speed, and range of the conveyance involved. The six chapters that follow highlight the different requirements of wayfinding maps that inevitably arise as the traveler moves from the sea, to roads and rails, and to the air. Tacitly—but more explicitly, in the case of Delano-Smith's chapter—they also ponder the sudden appearance and extensive reliance on travel and navigational maps during the nineteenth and twentieth centuries. Did modern transportation technologies fundamentally change the needs of travelers? Or was it—has it always been—that maps depicting paths of movement served other purposes than wayfinding? Indeed, what does it mean to use a map to navigate?

Mapping and Navigation

As a technical term, "navigation" generally refers to the act of operating, steering, or guiding a vehicle or vessel from one place to another. Historically, navigation referred to the operation of marine vessels, but, particularly since the advent of the airplane, it has come to refer to the direction and operation of other means of transportation as well.[3] We should note that additional meanings refer more broadly to the act of traversing from place to place or operating a vehicle, or to ship traffic and commerce. For our purposes, "navigation" is roughly synonymous with wayfinding—that is, the successful movement of a vehicle or one's person from one geographical location to another, be it over predetermined routes or routes of the navigator's choosing.

It follows that maps for navigation or wayfinding are those that assist movement along some route. By this definition, any map that gives the relative location of two or more places—that is, virtually any map—could be called a navigational map, but common sense limits us to maps produced and designed primarily to assist navigation. Even this narrowing of the field allows for a broad range of map types and levels of specificity. To simplify matters we may group all navigational, or wayfinding, maps into two broad categories—*network* maps, and *route-specific* maps—recognizing that these are ideal types that are frequently hybridized and combined within specific publications.

Network maps tend to be smaller in scale and concerned with depicting an entire system of routes and pathways (see, e.g., figures 4.14 and 5.12). Most modern road and railroad maps fall into this category, as do smaller scale aeronautical and sea charts. Depending on the situation, such maps may be used only for planning a route in advance of an actual trip, or also to verify and correct the course of travel during the journey. A motorist may use a map to plan a pleasure trip months in advance, thus navigating the terrain in her imagination. Encountering an unforeseen obstacle while on the road, she may refer to the map again to choose an alternate route.

Route-specific maps tend to be larger in scale, and as the name implies, are concerned with the navigation of specific routes or pathways. Strip road maps, for instance, or charts showing specific seaways and inland waterways, tend to be deployed when specific hazards, landmarks, decision points (such as crossroads), topography, and cultural features warrant a traveler's close attention. This class of navigational map is the closest descendent of the verbal itinerary or sailing direction, which literally narrate a route, or instruct a traveler how to get from place to place in relation to local obstacles and landmarks.[4]

Though we might associate verbal navigational guides most readily with the premodern itineraries discussed in this volume by Catherine Delano-Smith, or with sailing directions (*portolani,* rutters), verbal navigation remains a part of modern publications and products such as railroad timetables, published automobile route logs, and most forms of digital in-car navigation systems. Even aeronautical navigation relies to a considerable extent on the oral instructions transmitted by air traffic controllers advising pilots to adopt specific bearings or altitudes at strategic points along their routes. (On some flights it is now possible for passengers to listen in on these communications.)

Route-specific maps are most useful when navigation requires the careful attention of the traveler or vehicle operator. Network maps, on the other hand, have the advantage of providing the traveler with the general geographical context of his chosen route and how it stands in relation to other routes and geographical features some distance away. The relative strengths of the two ideal types have remained unchanged over time and have consequently engaged in a tug-of-war through the recent history of travel mapping. Andrew Cook proposes a typology of sea charts that range from small-scale general charts used for route planning, to charts designed to show specific tracks. Railroad maps range from the large-scale strip maps and atlases of single rail lines prepared for railroad operators (see figure 4.7), to small scale maps intended to show the rail network of an entire continent. Between these two extremes are American railroad maps that have elements of both ideal types. They show the rudiments of the national rail network, but, in the manner of railroad timetables, they emphasize the sequence of stops along a single line, and seem to arrange all surrounding geography, including connecting rail lines, in relation to that sequence of stops (see 4.17). Road maps in strip and network format coexisted in eighteenth-century Britain (see figures 2.16 and 2.18). Many early American automobile road maps simply showed a single route in strip format (see figure 5.2). Early motor clubs issued cards with maps or lists describing individual routes; later, the AAA developed custom-made atlases of route-specific maps called *Triptiks* to depict an itinerary specified by motor club members. Early aeronautical mapping experimented with both types as well (see figures 6.6 and 6.13).

Map characteristics and design also reflect differences in the identity of the navigator. The title of this volume signals the fact that in many contexts, wayfinding is only secondary to the other uses made of travel maps. Such maps serve as descriptive guides, souvenirs, and promotional publications intended for travelers who have nothing to do with the pilotage or operation of the vehicle they ride.

I have used the terms "traveler" and "navigator" interchangeably to this point, but we should be clear about what is implied by these two terms. Most navigators are travelers, since the navigation of a vehicle for some distance—historically, at least—requires that the navigator travel *with* the vehicle.[5] All travelers, on the other hand, do not seem to be navigators. Premodern travelers mostly moved on foot or on beasts, whose rate and direction of movement they controlled. In a broad sense, then, they were navigators as well as travelers. Modern automobile travelers include drivers, who are navigators as well, and their passengers, who may assist in the navigation of the vehicle by reading maps and shouting directions such as "turn here" or "you missed your exit." As a consequence, the maps and guides used by motorists generally must provide a relatively large amount of navigational detail. My essay argues that this was particularly true in the early automobile era, but Robert French shows that it was no less true at the end of the twentieth century. Improvements in route marking during the 1920s lessened the navigational burden of road map design but did not eliminate it altogether. In contrast, under normal circumstances, modern commercial air, rail, and sea passengers play no role in navigating the planes, trains, and boats they ride, and the maps designed for them will differ significantly from the maps used by the pilots, engineers, and captains who travel with them. Maps made for nineteenth-century American railroad passengers, as Gerald Musich notes, thus often provided only minimal navigational details, being essentially promotional documents.

That said, all travel maps retain a navigational component, because all travelers must navigate themselves, at the very least, on and off the vehicles of their choosing. A rail traveler wishing to journey from New York to San Francisco is not called upon to navigate the vehicle in a traditional sense, but he is required to sort out the possibilities: Should he travel via Pittsburgh, St. Louis, and Denver, or take a more northerly route via Buffalo, Chicago, and Salt Lake City? Does he wish to stop for a few days somewhere along the route to rest or to visit a national park? These decisions are navigational decisions. Modern travelers may not always determine the pathways they follow, but they do determine *which* pathways they follow. Network and route-specific maps persist alongside each other because travelers, to one degree or another, still need both to plan and to execute their itineraries.

Thus, with regard to travelers' maps, navigation embraces route planning, route promotion, as well as pilotage itself.

The Rise of Travel Mapping

The rise of transportation cartography in the decades before and after 1800 had much to do with factors that were then generic to the general expansion and diversification of map publishing. The application of lithography, steel engraving, cerography, and photolithography to cartography greatly simplified and reduced the cost of making maps. The rise of modern consumerism increased the demand for maps as material goods and stimulated the growth of new economic sectors that either encouraged or demanded map use.[6] The growth of public education and rising literacy rates fostered a corresponding growth in map literacy and interest in geographical topics. Hence, we find a general expansion in the production and use of school geographies, wall maps, and atlases throughout Europe and North America.[7] Likewise, periodicals and newspapers increased their use of maps, along with other forms of illustration, as literacy rates expanded and developments in printing technology eased the reproduction of simple graphics.[8] Over the course of the eighteenth century in Britain, we may also point to a significant rise in the opportunity and inclination to travel displayed by the emerging upper middle class and the aristocracy, which fostered the growth of modern tourism, both domestic and foreign, accounting for a significant upturn in the demand for travel guides of all sorts, including portable practical road atlases and maps.[9]

But, most fundamentally, the emergence of the travel cartography may be seen as a culminating effect of the transformation of travel technology itself. In his survey of the historical geography of modern transportation, James Vance has argued that Europeans underwent a transportation revolution from roughly 1500–1800. Advances in engineering and renewed dedication to bridge and canal building and wagon construction greatly improved inland movement just as the Great Discoveries quickened interest in travel and the movement of goods overseas. The mercantilist political philosophy of most European governments fostered the general growth of trade by investing in infrastructure and lifting or streamlining the legal impediments to travel.[10]

Though these transformations brought about a significant expansion of European horizons—itself marked by the wider circulation of maps of all sorts—it was a second, *vehicular* revolution (in Vance's words), which brought about the proliferation of specialized travel cartography. The emergence of artificially powered movement in the late eighteenth and early nineteenth-centuries greatly increased both the range and speed of transportation, requiring travelers and vehicle operators to absorb geographic information more quickly and over a broader area. Single-minded in their description of individual principal routes of travel, itineraries were poorly suited to the continental-sized spaces that trains and cars could traverse in days and airplanes in hours. General reference maps could be adapted for use by travelers, and often were, but they were poorly suited to the description of complex route choices made available during the nineteenth and twentieth centuries.

In addition to increasing the speed and range of travel, inland and overland transportation technologies of the industrial era required the creation of new pathways suited to their particular needs. Roads and trails that had grown organically for centuries in Europe and North America were straightened, paved over, obliterated, sidestepped, and overflown by modern railroads and superhighways.[11] There were continuities between old routes and new. Early American railroads closely followed the favorable pathways made by rivers and streams or traveled across passes preferred by existing roads. Early automobile highways often replicated railroad paths because they were the most direct and recognizable routes between major population centers. Early air routes often followed railroads or roads for much the same reason. Yet, in a broader sense, modern transportation technologies demanded a reimagining of the spaces they traversed. The labor- and capital-intensive grading, bridge building, and tunnel digging essential to railway construction required surveys and advance planning. This in turn required the publication of maps intended to convince capitalists of the profitability of specific rail projects, and later to convince potential travelers of the desirability of rail travel and the destinations the railroads served. To fulfill the needs of rail promoters and travelers alike, Gerald Musich argues in this volume, map publishers developed a cartographic style that contorted geography and scale in order to promote the interests of specific lines. Maps played a similar role in stimulating the imagination of twentieth-century automobile travelers. The earliest American road maps were simply concerned with transforming the ad hoc and disconnected nineteenth-century system of rural paths and wagon roads into automobile roads. (This took a great deal of imagination in the 1900s and 1910s.) Later, after automobile travel had been become practical for a majority of Americans, road map publishers and sponsors turned their attention to the creation of a national tourist's geography that privileged specific values, destinations, and even itineraries. The practical navigation of American air space required a radical rethinking of navigational needs so complete that no existing set of cartographic traditions or cartographic conventions served it well.

Each of these navigational worlds—of the ship, the train, the car, and the airplane—appeared to be mutually exclusive on maps. Travelers might easily transfer from one form of transportation to another as the need or mood struck them, but the most effective travelers' maps tended to be mode-specific, referring to other transportation systems only as points of reference. Sea charts rarely provided details about land more than a few miles inland, depicting only the veneer of the coastline for orientation. Maps made for American overland travelers up to the 1840s gave equal emphasis to roads, rails, and waterways. But as railroads came to dominate the travel landscape of the later nineteenth century, roads and canals disappeared from maps made for rail travelers. As the road system itself became a fully viable alternative, and then a dominant transcontinental transportation system, twentieth-century highway maps in turn excluded railroads. In like manner, aeronautical charts gradually abandoned their early interest in roads and railroads to emphasize the airfields and beacons that increasingly populated the emerging airspace of the United States.

Maps and the History of Travel

In exploring even this small portion of a huge subject the authors of the six chapters that follow have had relatively few guideposts to follow. The history of navigation, of course, has been central to maritime history, and has spawned a number of works that discuss the history of sea charts at length.[12] The portolan charts produced in the Mediterranean world from the thirteenth to the sixteenth centuries have long enjoyed a privileged place in the cartographic canon, as have the charts that document the rapid expansion of European knowledge of the wider world in the first few decades after Columbus. Accordingly, map historians and librarians have produced a number of specialized books with high quality images or early sea charts and critical facsimiles of early sea atlases.[13] Many of the more popular one-volume histories of cartography have devoted entire chapters to this specialized topic, while treating the rest of the subject chronologically or geographically.[14] This persistent interest among cartographic scholars has been sustained partly by the history of cartography's particular fascination with the "age of discovery," in which maritime navigation played an obvious part.[15] Sheer numbers also play a part; sea charts constitute a sizeable portion of the cartographic record before 1550, and they are consequently hard to ignore. Significantly, the far more ubiquitous and more rigorously mathematical charts produced from the late eighteenth century to the present day, which are the concern of Andrew Cook in chapter 3, have not received nearly as much attention.

Mapping for overland navigation and travel, in contrast to maritime mapping, has been woefully neglected. Of the four hundred and fifty feature articles published in the field's principal journal, *Imago Mundi,* from 1937–2001, only fourteen concerned road maps, road or route description, or road navigation;[16] and four concerned the mapping of rivers or inland water routes. Astonishingly, only one article discusses developments since 1800. No articles have been published in this journal on the history of aeronautical mapping or of railroad cartography. Sea charts and other guides to oceanic navigation have fared rather better in *Imago Mundi,* accounting for sixty-four, or 14 percent of all articles; but these, too, are overwhelmingly concerned with developments before 1800. It should be noted as well that about half of these articles on sea charts are primarily concerned with them as records of evolving geographical knowledge of the world rather than with their characteristics and use as navigational tools.

Synthetic works on overland travel mapping are almost nonexistent. The huge legacy of railroad cartography in North America has been treated by a single work, Andrew Modelski's well-researched and well-illustrated but essentially anecdotal *Railroad Maps of North America.*[17] In our volume, Ralph Ehrenberg cites only one article-length bibliographic study of the history of aeronautical charts.[18] Tim Nicholson's fine study of bicycle and early automobile maps in Britain, and Yorke, Margolies, and Baker's album of American automobile map cover art, are the only longer works on automobile road maps published in English.[19]

General histories of cartography tend to mention only the scattered monuments of overland route mapping—The Peutinger Map, Matthew Paris's graphic itinerary, Erhard Etzlaub's *Rom Weg* map, and John Ogilby's *Britannia.* In part this is due to the tendency to organize the history of cartography into a single narrative culminating in the Enlightenment and the creation of national topographic mapping agencies. A quarter-century ago Michael Blakemore and J. B. Harley documented a pattern of scholarly aversion to topics relating to the history of cartography since 1800, and Mark Monmonier recently noted the persistence of this pattern.[20] Lloyd Brown's *The Story of Maps* and Leo Bagrow's *History of Cartography* devote entire chapters to medieval and early modern sea charts, but make only a few scattered references to road maps en route to an eighteenth-century climax. Norman Thrower's *Maps and Civilization,* in refreshing contrast, devotes three of its nine chapters to later modern developments, and thus has made room for discussions of more recent road and railroad maps.[21]

Alan Hodgkiss organizes his *Understanding Maps* according to the uses of maps, providing the only existing synthesis of the history of what he calls "route maps" that is inclusive of road, rail, waterway, and aeronautical mapping.[22]

Several regional histories of cartography, freed of the need to be comprehensive, do a rather better job of addressing the subject. Catherine Delano-Smith and Roger Kain's survey of *English Maps* includes a fine chapter called "Maps and Travel," which, regrettably, does not venture far into the nineteenth century.[23] Walter Ristow's *American Maps and Mapmakers* offers the publishing histories of several important producers of nineteenth-century travelers' maps, including Samuel Augustus Mitchell, J. H. Colton, and Rand McNally & Co.[24] David Buisseret's collection of essays on maps as documents for the study of North American history, providing a mostly thorough survey of nineteenth and twentieth-century mapping genres, includes a fine chapter on twentieth-century highway maps, but does not consider other genres of travel mapping, most notably excluding railroad maps.[25]

Much of the historic work on travel and navigational mapping has turned up in the professional cartographic and technical journals. Walter Ristow's classic 1964 survey of American road mapping, for example, appeared in *Surveying and Mapping,* the organ of the American Congress on Surveying and Mapping, and MacEachren and Johnson's brief history of strip-format travel maps appeared in the *Cartographic Journal.*[26] The recent history of navigational tools is covered occasionally by the specialist journal *Navigation: The Journal of the Institute of Navigation.* Collectors of travel ephemera have also developed their own specialist literature that has yet to be fully explored by mainstream cartographic scholars. *The Legend,* the newsletter of the North American–based Road Map Collectors' Association (RMCA), and *Check the Oil!* a newsletter for collectors of "petroliana," have regularly offered columns on American automobile road maps since the 1990s. The RMCA operates an informative website that includes a number of bibliographic and feature articles, and links to other sites with similar interests. Likewise, collector-oriented magazines with more antiquarian interests, including *The Map Collector,* its recently demised successor *Mercator's World,* and Britain-based *IMCoS Journal,* have been far more willing to publish articles on recent, popular, and informal types of cartography, including travel cartography, than more scholarly journals.

For their part, historians of transportation have scarcely done any better than historians of cartography. Despite the essential role mapping played in the development of railroad networks, American railroad historians have barely acknowledged it. The most intensely geographical of all railroad histories, James Vance's magisterial historical geography of *The North American Railroad,* is illustrated with a handful of historic maps, but, with the notable exception of the Pacific Railroad Surveys, does not treat them as a historical subject; "map" itself is absent from the index. James Stover's authoritative history of American railroads does not mention railroad maps at all. George Douglas's *All Aboard* likewise avoids the topic, despite Douglas's broader interest in the projection of the railroad into popular culture.[27] The standard histories of the automobile written by James Flink and John B. Rae do not mention maps.[28]

There is as yet but slight interest in mapping among historians of travel and tourism. John Jakle's study of early twentieth-century American tourists frequently refers to motorists' use of maps. Valerie Fifer's survey of the "growth of the transport, tourist, and information industries in the nineteenth-century West" discusses travel publications at length, but makes only scattered and unindexed references to maps.[29] Splendid recent studies by historians of tourism of Baedeker and Michelin guides address textual but not cartographic content.[30] The essays assembled by Stephen Hanna and Vincent Del Casino in the recent volume *Mapping Tourism* provide the first critical examination of the way in which contemporary maps made for tourists construct identities and reconstruct history.[31] Very recently, historian Jordana Dym has examined the maps produced by and for travelers to Central America from the nineteenth through the early twentieth century, reflecting on the interplay between the maps and politics, economics, attitudes toward Central Americans, and the nature of travel itself.[32]

These six chapters, then, can hope to offer only a part of a still largely unwritten narrative. We hope nevertheless that the present volume outlines many issues that future work on the modern history of travel mapping should address. The biases of the volume are plain enough. The essays unabashedly focus primarily on British and American cartography; and while it may be possible to extrapolate from these essays to other geographical and chronological contexts, we have sought to provide breadth by surveying the cartography related to different modes of travel (by road, on the sea, on rails, and in the air), while holding the geographical context—or shall we say cultural context—more or less constant.

Catherine Delano-Smith sets the stage with a comprehensive survey of European overland travel, travel guides, and mapping in medieval Europe and early modern times. She begins with a profile of the medieval European traveler, dashing the commonly held assumption that medieval Europeans were not mobile. Geographical mobility, she finds, was neither restricted by class, nor prohibited by the poor quality of roads, nor confined to pilgrimages. The demand for information about routes, therefore, was considerable, but was almost always supplied in the form of written itineraries. Indeed, she notes that most celebrated early "graphic itineraries," by Matthew Paris and John Ogilby, were not only separated by several centuries, but appear to have had more to do with vicarious travel than practical guidance. The most celebrated early maps of route networks, the Peutinger Map and the Gough map of Britain, are similarly isolated examples of uncertain utility to travelers. Despite the increasing circulation of maps in Europe after the invention of map printing, specialized maps for travelers were slow to develop. Maps were not attached to practical travel guides with any frequency until the end of the seventeenth century. This change, Delano-Smith suggests, had more to do with the increase in commercial traffic brought about by industrialization than with the needs of private travelers.

Andrew Cook's chapter shifts focus to the other ancient travel venue, the sea and the sailing ship. Rather than attempt a survey of the well-trodden ground of the broad history of sea charts, Cook chooses to focus on what he sees as a turning point in the history of British hydrography, namely the efforts begun by Alexander Dalrymple to consolidate and standardize the compilation and production of charts used by British merchant and naval vessels. As unofficial hydrographer to the British East India Company, and then by formal appointment to the Admiralty, Dalrymple campaigned to increase administrative oversight of the gathering, recording, and cartographic representation of navigational information. Dalrymple's lifelong campaign for standardization was hindered by naval tradition and the private rights and motivations both of naval officers and commercial chart publishers. As on land, sea charts were only one set of tools available to navigators, who depended as much, if not more, on their instruments and experience. Increasing commercial traffic during the eighteenth century generated the need for greater predictability in traveling along the established routes of the British Empire, tilting the balance in favor of greater governmental control of navigational information and cartographic standards. In this respect, modern maritime cartography differed profoundly from modern navigational cartography on land, which—reflecting the less regulated and more flexible nature of overland travel—is neither standardized nor dominated by state agency.

The next four chapters shift their field of vision primarily to the United States and the nineteenth and twentieth centuries, examining the mapping associated with train, automobile, and airplane travel, each in their rough chronological turn. Gerald Musich surveys the universe of early railroad mapping and then focuses on the history of mapping for rail passengers up to the early twentieth century. Musich's four-stage narrative argues that the history of railroad mapping is inextricable from the history of the rail network it portrays. For the first time we see on a grand scale how mapping was used as a developmental tool, promoting the creation and patronage of specific rail lines as well as the landscapes served by those lines. The character of maps used by rail travelers was very much conditioned as well by the geographic extent and economic status of American railroads at particular times.

My contribution argues that a similar symbiosis existed between road maps and the development of American automobile highways. Particularly in the early decades of the twentieth century, automobile maps and mapmakers played a significant role in promoting and developing automobile highways, even as the navigational character of road maps was conditioned by the state of the highways themselves. The completion of an effective highway network subsequently allowed highway mapmakers and sponsors to turn to more fully promotional agendas. Chief among these agendas were those designed to equate automobile travel with national citizenship.

Ralph Ehrenberg addresses an entirely different set of navigational problems in his study of early aeronautical mapping. Early air pilots had many of the same needs as maritime navigators—an awareness of bearing and speed of travel, knowledge of landmarks and safe havens and the difficulties of approaching them. Yet since pilots operated almost entirely over land, the charts they developed were initially hybrids of maps intended for ground and maritime travelers. Here, as with maritime mapping, the production of aeronautical charts was originally undertaken a mixture of private firms, individuals, and governmental agencies. And, as on the sea, the increasing traffic that followed the commercialization of freight and passenger travel necessitated a shift toward greater governmental control of map standards and production.

Robert French provides a coda to the history of travel mapping that reminds us once again of the persistent utility of nongraphic solutions to navigational problems. Setting the stage with an overview of the history of mechanical devices used for road navigation, French then explores the relatively brief history of electronic navigational systems, which debuted in specific military and commercial applications ranging from jeep navigation to newspaper delivery. French envisions a future of automobile navigation in which maps might once again play a relatively minor role as navigational tools. At the turn of the twenty-first century, this vision remains largely unrealized, in part because of the significant cost of digital navigation systems, but also perhaps because of the lingering sentimental hold exerted by paper maps.

Within the limited scope of the present volume we have necessarily omitted modes of transportation and venues that are sorely in need of attention. The mapping of inland waterways, including navigable rivers and canals, for navigators and travelers, particularly in an American context, is a subject in search of a scholar. In the United States, the most widely distributed travel and navigational mapping during the first three decades of the nineteenth century appeared in the popular guides to the Ohio and Mississippi Rivers, particularly Zadok Cramer's *The Navigator* and Samuel Cuming's *Western Pilot.*[33] But, despite the occasional flourishes of activity associated with the construction of particular canals, the commercial publication of inland and coastal waterway maps in the United States never approached the level of activity associated with railroads. After the Civil War, inland waterway navigational mapping passed primarily into the hands of governmental agencies.[34] The higher use of the comparatively denser network of navigable waterways and canals in by travelers in Britain and Western Europe, by way of contrast, has sustained a more vigorous and varied tradition of waterway mapping. A parallel contrast between the European and American experience with travel cartography, influenced both by geographical and cultural differences, is evident in the greater importance of hiking and bicycling map publication in Europe both by governmental agencies, such as the Ordnance Survey, and commercial firms. The gap between American and European output of these two types of transportation maps, however, has narrowed considerably in recent decades as recreational interest in cycling and hiking has increased.

With the exception of Gerald Musich's discussion of interurban mapping, the present volume has also passed over the considerable mapping devoted to the navigation of more localized and urban transportation networks, including street maps and public transportation maps. The celebrated map of the London Underground designed by Harry Beck in 1933 now enjoys an almost iconic status in popular culture and has been the subject of an admiring monograph.[35] But despite their widespread publication around the world throughout the twentieth century, the broader history of public transportation mapping has been largely overlooked by scholars.

For most scholars, mapping remains a footnote in the history of travel and movement. The goal of the Kenneth Nebenzahl, Jr., Lectures has always been to call attention to and address relatively undeveloped themes in the history of cartography. Some years ago, another volume in this series, *Monarchs, Ministers, and Maps,* identified an early modern watershed in the use of maps by the state, helping to elevate historians' appreciation of the importance of maps to the modern exercise of political power.[36] Travel, like the exercise of political power, is by its very nature a geographical exercise. As a consequence, the use of maps to appeal to and guide modern travelers experienced a similar, if later, watershed. Readers of this volume will come to appreciate that the planners and operators of modern transportation systems and the travelers who used them valued their maps on the move both as navigational tools and as representations of radically new forms of mobility.

2

Milieus of Mobility

ITINERARIES, ROUTE MAPS, AND ROAD MAPS

CATHERINE DELANO-SMITH

Implicit in the theme of the Twelfth Nebenzahl Lectures is the idea that maps are used to help people find their way about. This is certainly so today. Was it always the case? Were maps used for wayfinding in the European Middle Ages? Are maps really the best form of travel aid? The answer to all three questions is no. On the contrary, the paradox is that what today constitutes by far the commonest use of the commonest types of maps (the printed topographical sheet of middling scale and the road map or road atlas) is the one purpose for which maps were not used in premodern times. To examine the changing place of maps in travel without first considering the needs of travelers—the potential map users—would be like describing the mechanics of an engine without considering the fuel. We need to take a long look at exactly who was on the road in medieval and premodern times and why. Our findings may lead us to recast some of our preconceptions about early road conditions, and the "difficulties" of travel in the past, if we are to improve our understanding of the place of maps in early mobility.[1]

No less than in ancient times, the roads of medieval Europe were busy, "full of movement," as Norbert Ohler puts it.[2] Ohler describes how rich and poor, churchmen and laity, men and women, young and old, all used roads. There was no lack of desire or opportunity for people to travel. Their mobility was no more prevented in the Middle Ages by the "poor quality" of roads than in the sixteenth century, when nonessential travel so expanded as to justify publication of a variety of travel aids and other travel-related writing. As today, medieval travelers made their journeys with a clear knowledge of the dangers.[3] If the documents are silent on either the state of contemporary roads or on the process of travel, it can only be because neither called for remark, not because nobody moved.

Nor was all movement local. When Fernand Braudel remarked that distance was "public enemy number one," he was thinking of the entire Mediterranean world and its external relations.[4] When Geoffrey Parker echoed Braudel's phrase, he was writing about Spain in Flanders, another situation involving transcontinental movement. In other circumstances, however, the distances involved are much more modest. Villagers moved about their fields in a daily routine, which for most rarely extended much beyond the nearest market or fair town.[5] For such people, distance was no "enemy," although it may sometimes have been an inconvenience. Their destinations were close to hand and the way there well known.[6]

Between the two poles of local movement on the one hand and continental movement on the other exists a wide range of public and private mobility. Medieval villagers occasionally had to make long journeys. Feudal obligations, for instance, could mean that a villein was called upon to run an errand for the lord of the manor. In the 1230s, villeins living on the manor of Combe, 10 miles (16 km) distant from Highclere in Hampshire (a manor belonging to the monastery of Bec in Normandy), were expected to be ready to go to market for provisions when the abbot was visiting, to take cheese to Southampton (35 miles/56 km away) for export to Bec, or to take wood to Quanly, another of Bec's Hampshire manors.[7] In 1301–2, a certain Duncan from Farnham manor, Surrey, was sent to York, 250 miles (400 km) away, "bearing there 1 letter on the bishop's business."[8]

The rate of travel was normally slow. Most people travelled on foot or on horseback. They had an intimacy with the road that has been largely lost today in the Western world and time to observe landmarks and make inquiries. Even for carriage users, contact with local folk and fellow travelers could not be entirely avoided. Waits to cross a river, or halts to water, feed, rest, or exchange mounts at the inn were not only frequent but could be prolonged. By and large, a traveler on foot might aspire to cover some 25 miles (40 kilometres) a day over undemanding terrain and an ordinary rider might travel some 30 miles (50 kilometres) in a day without punishing his mount.[9] Wheeled transport was slower. When the emperor Charlemagne travelled in an ordinary rustic wagon, he was unlikely to move more than 20 miles (30 kilometres) in a day, especially if a train of retainers and baggage accompanied him. We hear how in early thirteenth-century England "the essentials of government, the *hospitum regis*... followed the court; a train of from ten to twenty carts and waggons."[10] Only the privileged—those who could pay for themselves, their baggage, and a guide—would hire a cart or wagon for individual sections of a journey. Their luggage followed behind in separate carts. The advantage of such carts was the local drivers who accompanied them and who were expected to know the way over their particular stretch of road. The heaviest vehicles, four-wheeled wagons like the *longa caretta* used in France in the fourteenth century, were of little use on rough terrain in forested districts. According to Henry Earl of Derby's expense accounts, he and his companions had to transfer on more than one occasion "in le wylderness" from cart to horseback, while their luggage was unloaded from the broken down cart onto pack animals.[11]

Travelers

In the older literature, travelers are usually described according to their personal occupation. The still much-cited French writer J. J. Jusserand has, for example, one chapter on "Lay Wayfarers" and another chapter on "Religious Wayfarers."[12] The first chapter includes a motley collection of herbalists, charlatans, minstrels, jugglers, tumblers, messengers, itinerant merchants, pedlars, outlaws, wandering workmen, and peasants out of bond. The second chapter deals with wandering preachers, friars, pardoners, and pilgrims. Jusserand's choice of the word "wayfaring" is unfortunate, for it has a casual, almost indolent ring that belies the facts. When one considers the reasons for the vast majority of early journeys, a quite different impression emerges. The words that then come to mind are purposefulness, professionalism, and organization.

REGULAR PROFESSIONAL TRAVELERS

Professional travelers are those obliged to travel by virtue of status, occupation, or employment. They travel on behalf of an institution or association that has invested a certain amount of accumulated travel information on which the traveler can draw prior to setting out and to which he adds on return. A distinction can be made between those professionals whose commitment to the road is long-term (the regular professional traveler) and those who are on the road only intermittently (the occasional professional traveler).

Early regular travelers included those responsible for government and administration, both lay and ecclesiastical. Those with the highest profile were the kings and emperors of Europe for whom travelling was virtually synonymous with ruling. Maps of royal progresses, such as those showing the journeys made by the emperor Charlemagne or of the English kings John or Edward I, give an idea of the burden laid on medieval rulers.[13] They moved about the territory under their control to hold court and administer justice. Later, when this personal involvement of the crown was replaced by increasingly bureaucratic forms of administration, government agents, messengers, and emissaries were sent out all over the country carrying letters and conveying orders and edicts, and the royal progress was primarily a propaganda effort. In thirteenth-century England, however, both royal and ecclesiastical dignitaries still lived what has been described as "a largely nomadic life," travelling from manor to manor.[14] In fact, most English medieval kings spent much of their life on the road, either on such progresses or on military campaigns against the Scots, Welsh, or the French.

Princes and knights served in similar fashion. Between 1390 and 1392, Henry, earl of Derby (later King Henry IV), spent two years travelling on the continent of Europe.[15] The first of his two main journeys was a military expedition in support of the Teutonic Knights in Prussia in order to protect England's vital Hanseatic trade. The second, motivated primarily by political considerations at home (his desire to be out of the way of domestic troubles), took Henry and his knights from a second visit to Prussia on to the Holy Land. Bishops (good bishops, that is to say) also went on regular progresses about their diocese in order to get to know their clergy. In the first seven months of his reign, John le Romeyn, archbishop of York from 1286 to 1296, visited eighteen parishes, one abbey, twelve rural deaneries, the city of York, three jurisdictions (Otley, Ripon, Beverley) and attended two ordinations and one convocation.[16] In the 296 days covered by an extant fragment of the roll recording the domestic expenses of Richard de Swinfield, Bishop of Hereford (d. 1317), Richard moved his household no less than eighty-one times and slept in thirty-eight different places in the fifty-one days between April 10 and June 5, 1290.[17] Nor were these moves the simple journeys of a small band of horsemen; like kings and archbishops, bishops travelled with their retinues. Bishop Swinfield travelled with part of his household—thirty riders—and was preceded by wagon-loads of baggage, which would have included his library.

Good government, like efficient administration, rests on up-to-date information and an efficient communications infrastructure. The Romans set up the *cursus publicus,* a system ensuring the availability of the means of transport "at particular places along certain routes."[18] Evidence for the use of relay runners date mostly from the fifteenth century when "war, with its handmaid diplomacy" proved once more to be "the decisive factor for

FIGURE 2.1 A professional messenger. Marginal drawing in an early fourteenth-century manuscript, the Decretals of Gregory IX. B.L., Royal MS 10.E.IV, fol. 302v. By permission of The British Library.

innovating central institutions, and produced in the relay system a new instrument of absolutism."[19] Professional messengers had always been used when needed. No major medieval institution, whether a trading or banking company, an urban guild, or a monastic, ecclesiastical or university community, could function without some sort of messenger system (figure 2.1). The foundation charter of the Benedictine monastery of Limburg, Flanders (1035), stipulates that "every member of the monastery was bound to ride out each day wherever the abbot sent him."[20] In France the monastery of Cluny had established its own system of emissaries to maintain links with daughter houses all over Western Europe, and Rome was considered "the best informed city in Christendom."[21] Traders and financiers made extensive use of agents and couriers, carters and carriers. Merchants and governments established their own networks of informants. In 1357 in Florence, seventeen trading companies combined to found a joint messenger service, the *Scarsella dei Mercatani Fiorentini* (The Messenger Packets of the Florentine Merchants).[22] Intelligence gathering was a major activity, whether carried out by special emissaries or more informally through the firsthand observation of any returning traveler. In 1418, Henry VI sent John Harding to Scotland "to spy out with all kinds of diligence" any evidence for Henry's claim to the country and to send back information as to the best way Henry might invade it.[23] A few years later, in 1421, Gilbert de Lannoy, Philip of Burgundy's Flemish chamberlain, "undertook the Jerusalem journey by land at the request of the king of England, and of the king of France, and of Monseigneur Duke Philip, chief mover" with express orders to supply information on the military strength and strategic layout of Egypt and Syria.[24] In 1521 Henry VIII paid Sebastian Cabot for his map of Gascony, at the time still in English hands: one wonders what other information Cabot brought back for his monarch.[25]

Traders kept the wheels of commerce turning busily in the Middle Ages as they would in the sixteenth century.[26] Even in a predominantly rural world, trade was important. Rural industries such as mining, quarrying, metal working, charcoal burning, and glass making involved mobility, and most forms of agriculture generated a transport industry of some sort for the exportation of products.[27] Medieval commerce was far from only local or regional. Surviving records yield glimpses of the life of wool traders as they made their journeys across Europe with raw wool one way and woven cloth on their return from one of the great cloth-producing regions of western Europe (the Netherlands, Catalonia, or Lombardy).[28] Take the Genovese merchant Lafranco Lecari for instance. In February 1290, Lecari arrived with his carter (*voiturier*) and two bales of merchandise at the recently constructed port of Aigues-Mortes.[29] He had intended to make his way as usual around the head of the Rhone delta, across the two main branches of the Rhone at Arles, in order to take the easier gradient of the king's public highway (*iter publicum domini regis*) northwards along the Rhone valley. But no sooner out of Aigues-Mortes at St Gilles, Lecari found that officials had barred the way ahead. Offers to pay the usual toll were of no avail: the route, he was told, had been closed to merchants and traders on the orders of the French king. All long-distance travelers were to use the upland route across the high plateaux of the Central Massif. Lacari was obliged to comply; indeed, two years later, he was still using the Central Massif route.[30] Expense accounts kept by wool couriers also serve as a record of routes taken. One such account relates to the transport of a sack of wool (packed into two bales to be carried by one pack horse) being escorted from London to Lombardy sometime between 1310 and 1340.[31] The "swain" responsible made his journey in three stages: first, by sea from London to Libourne on the Gironde, then overland to Aigues-Mortes, then by sea again to northern Italy. Another account for the transport of wool is equally detailed. Dated 1390, it relates the arduous transalpine journey from Constance, Switzerland, to Bellinzona in Italy.[32]

Political hostilities generated professional travel for all sorts of people. Not only were rulers and commanders involved; so too were a multitude of spies, scouts, guides, troop escorts, and quartermasters. Military corridors allowed an army on active service to connect with its distant recruiting ground.[33] When in the early summer of 1572 the Duke of Alva sent reinforcements to the Spanish army in Flanders to raise the permanent defence corps of 13,000 to a full-time war establishment of 67,000 men, the extra troops travelled by sea from Spain to Lombardy and then continued north on foot. They followed a preselected sequence of roads, allowing the troops to travel almost entirely within territory controlled by Spain and to avoid the hostile cities of Geneva, Besançon, and Metz.[34] There was nothing special about this "Spanish Road," as it was then known. Traders and troops shared whatever local tracks were available, the troops receiving instructions as to which route they were to follow, which bridges, fords and ferries to use, and in which towns and villages to seek overnight accommodation. One commander, Don Lope de Acuña, provided himself with simple but effective sketch maps showing everything an army on the march needed to know: the route, major rivers and forests, alternative itineraries, and the nearest towns (figure 2.2). "It is difficult to see," says Geoffrey Parker, "how, in the sixteenth century, better summary guides could have been produced at short notice."[35]

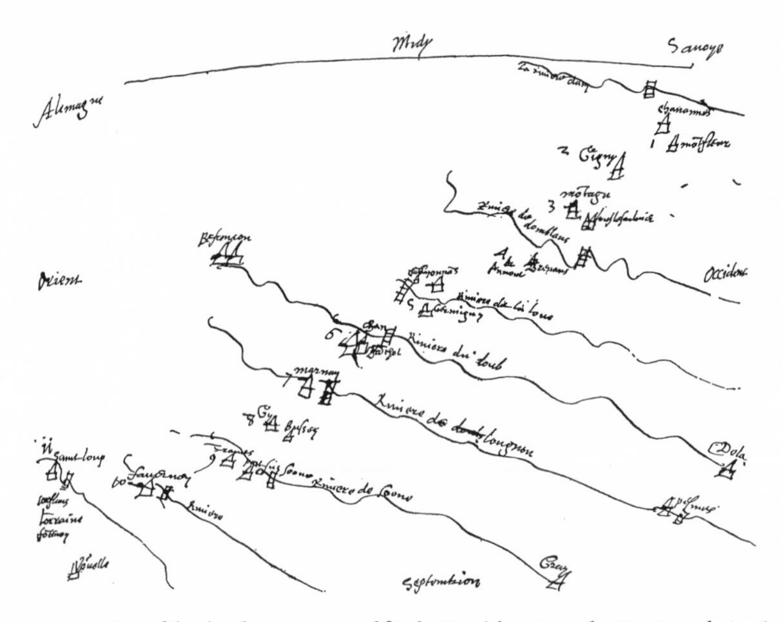

FIGURE 2.2 One of the sketch maps prepared for the Spanish commander Don Lope de Acuña as an aide-memoire, showing river crossings, the *étapes* (supply points), and towns along or near the routes the troops were to follow across eastern France on their way to reinforce the Spanish army in Flanders in 1573. From Geoffrey Parker, *The Army of Flanders and the Spanish Road, 1567–1659* (Cambridge, Cambridge University Press, 1972), 102–3. Reproduced with permission from the Syndicates of Cambridge University Press, 1972.

The frequency with which the professional set out on a journey differentiates a regular and an occasional professional traveler. Occasional professional travelers in the early Middle Ages would have included feudal peasants dispatched on manorial errands, ecclesiastics and monks visiting a parent church or daughter house, and abbots taking up a new appointment.[36] The road to Rome was particularly well trodden at all times by those on occasional journeys as well as by those frequenting Rome regularly. In the seventh century Benedict Biscop, founder of the monasteries at Jarrow and Wearmouth, Northumbria, made the journey to Rome at least six times, returning with books for the Jarrow library and *scriptorium.*[37] Biscop's successor, Abbot Ceolfrith, visited twice, also returning with books. The London priest Nothelm went to copy whatever documents he could find in the papal archives relating to St Augustine's mission to England in the time of Gregory I.[38] In 990, Sigeric, newly-appointed archbishop of Canterbury, went to Rome to receive his *pallium.*[39]

Other monks made the occasional journey for purposes of scholarship. William of Malmesbury has been called one of the "first Englishman to travel to get his evidence [and] to write history from information collected on the spot."[40] Malmesbury's writing is packed with first-hand descriptions of castle and church architecture and his ecclesiastical history, the *Gesta Pontificum* (1125), has been described as "virtually a gazetteer of ecclesiastical England."[41] Another twelfth-century traveler, the anonymous author of the *Gesta Stephani,* described the castles of Exeter, Bristol, Bath, Oxford, Cricklade, and Farringdon. William of Newburgh (1136–98) gave a description of Scarsbrough castle in his *Historia Rerum Anglicarum.* William FitzStephen, a clerk in the service of Thomas Becket, described the layout of the city of London (1173–75). Gerald of Wales travelled in Ireland with Prince John in 1185 and toured Wales with Archbishop Baldwin in 1188. He compiled detailed written accounts of both countries and left sketch maps, which he may also have used for his Oxford lectures.[42] In the thirteenth century, Matthew Paris visited London and Winchester, and in 1248 he went to Norway.[43] Not every travelling scholar was a monk. William Worcestre (d. 1482) was a "gentleman bureaucrat" who had spent his working life as secretary to Sir John Falstaff. Finding himself free after his master's death, he spent his last years indulging antiquarian interests, travelling indefatigably in southern England, observing, writing notes, and sketching.[44] In 1478 he rode from Norfolk to London and on to St Michael's Mount, Cornwall; in 1479 he travelled within the county of Norfolk; and in 1480 (aged sixty-five) he rode from London to Glastonbury.

Skilled workers, such as master craftsmen, stonecutters, and masons, travelled from site to site, even country to country, as opportunities for work arose. Jean Gimpel cites the way English kings were able to conscript workers for "the site of a fortified castle which might be several hundred miles away."[45] The search for minerals could also prompt an occasional journey. Two Silesian documents from the 1470s record the route some prospectors were to follow in order to reach mineral-bearing veins in the Riesengebirge.[46] In mid sixteenth-century Italy, the search for employment drove master iron workers from the Bergamo district to other parts of the Italian peninsula, as far as Reggio Calabria and Sicily in the south, Savoy to the northwest, and Austria to the northeast.[47] In 1661, another Bergamesque iron master sent Costanzo Gervasone on a six-day round trip across the mountains and lakes of northern Italy with a brief to persuade another ironmaster to purchase some redundant metal workings.[48]

Those pressed into the king's service as soldiers can also be classified as occasional travelers. Between April 20, and May 24, 1295, a group of 234 soldiers set out with an escort from Languedoc for northern France to take part in operations against the English.[49] Departing from Aigues-Mortes, they would have followed much the same road across the Central Massif as Lecari had been obliged to take a few years earlier.[50] Seventeen night-stops later, the group reached the River Loire at Nevers. Here, most of the group were put on boats to travel downstream to Bonny-sur-Loire while the others escorted the group's mounts overland. Meeting up two days later, the group continued northwards to Mantes, where a similar separation took place, the main body of men taking boats down the river Seine to Rouen and the other twelve riding overland with the horses.

INDEPENDENT TRAVELERS

Independent travelers are those who undertook their journeys as individuals, on behalf of no institution and responsible only to themselves and their families. They were amateurs on the road, and their reasons for setting out were usually highly personal. Despite an overeagerness in the traditional and general travel literature to see a pilgrim in every early traveler (and a "pilgrim's road" in every track), medieval and early modern pilgrims undoubtedly comprised the largest single group of independent travelers, especially in millennial and other "holy years."[51] The need to avoid punishment or religious persecution, especially during the years of the Reformation, or to plead a cause with the pope, sent others more reluctantly on the road. A few individuals travelled simply because they had the leisure, the financial resources, and the intellectual curiosity to visit other places for pleasure, or self-education, or social advancement.[52]

Travelling for pleasure was not an invention of the modern period. Some of the poet Petrarch's recorded journeys were imaginary, but he really did climb to the summit of Mount Ventoux in 1336.[53] William Worcestre travelled to occupy himself in retirement. Another late fifteenth-century independent traveler was the German knight Arnold von Harff, who started out from his home in Cologne in 1496 on what turned out to be a three-year pilgrimage to Rome and the Holy Land. Von Harff travelled much of the time in the company of merchants, first to Venice for Palestine; then from Venice to Santiago di Compostella, Spain; then home by way of Britanny and Mont St Michel.[54] A little earlier, a young Gascon noble, Jacques Nompar de Caumont, had completed a similar double pilgrimage but in the reverse direction and with a brief rest at home between the two journeys.[55] Like many fourteenth- and fifteenth-century pilgrims, de Caumont had planned to embark at Venice, but had scarcely left home when he met his uncle, who urged him to change routes to avoid the troubles on the Franco-Italian frontier resulting from Charles VIII of France's incursion into northern Italy. De Caumont headed instead for Barcelona, which meant he had an opportunity to explore Sardinia and Sicily before catching up with a group of pilgrims who had reached Venice from other directions and who were already in the Ionian Islands on their way to the Holy Land.

By the end of the sixteenth century, not pilgrims but tourists contributed to the increasing traffic on the roads of Europe. Fynes Moryson was a highly educated and wealthy young lawyer. Fresh from attending both Oxford and Cambridge universities, he toured Europe in 1591, impatient to "see forraigne countries" and sufficiently monied to hire a coach with driver.[56] On his return, he published his experiences for the benefit of those who might be tempted to follow his example. Besides a wealth of practical hints (such as the type of horseshoe to use for mounts crossing the Alps), Moryson urged tourists to "observe the underwritten things" and to look out for signs of the relative prosperity of each region through which they travelled so as to report on such matters as the abundance of mines, industries, bridges, outstanding buildings (including libraries "with the most rare books"), the layout of towns and the nature of their fortifications, and a host of other things.[57] Moryson also urged travelers to ask the local doctor or "man of principall account" to take them around the city in which they had arrived. When a tourist had the opportunity to "observe the situation of any City," he should, "(if he may without jealousy of the Inhabitants,) first climb one of the highest steeples, where having taken the general situation of the City, he shall better remember in order the particular things to be seen in the city. To which end, let him carry about him a Dyall, which may shew him the North, South, East and West . . . after, being returned to his Inne, [he] may draw it on paper, if he thinks good."[58] Such sideline interests could be perilous, though. Thomas Coryate warned of the dangers in prying "very curiously into State matters" and alluded to "a most tragicall" experience that cost some unfortunate traveler "dear" in Strasbourg not long before Coryate's arrival there in September 1608.[59]

Moryson was not the first to write up his experiences. Some medieval pilgrims had recorded their itineraries or described their travels, giving details about the places visited as well as their location.[60] From these we may also learn something of the travelers' attitude to foreigners and their lands. We learn nothing, however, about the way and the extent to which a traveler prepared himself mentally (or even spiritually) for the journey, or which books he read or maps he consulted.[61] The outstanding exception is Felix Fabri, a Dominican monk from Ulm in southern Germany who made two journeys to the Holy Land between 1480 and 1483. After his second journey, Fabri compiled a detailed, perceptive, and entertaining account of his experiences for his fellow monks, confessing how, on his return from his first visit, he suddenly realised what he had missed through sheer ignorance.[62] Before setting off again, he "worked harder" he tells us, "in running from book to book, in copying, correcting, collating what I had written, than I did in journeying from place to place on my pilgrimage."[63] More to the point, he evidently studied maps, although we only hear about those from the edition of Ptolemy that had just been published in Fabri's home town and rather more vaguely about "the new maps of the world" that showed "the regions of the east so far distant from us that according to modern geometers and mathematicians they who live there are antipodes, which however the ancients, such as Aristotle, Ptolemy, and Augustine could not admit."[64] But Fabri says nothing at all about maps in relation to his journeying, or about how he or the merchants he travelled with found their way. The only other map we hear about (apart from Fra Mauro's map Fabri reported seeing on the island of Murano in the Venetian lagoon) is the navigational chart he saw on board his ship, a "chart, which is all of an ell long and an ell broad, whereon the whole sea is drawn with thousands and thousands of lines, and countries are marked with dots and miles by figures. In this chart they see where they are, even when they can see no land, and when the stars themselves are hid by clouds. This they find out on the chart by drawing a curve from one point to another."[65] Moryson had even less to say than Fabri about maps. He merely commented that a traveler should acquire "some skill (at least superficially) in the Art of

Cosmography" in order to know something about the "Kingdoms which he is to pass" through.[66]

Moryson represented a new class of private traveler with different reading needs from those of a pious and theologically-minded medieval pilgrim like Felix Fabri, and by the middle of the sixteenth century, a new form of literature began to make its appearance: the popular guide book. In 1549 Anton Franceso Doni published a letter in which he made suggestions about lodgings and itineraries in Florence.[67] Later in the century another genre was added to the travel literature. First, in Germany, came Hieronymous [Jerome] Turler's *De peregrinatione* (Strasbourg 1575), translated the same year into English as *The Traveiler.*[68] Turler's definition of travel reveals an attitude setting the new kind of traveler well apart from the traditional run of professional road users. For Turler, travelling was "nothing else but a painstaking to see and search foreign lands, not to be taken into hand by all sorts of persons, or unadvisedly, but by such as are meete therto, either to the end that they may attain to such arts and knowledge as they are desirous to learn or exercise; or else to see, learn, and diligently to marke such things in strange countries, as they shall have need to use in the common trade of life, whereby they may profit themselves, their friends, and Countrey if need require."[69] Turler's book is a philosophical treatise on the "art of travel," the aim of which was to impart an intellectual framework for knowledge acquired through direct observation and the exercise of reason.[70] If maps were mentioned at all in this kind of book, it was to place them in a formal, Ramist, ordering of knowledge under headings such as "How to use reason in travelling."[71]

Tourist travel accounts from the late sixteenth century onwards tend to be highly personal. The authors complained about physical discomforts they had to endure on the journey: the cold and wet, the dizzyingly high mountains, and the treacherous passes. In contrast, the physical environment was all but ignored in the itineraries used by professional travelers. Rivers, major forests, even mountains (other than as passes) rarely receive mention.[72] What is emphasised are political frontiers and conditions (table 2.1). The Titchfield Abbey itinerary of c.1400 to Rome specified an alternative route in northern Italy "if there is war in Lombardy."[73] Despite the silence in contemporary itineraries on physical features, so much is made in some modern literature of the hardships of early land travel that one wonders how anybody dared set foot across their own threshold. Of course travel was dangerous—it still is—but much of the emphasis on physical discomfort is misplaced. So-called regions of difficulty or hardship are defined not by inhabitants but by outsiders. Local people, as Eugene Weber points out, have "always ignored difficult terrain and

Table 2.1 Extract from Arnold von Harff's itinerary from Cologne to Rome and the Holy Land, 1496.

EIN STIJFT VAN COELLEN [ARCHBISHOPRIC OF COLOGNE]	Miljen
Van Coellen [from Cologne]	iiij
Bonne [Bonn]	ij
Winteren [Oberwinter]	j
Remagen [Remagen]	ij
Prijsac [Breisig]	ij
Andernach [Andernach]	iij
EIN STIFT VAN TRIERE [ARCHBISHOPRIC OF TRÈVES]	
Covelenz [Koblenz]	ij
Rense [Rhens]	ij
Borardenc [Boppard]	j
Histzenauwe [Hirznach]	j
Sent Gewer [St-Goar]	j
Wesel [Oberwesel]	j
Trecks husen [Trechingshausen]	j
Bingen [Bingen]	ij
EIN STIFT VAN MENS [ARCHBISHOPRIC OF MAINZ]	Mijlen
Ingelhusen [Ingelheim]	ij
Mentz [Mainz]	iij
Oppenheim [Oppenheim]	iij
Worms [Worms]	vj
Spire [Speyer]	iij
Broessel [Brucksal]	ij
SWAEBEN LANT WIRTENBERCH [SWABIA, WÜRTTEMBERG]	
Breten [Bretten]	j
Smeen [Schmie]	j
Feygengen [Vaihingen]	j
Swepertingen [Schwieberdingen]	j
Canstatt [Cannstatt]	j
Esslinghen [Esslingen]	iij
Gyspingen [Göppingen]	ij
Gislingen [Geislingen]	iij
Vlm [Ulm]	vj
Memmingen [Memmingen]	iiij

Kempten [Kempten]	ij
Messelbanck [Nesselwang]	j
Fyltz [Vils]	j
Rute [Reutte]	ij
HERTOCH SEGEMONYS LANT [LANDS OF DUKE SIGMUND]	
Lermois [Lermoos]	ij
DER VERNER EYN BERG [MOUNT VERNER = FERN PASS]	Mijlen
Nasareth [Nassereith]	j
Eyms [Imst]	ij

Notes: The political units (archbishoprics or dukedoms) are identified, but there is no mention of the Rhine, which had to be crossed at Speyer. Transcription taken from E.-T. Hamy, *Le livre de la description des pays de Gilles le Bouvier, dit Berry*. Receuil de voyages et de documents pour servir à l'histoire de la géographie, depuis le IIIe jusqu'à la fin du XVIe siècle, no. 22 (Paris, 1908), 220–21.

stalked over it regardless," resenting bad roads only when better ones became available.[74] It is worth bearing in mind, moreover, just how much of Western Europe would have offered even the medieval traveler an "easy going" landscape of "fertile, well-farmed countryside with good paths and bridges and settlements not too far apart."[75] Even Braudel admitted that sixteenth-century travel could have been "positively pleasant where towns and villages were close together."[76]

Roads and Routes

"Roads, roads, and still more roads": Thus Eugene Weber heads the chapter of his book on the state of France in the early nineteenth century.[77] But what is a road? And what is the difference between a road and a route? In modern writing, loose use of vocabulary has given rise to simplistic notions about early roads and road travel. What exactly did Jusserand have in mind when he asserted that in the Middle Ages "the roads in England would have been entirely impassable," and whom was that alleged impassability supposed to have prevented from travelling?[78] Certainly not the merchants and other professional travelers who had crowded the roads since early antiquity. Likewise, what did H. G. Fordham mean when he remarked (self-contradictorily) that in the third decade of the sixteenth-century in

England "roads hardly existed" when Leland was travelling throughout the country?[79] The myth of a blanket "impassability" dies hard; in 1980 we were still being told that a "lack of desire or opportunity for people to travel" was associated with the fact that the roads were of "too poor quality" to permit them to do so.[80] The point has been missed yet again: it was not the professional merchants and traders who were deterred from venturing forth "because of the state of the road," but the independent traveler, whose journey was entirely optional. Complaints about the state of roads abounded at all times, but business and commerce carried on nonetheless.

Long before the arrival of John Macadam's solid, drained, and dry road surfaces in the early decades of the nineteenth century, and their eventual waterproofing with asphalt or bitumen, there existed an intricate hierarchy of lines of movement for which modern English has a variety of terms (way, path, track, lane, road, highway, etc.). The difficulty for the historian is to make the appropriate match between the term in a document, the line on the map, and the feature on the ground. The commonest confusion is between "road" and "route," words with different etymologies and, although commonly used interchangeably, different meanings. The word "road" is well grounded, coming as it does from the Old English *rad* (and the cognate Middle Dutch and Old Frisian form *red*) and associated verbs such as *ridan,* meaning "to ride" on horseback).[81] In contrast, "route" has always had an abstract dimension. It derives from the Latin adjective *rupta* as in *via rupta,* a line cleared through woodland, thus permitting access or a way through. Within the area of the Roman Empire, Latin spawned a hierarchy of words, such as la*voie* (from *via*), le *chemin* (from *caminum*), and l'*estrade* (from*strata*). The picture is complicated by regional as well as chronological factors. In twelfth-century France, for example, *route* was standard in the north but scarcely known in the south.[82]

The word "road" thus usually signifies in the first instance a physical feature in the landscape. At the same time, the existence of the physical feature conveys intangible aspects, notably the right of people—either specified individuals or the public at large—to traverse the land from one point to another.[83] It also implies that such movement along the road is normally feasible, apart from constraints such as seasonal flooding. In describing the shortest route between Exeter and Bristol, which lies across the water-logged Somerset Fens, William Smith warned that "no man can travell it well, except it be in somer tyme, or ells when it is a great frost."[84] On the other hand, the existence of a road does not always imply its current use. Medieval roads did not necessarily follow the local section of

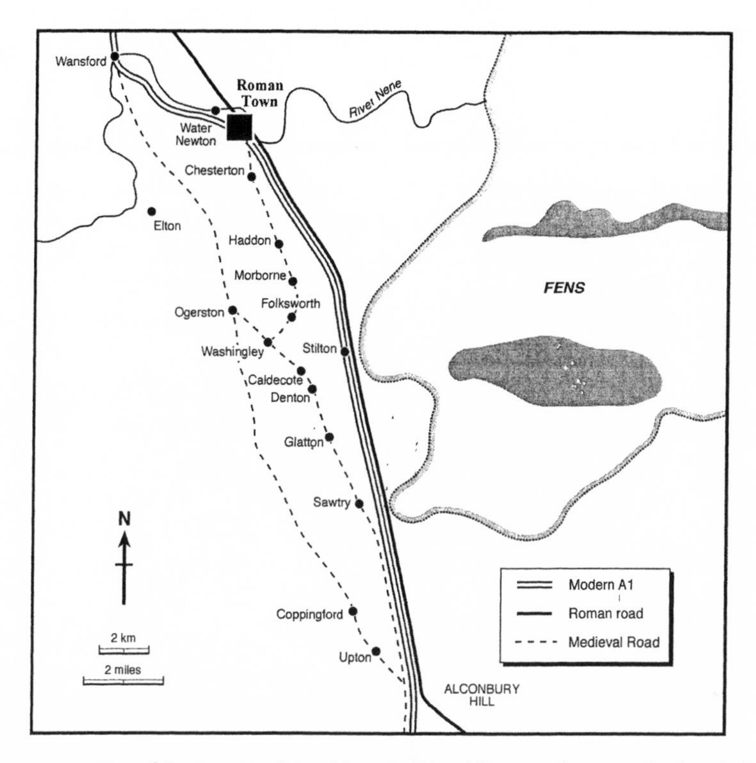

FIGURE 2.3 Part of the Great North Road (Ermine Street) from London to Scotland. Only the Roman and the modern alignments coincide in this Northamptonshire section. The tracks to the west that were used in the Middle Ages are today minor roads. Adapted from Christopher Taylor, *Roads and Tracks of Britain* (London: Dent, 1979), fig. 57.

Roman roads, despite the latter's better construction, when adjacent places proved more attractive or some other factor deflected travel to a different alignment (figure 2.3). The width of a road could also be an issue. In open country and on steep slopes on soft soils and damp clays, the trodden surface was liable to encroach on to the adjoining land as riders, carriages, and wagons attempted to bypass the muddiest and most deeply rutted patches (figure 2.4).[85] In bocage regions, where the hedges of the fields prevented lateral encroachment, the track remained narrow and deepened over time into "hollow ways," which can be impassable even today in times of extreme wet, deep snow or when blocked by a fallen tree (figure 2.5). In these lanes, travelers had restricted visibility, and the risk of ambushes from highwaymen was a constant fear. The difference between closed and open roads was important enough to the early traveler for John Ogilby to make the distinction clear on his post road maps in *Britannia* (1675).

FIGURE 2.4 An open road: the Oxfordshire section of the Icknield Way, an unmetalled prehistoric trackway along the crest of the North Downs, now partly fenced off. The full width of the trackway is not normally used but tracks in the grass show where traffic has attempted to avoid the wettest parts. Author's photograph.

A route is not a road, nor in itself a physical feature, but a direction, an imaginary line linking a point of departure with a destination. Only its description gives it tangible form in speech or gesture, writing or image. A route may relate to a journey already undertaken or to one in the future. It may be the personal creation of an individual traveler or group of travelers or it may be official, like the Roman *cursus publicus* and the later postal relay systems. It may be held privately in a library or displayed publicly for all to see, like the Roman *tabellaria,* which were erected by road builders at strategic points or at intervals along a road and on which were listed, in sequence and usually with distances, a selection of the places along that road in both directions.[86] The most impressive example in its day may have been the *tabellaria* at Gades, Spain, which appears to have given the route (in individual stages and with distances in miles) from Gades through southern France to Rome in lists on four plaques or panels, but a recently reconstituted example from Patera, southern Turkey, shows how many of these directional monuments were likely to have been located at the gates of towns.[87] The Patera *tabellaria* has been reassembled from fifty-three fragments recovered from the marshy ruins of the Roman town walls. It appears to have been a rectangular monument some five and a half metres tall, surmounted by a statue, on the flanks of which were detailed the places (with distances between them in *stades*) to which the three highways which radiated from the gate led. Itineraries, such monuments remind us, describe places between a starting point and a terminal point, and it is only when a route has been decided on that the appropriate road can be identified.

FIGURE 2.5 A closed road: a sunken unmetalled lane in Devonshire. Author's photograph.

Written Itineraries

The normal guide to wayfinding, then, has always been the itinerary. An itinerary comprises little more than a list of names in geographical order of places to be passed through on the journey to a specified destination. In addition, certain places may be designated as suitable overnight stops. Distances may be given: between each place, for each day's travel, or for subdivisions of the itinerary (within a particular political unit, for example). Sometimes additional brief notes indicate special points for the traveler to anticipate, such as the frontier, a special toll post, or the advisability to take a boat. Many such details were included in the Titchfield Abbey itinerary to Rome, compiled from a journey made between 1400 and 1405 by a group of the abbey's canons (figure 2.6).[88] Additional details include a note at the St. Gotthard pass indicating, encouragingly, the half-way mark in the journey,

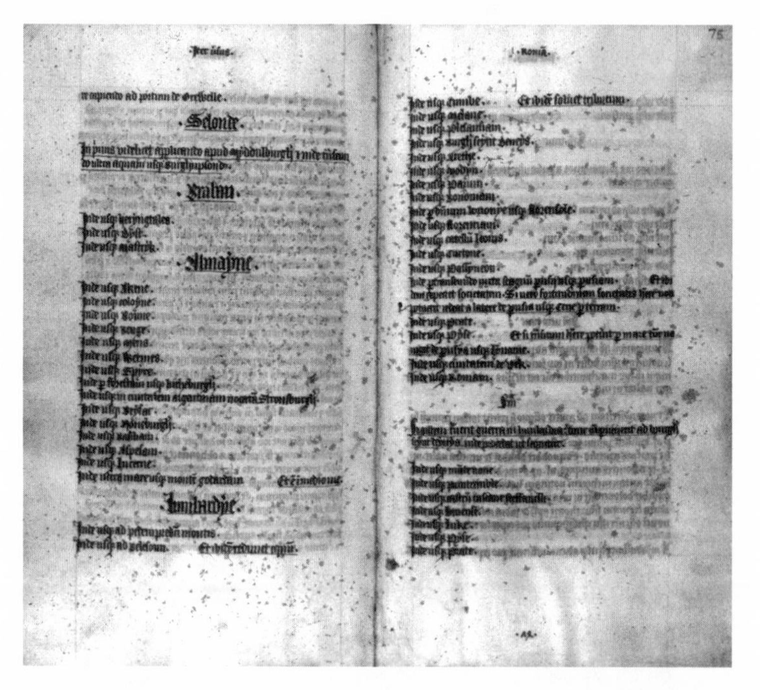

FIGURE 2.6 Extract from an itinerary from England to Rome (c. 1400) from Titchfield Abbey, Southampton. The itinerary is set out according to political divisions. One of the notes on the right side of the column on the left page marks the half-way point of the journey (*Et est in medio [it]ine[re]*), another indicates the stage at which the horses have to be paid for (*Et ibidem redimet equum*), and the last main heading on the right page indicates the alternative route to be followed if there is war in Lombardy (*Si autem fuerit guerra in Lombardia, tunc cum pervenerit ad bourgh' seynet deyneys* [Borgo S Denis, now Fidenza] *inde procedat ut sequitur*). B.L., MS Add. 70507, fols. 74–75v. By permission of The British Library.

and another note at Perugia, advising travelers to join merchants for the next section of the journey (across the infamous Tuscan Maremma).[89]

Itineraries are essentially written documents. Little is to be gained in a strictly utilitarian sense from attempting to portray a list of places graphically, or pictorially. A line of place-names in geographical sequence is in itself a topological map, and where alternative routes are added there may be some advantage in setting out the itinerary as a sketch map, as William Bowles did in 1578 in planning Queen Elizabeth I's Norfolk progress (figure 2.7).[90] To portray such a prosaic, practical travel aid as a work of art, though, seems pointless at first sight. A different interpretation is therefore needed to explain the function of the only known medieval graphic itinerary, the one compiled by Matthew Paris in the 1250s at St Albans, Hertfordshire, and the collection of graphic itineraries in John Ogilby's *Britannia* (1675).

The format and context of surviving itineraries can be a useful guide to the status of the traveler or travelers involved. One early reason for keeping records of itineraries, apart from the obvious need to store route information in the event of a journey, was the need of officials to check distances before disbursing payment for mileage claimed.[91] The most business-like format is also the most common: a simple list of places set out down the page. Sometimes it is obvious that the list represents a hasty jotting down of a route the writer had just heard about. A couple of lines of faded ink and all but illegible writing crammed on to the bottom of a letter gives the route from Boulogne to Orleans.[92] The blank folios at the beginning or end of a codex provided a more convenient place to store a note of itinerary, especially when writing materials were scarce and expensive. One list of the ports between Marseille and Jerusalem, with distances, is written in ink on the end paper of a miscellany of scientific treatises dating from about 1300.[93] A fifteenth-century collection of tracts contains a note about the six stages of the sea route from Venice to Jaffa on a spare page.[94] The flyleaf of a fifteenth-century English translation of the travels of Sir John de Mandeville carries a note written in English of a route from somewhere in northern Europe (the page has been torn) to Florence.[95] In the same century, a group of itineraries was written down, in German and unusually neatly, at the end of a lavishly illustrated account of Gabriel Maffre's journey to Jerusalem in 1465.[96] In fact, any surface at hand could serve to record an itinerary. When in the 1550s William Cecil wanted to note down the routes from Ghent to Bruges and Antwerp to St Quentin, he used the back of a piece of folded paper that happened to be handy, Laurence Nowell's map of England and Wales.[97] Cecil continued to collect itineraries when in the 1570s he became Lord Burghley and Elizabeth I's secretary

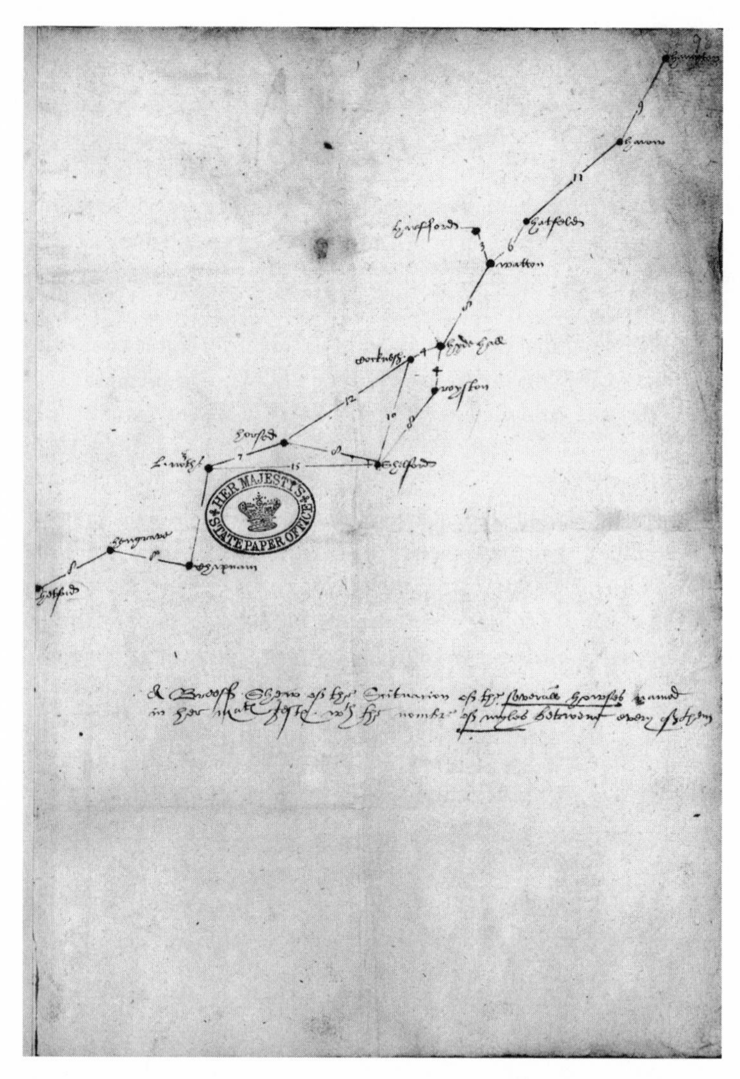

FIGURE 2.7 William Bowles's sketch map setting out the possibilities for Queen Elizabeth I's route to Norfolk in 1578. Reproduced with permission from the National Archives of the United Kingdom: SP 12/125, p. 98.

of state (figure 2.8).[98] Other routes (and notes on the cost of upkeep of certain trunk roads) are noted in his or a secretary's hand on folios kept in the same folder as his much-used proof copies of Saxton's maps.[99]

Those whose business it was to direct others, like Cecil Lord Burghley, made a point of collecting potentially useful itineraries. Military commanders also needed routes. Mention has already been made of Don Lope de Acuña's sketch maps for troop movements along the Spanish Road.

In the eighteenth century, the duke of Marlborough, on campaign in the Netherlands, noted an itinerary on a page that a quartermaster had already (or later) used to record the arrival of luggage and artillery.[100] It was in the interests of the institutions, too, to collect relevant itineraries so that these could be made readily available when one of its members was about to make a journey. Religious houses maintained regular contact with others of the same order and the Titchfield Abbey volume also contains itineraries to each of the twenty-nine daughter houses in England.[101] All of the Titchfield itineraries were neatly written out and bound up with other

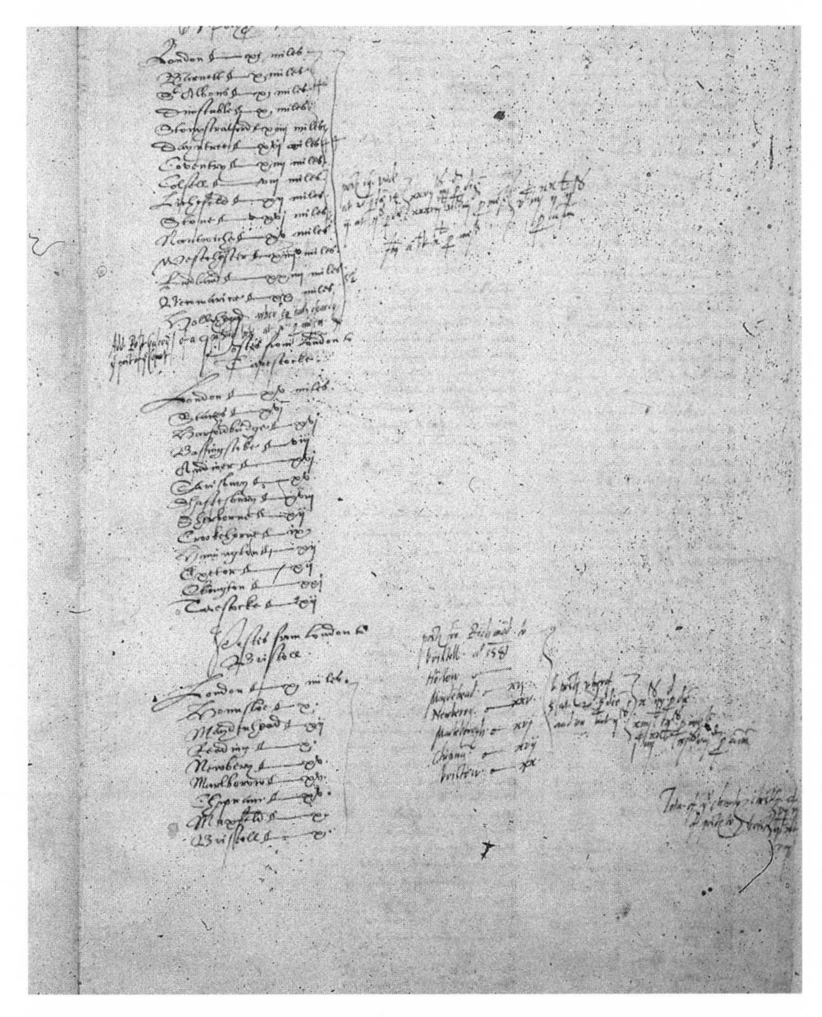

FIGURE 2.8 William Cecil's post road notes. Three itineraries are listed: London to Holyhead, London to Tavistock, and London to Bristol. At the bottom, to the right, is the "Post fro[m] Richmond to Bristol." B.L., MS Royal 18. D. III, fol. 4. By permission of The British Library.

important documents relating to the monastery's affairs to form a stalwart codex protected by vellum-covered oak boards with fastenings and leather page-markers. Merchant houses also took good care of the route knowledge they amassed. In the Flemish city of Bruges, a group of itineraries compiled around 1500 from various sources and kept for use by Hansa traders were bound into a large and apparently "sumptuous" volume.[102]

The itineraries mentioned so far represent functional route descriptions that were recycled over and over again. Early itineraries are also found in other formats and contexts. The itinerary of Archbishop Sigeric's return journey from Rome in the tenth century is given in an account of his life.[103] William Worcestre tabulated features drawn from his itinerary separately by way of topographical analysis: "The bridges of Cornwall, from Exeter to the Mount" and "The islands in the Severn towards Bristol," for example. Often the exact route followed by the author is lost in one of these "expanded itineraries" and indeed the modern editor of Worcestre's *Itinerarium* comments that the many digressions make it "difficult to get a clear picture of his average day's journey."[104] Many travel accounts identified by their authors as itineraries give no indication of the actual route taken. A case in point is the fourteenth-century copy of three separate narratives relating the travels of William of Rubrick (1253) and of Symeonis Semeonis (1332) and Friar Odoric (who wrote in 1330), each called *Itinerarium.*[105] William Wey's account of his pilgrimage to the Holy Land in 1462, like the sixteenth-century antiquarian John Leland's diary, is also an example of the genre.[106]

The arrival of printing did little to alter the way itineraries were recorded, but it did make route knowledge available more generally. Buried in a compendium of customs and charters relating to the city of London that was printed in 1503 is a note of "The way from Calais to Rome through France" and on to Naples.[107] Then in 1542 the stationer John Judson printed a small (octavo) booklet containing a chronicle continuation that described eleven routes in England and Wales.[108] Judson's book was followed the next year by William Middleton's almanac, the title page of which announces that he has added a description of "the ways leading to the notable places and the distance between the same."[109] The routes are those described by Judson. The longest (Berwick to London via York) has twenty-one stages, the shortest (Dover to London) has six stages (figure 2.9). These itineraries are clearly part of a recognized corpus of trunk routes, most of which would have dated back to Roman times and earlier (figure 2.10). The itineraries would not have changed much except to accommodate new towns or because national politics altered the relative importance of one route.[110]

March of Scotlande, to Totnes in Devonshire is. iiii.C.myles.
The bredth fro saynt Dauids in Wales unto Dower is iii.C.myles.
And the compasse of Englande rounde aboute, is iiii.M.iii C.lx.myles.
There bene in Englande of parish Churches, to the nombre of.xlviii.M.viii.C.xxii.
Also there bene in Englande of Byschopryckes, to the nombre of.xxv.
Also there bene in Englande of shyres, or coūties, to the nombre of.xxxvii.
Also there bene in Englande of Townes besyde Cyties, and Castels, to the nombre of.lii.M.lxxx.

Here is shewed how the shyres lye, with the most and greatest notable townes: as shyretownes, Cyties, Borowes, and market Townes, and what distaūce they be a sonder.

Here foloweth the waye from Walsyngam to London.

From Walsyngam to Piknam.	xii.myle.
From Piknam to Brandonfere.	x.myle.
From Brandonfere to Newemarket.	x.myle.
From Newemarket to Brabram.	xii.myle.
From Brabram to Barkway.	x.myle.
From Barkway to Pukrych.	vii.mile.
From Pukryche to Ware.	v.mile.
From Ware to Waltham.	viii.myle
From Waltham to London.	xii.myle.

Here foloweth the waye from Berwycke to Yorke, and so to London.

From Berwicke to Anwik.	xii.myle.
From Anwik to Morpit.	xii.myle.
From Morpit to Newecastell.	xii.myle.
From Newecastell to Doram.	xii.myle.
From Doram to Darryngton.	xiiii.myle.
From Darryngton to Northalerton.	xiiii.myle.
From Northalerton to Topclif.	viii.myle.
From Topclif to Yorke.	xvi.myle.
From Yorke to Tadcaster.	viii.myle.
From Tadcaster to Wentbrydge.	x.myle.
From Wentbryg to Dankastre.	viii.myle.
From Dankastre to Toxforde.	xviii.myle.
From Toxforde to Newarke.	x.myle.
From Newarke to Grantham.	x.myle.
From Grantham to Staunforde.	xv.myle.
From Staunford to Stylton.	xii.myle.
From Stylton to Huntyngton.	ix.myle.
From Huntyngton to Royston.	xv.myle.
From Royston to Ware.	xiii.myle.
From Ware to Waltham.	viii.myle.
From Waltham to London.	xii.myle.

Here foloweth the waye from Carnaruan to Chestre and so to London.

From Carnaruan to Conway.	xxiiii.myle.
From Conway to Dynbigh.	xii.myle.
From Dynbigh to Flynt.	xii.myle.
From Flynt to Chestre.	x.myle.
From Chestre to Wyche.	xiiii.myle.
From Wyche to Stone.	xv.myle.
From Stone to Lichfelde.	xv.myle.
From Lichfelde to Colsill.	xii.myle.
From Colsill to Couentre.	viii.myle.

And so from Couentre to London as hereafter.

FIGURE 2.9 Itineraries in a printed almanac. The headnote introduces the information on "how the ways lye, with the most and greatest notable townes: as the shyre towne, cyties, borowes, and market townes, and what distance they be a sonder." Two pages from William Middleton's almanac of 1544, B.L., G. 5892. By permission of The British Library.

Road books were something new in the first half of the sixteenth century, but thereafter they were issued with impressive frequency up to the nineteenth century. The (usually) leather-bound pocket-sized road book became the first place to turn to for an itinerary. Later, especially in the eighteenth and early nineteenth centuries, some books dealt with a single road, some with a specified category (post roads) for either the country as a whole or an individual county. Typically, the routes are set out in columns, starting at the top of the page, and distances are given in miles and furlongs. A separate column may indicate the more important crossroads. The books supplied a new source of demand and a new class of road users, people who had time to take an interest in the surrounding landscape as well as the need to confirm a town's market day or check on the distance between one place and another on a national highway. The road books came to include, as in William Worcestre's day, brief historical anecdotes or comments on a particularly fine bridge or other structure. In a book of English roads published in France by Jean Bernard, the *Guide des chemins d'Angleterre* (Paris, 1579), the adventitious content was advertised on the title page: *J'ai aussi*

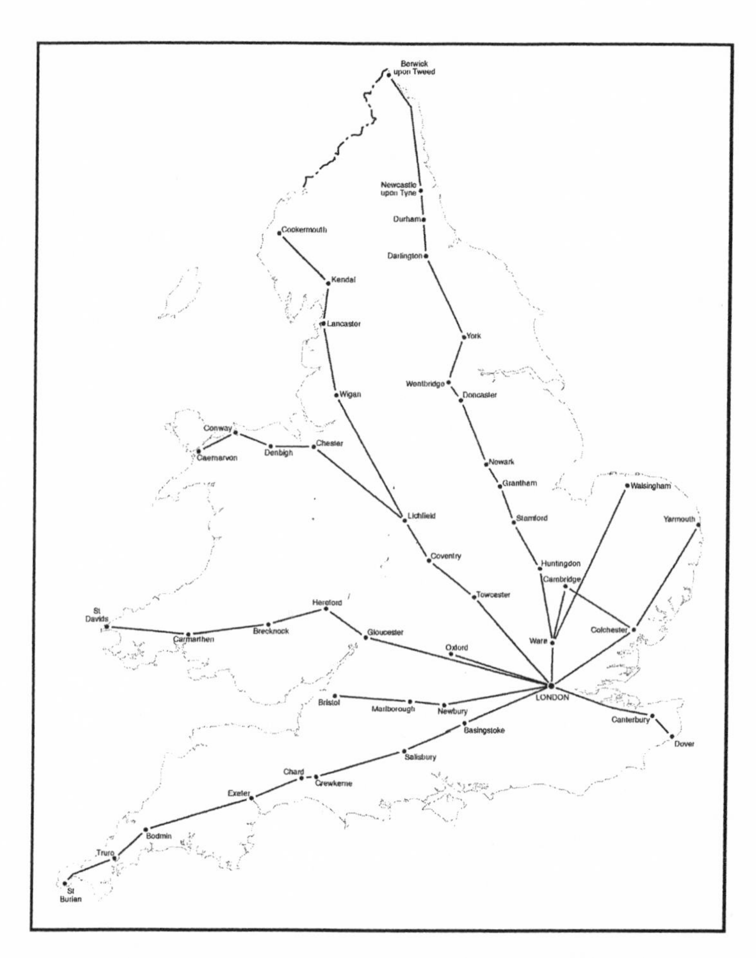

FIGURE 2.10 The trunk roads of England and Wales in the second half of the sixteenth century. A simplification of the route network in England and Wales, emphasizing the key routes. The pattern was in fact much more intricate, there were cross routes—from Oxford, via Northampton and Nottingham to Doncaster, for example, and from Southampton to York by way of Bristol and Northwich (near Chester)—and subsidiary routes, such as from London to Rye on the south coast or London to Norwich, west of Yarmouth on the east coast. Based on the forty-one itineraries in Richard Grafton, *Abridgement of the Chronicles of England* (1571).

rapporté certaines particularitéz dignes d'estre cogneuës à ceux qui passeront de ville en ville . . . tels le nombre des parroisses, Eglises, villes & Euechéz.[111] It made good sense for visitors from France to be provided with itineraries for the roads they were most likely to be using in England, especially the road from Dover to London and perhaps thence in various directions to the provinces.

Losses of heavily used pocket books must have been high and our picture of their production and dissemination must be considered incomplete, especially for the early part of the modern period. There appears to have been much borrowing from one edition to another, or from common sources that are no longer extant or identifiable, despite Charles Estienne's assurance in the *La guide des chemins de France* (Paris, 1552) that this "novel and original" work incorporated firsthand information supplied by "messengers, traders and pilgrims."[112] Etienne offered his readers a real choice of alternative routes—"by Orleans" as opposed to "by Burgundy" (Paris to Lyons); "the most direct" and "the easiest" (Paris to Grenoble); the "most direct," "shortest," and "longest" (Paris to Lusson); the "main road" (*le grand chemin*), "the nicest and shortest"—and occasionally precise instructions: "Passe par le bout des hayes du village et laisse à gauche." Bigger works, however, tended to keep to the point, repeating the main routes. The substance of the thirty-nine "High Wais, from any notable towne in England to the Cittie of London, and lykewyse from one notable towne to another," for example, which was printed in Richard Grafton's fat little *Abridgement of the Chronicles of England* (1571), is found in Raphael Holinshed and William Harrison's *Chronicles of England, Scotland, and Ireland* (1577), Richard Rowland's *The Post of the World* (1576), and William Smith's manuscript of *The Particular Description of England* (c. 1588). In these and others, the exact number of routes described and their details may differ from one to another, but in all essentials they are identical.[113] Like travel accounts with the word "itinerary" in the title, not every book that might be thought a road book relates to the practical aspects of travel. The *Direction for Travaillers* (London, 1592), for instance, turns out to be a homily on the proper attitude to travel for "the right honourable Lord, the young Earl of Bedford, being now ready for travell."[114]

Maps were only occasionally included in almanacs and road books and then only from the late seventeenth century onwards. There is no known sixteenth-century or early seventeenth-century road book with a map of a road or a road network. In 1676, however, the publishers of a new edition of John Speed's *Theatre of the Empire of Great Britaine* added five itinerary or route maps. The maps are thematic, showing only the place names and "computed distances" of each of the "Principal Roads and their Branches" in a "new and accurate method," as advertised on the title page (figure 2.11). Some of the maps also appeared in the same year in an anonymous road book called *The English Traveller's Companion,* much folded to fit the pocketbook's distinctive format (20 × 8 cms; 8 × 3 ins).[115] Soon after, Abell Swall included a similar type of map in *The Traveller's Guide: or, a most exact description of the ROADS of ENGLAND* (1699).[116] More maps appeared in eighteenth century road books. An unattributed map, possibly copied from one published in 1737 and showing the usual London-based route system,

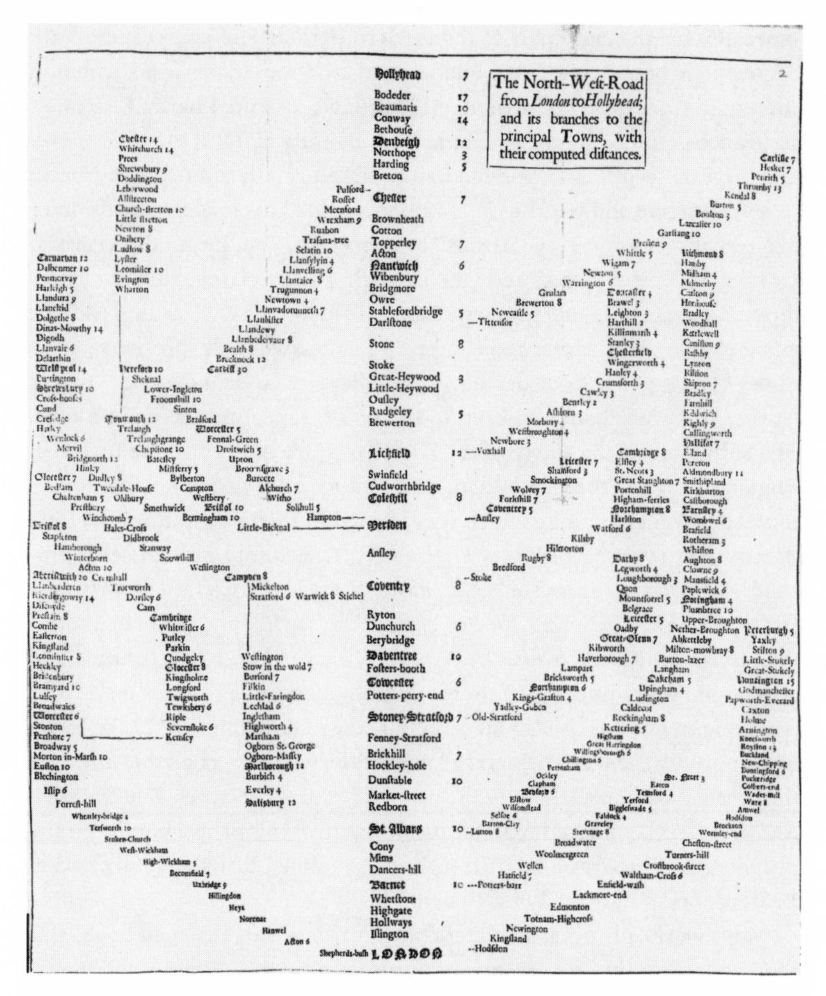

FIGURE 2.11 "The North-West-Road from London to Hollyhead; and its branches to the principal Towns, with their computed distances." One of five itinerary maps inserted into the 1676 edition of John Speed's *Theatre of the Empire of Great Britaine.* The system starts from London (at the bottom) and is made up of seventeen separate itineraries. The maps were also included in a road book, *The English Travellers Companion,* published in the same year. B.L., Maps C.21.b.16. By permission of The British Library.

was included in the *Traveller's Pocket Companion thro' England and Wales containing a Map of all the Direct and Principal Cross Roads laid down from Mr Ogilby's Survey* (1741).[117] *The Kentish Traveller's Companion* of 1776 presented the route from London to "three Miles beyond Rochester" in a different way, portraying the road in a series of strips laid across the page.[118] John Cary arranged his presentation of the road from London to Hampton Court

(1790) vertically, two columns to the page. His maps include lines from the road to "the points of sight from where the Houses are seen," reminding us of the nature of the clientele to whom Cary hoped his road book would appeal, namely the social traveler rather than the hard-pressed tradesman (figures 2.12 and 2.13).

Another travel aid appeared on the market in the sixteenth century. This was the printed, pocket-sized, multilingual phrase book for commercial travelers. Our interest in these book lies in the light they can shed on how travelers communicated, and, above all, on how they found their way. The tradition of supplying ready-made phrases in another language for the use of travelers was not new in the sixteenth century—Ohler alludes to the "Conversations in Old German" of the ninth and tenth centuries—but printing ensured easy accessibility and wide dissemination.[119] The number of languages given in each book varied from one edition to another, but the standard format was to set phrases down in the form of a dialogue, with a separate column for each language. Separate chapters deal with different situations in which a merchant might expect to find himself while travelling. Noel Berlaimont's seven-language phrase book, published in Antwerp in 1586, contains chapters on, among other matters, "Buying and selling," "Demanding one's debts," "Merchandising," "A dinner for 10 persons," "Obligations," "Letters, contracts, quittances," "In the inn," and—not of least interest in the context of the present essay—"To ask the way."[120] In the latter, the dialogue takes place between two travelers (A and B) on the road to Antwerp and a local shepherd (G). The key part is reproduced here:

A. I am afraid that we be out of our way . . . it is good to ask it.
B. Ask of that . . . shepherd.
A. My . . . friend, where is the right way from hence to Antwerp?
G. Right before you, turning neither on the right nor on to the left hand till you come to an high elm tree, then turn of the left hand.
A. How many miles have we from hence to the next village?
G. Two miles and a half and a little more.
. . .
A. I see the tree whereof he told us . . . the sun goeth down; I am afraid we shall not come by day-light to the town.
. . .
A. I see the steeple of the town . . .
B. Truly it will be late before we come there; I doubt that we shall not get in.

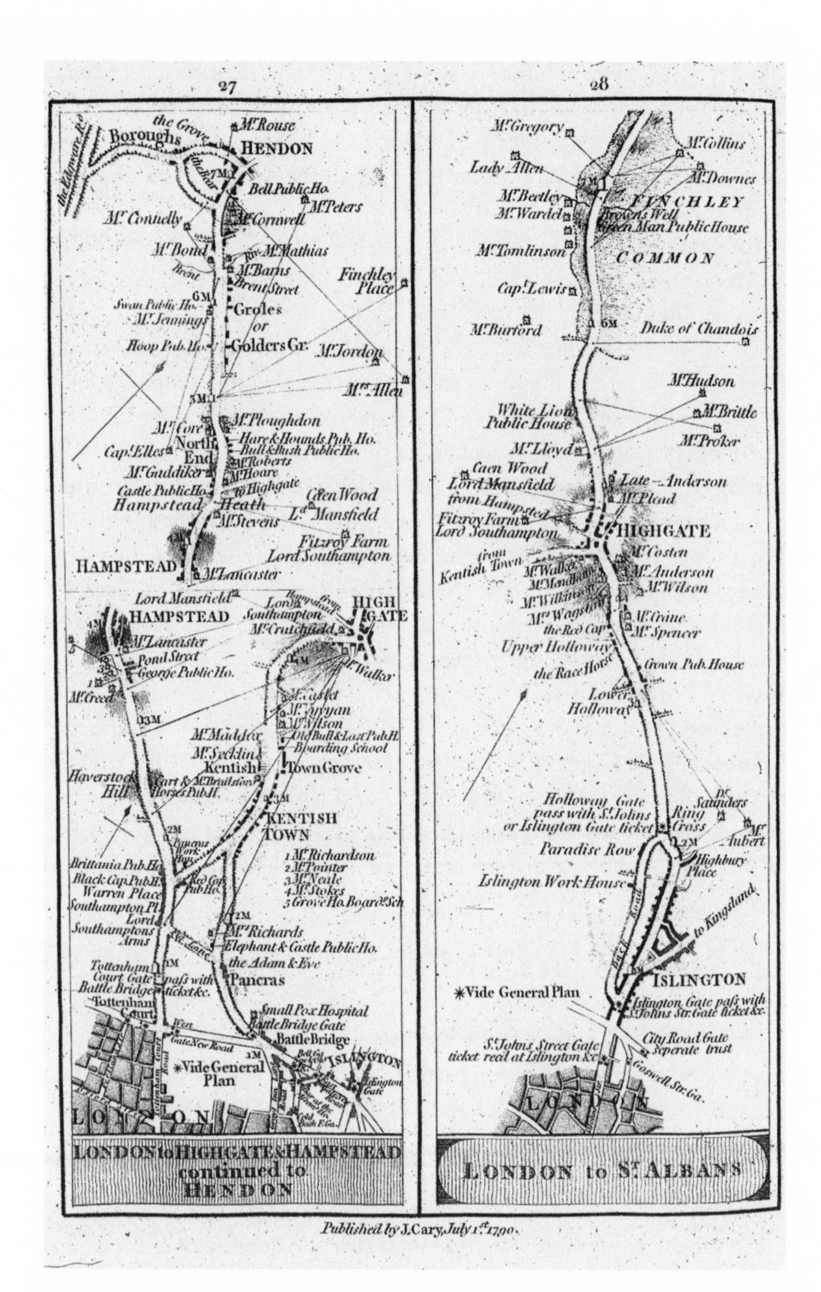

FIGURE 2.12 Maps in road books. John Cary's *Survey of the Roads from London to Hampton Court* (1790). Cary's ingenious sight lines indicate the points on the road from which the country houses of the gentry could be viewed (see fig. 2.14) and highlights the social function of road books. Some of these houses were newly built and many were having their estates professionally landscaped. The inclusion on the map of a particular house was also a way of attracting subscribers to underwrite the costs of production. Courtesy of the Newberry Library.

A. Yes forsooth they do not shut the gates before nine of the clock.
B. It is better, for I would not be gladly in the suburbs . . . Let us aske of these folks for the best inn.[121]

This simple conversation documents what common sense has long insisted on; that the chief mechanism of wayfinding has always been oral. Once the traveler was on the road, all maps were redundant so long as he had his itinerary. Regular checks with other travelers and local people at meeting places such as inns, relay and coaching stations, and river crossings ensured that the traveler was following the right road or knew which branch of the track to follow at a bifurcation. There may have been signposts, or markers

FIGURE 2.13 The view from the road. The landscape architect Humphrey Repton suggested that the hill to the right should be cut away to open up the "prospect" of Sheringham House, Norfolk, from the road. Detail from one of two drawings in Repton's *Red Book for Sheringham* (1812) showing the approach to the house. Reproduced with permission from the RIBA Library Drawings Collection, London.

of some sort (as there still are on many mountain paths in the Alps and elsewhere), but, as the above extract makes clear, the traveler was also told which landmarks to look out for. Even today, wayfinding remains largely a matter of reading the road signs that furnish all major roads in the course of following an itinerary.

Graphic Itineraries

From the traveler's practical point of view there is no need to translate a simple written list into graphic form. It could be useful to emphasise the sequence of places by writing them out one by one down the page and ruling a line between them. Sometimes it is helpful, where there are alternatives, to set out the places on the page roughly according to their relative geographical position, as William Bowles did in 1578 (see figure 2.7). As a matter of principle, though, such embellishments make no significant difference to what is already a topological map of a route. Why, then, anybody should have wanted to spend the time and energy deployed by Matthew Paris in the mid-thirteenth century, and John Ogilby in the later seventeenth century, in transforming words into images and a list into a work of art is at first sight puzzling. Tellingly, nobody else in the intervening four centuries seems to have attempted a parallel presentation. If reasons other than practical travel can be found for the portrayal of a route in graphic form, however, all becomes clear.

MATTHEW PARIS, C. 1200–1259

Matthew Paris entered the Benedictine monastery of St Albans, Hertfordshire in 1217 and remained there until his death in 1259. By all accounts, he was a remarkable scholar, artistically gifted, voracious for gossip, and indefatigably stimulated by the events of his day, which he dutifully recorded in the monastery's *Chronica maiora.* Paris's visual-mindedness was not all that unusual, for many medieval writers on scientific, exegetic, and philosophical matter used diagrams, maps, and plans as visual aids.[122] As we see it today, though, his exceptional gifts were expressed in both the range of his cartographic vision and the quality of his output. He has left us fifteen maps and diagrams on seven subjects in the various volumes of the *Chronica maiora.* Amongst these is an itinerary for the journey from London via Rome to the Apulian ports of Barletta, Trani, and Brindisi. The itinerary is extant in four versions, all in Paris's own hand but each sufficiently different to make it difficult to describe the itinerary as a single map.[123]

As befits an itinerary, Paris presented the route in columns (two, three, or four to the page, depending on the version) to be read from the bottom upwards. Towns, some located on major rivers, and important features such as the English Channel and the Alps are portrayed in geographical order in vignettes that are varied and detailed, some with identifiable features. Each section between the towns indicates one or sometimes one and a half day's travelling and are labelled in French as *j[o]urnee* or *j[o]urnee e demie*. The words are written between parallel lines ruled between the relevant places to form a visual link. On one manuscript (CCCC MS 26) the route is identified between London and Dover, *Le chemin a Roucestre* (Rochester) and *le chemin a Canterbire* (Canterbury), elsewhere as *Le chemin vers l'Orient*. The itinerary format is adhered to as far as Rome, beyond which Apulia and the rest of southern Italy is shown in map form. On two manuscripts (CCCC, MS 26 and BL, MS Royal 14, C VII) small flaps glued to the edge of the page open up to reveal a plan of Rome and an outline map of Sicily (figure 2.14). Where the itinerary was intended to end has been much debated, not least because the question is relevant to the interpretation of its function. Did Paris intend the itinerary to terminate in southern Italy, or does the fact that the next map in the codex shows the Holy Land, with the port of Acre and the city of Jerusalem exaggerated, indicate that he was linking London with Jerusalem? On each version of the map, the itinerary is treated differently when it reaches the Adriatic coast in Apulia. On one manuscript (Royal MS. 14.c. VII) a note at Rome explains that "many itineraries end here" (*terminus itineris multis*), an unsurprising observation.[124] However, on two manuscripts (CCCC MS 26), Paris made a clear connection between the *le chemin* in Italy and the Palestinian port of Acre. A note at the top of the first column on folio 4 of the Royal manuscript is linked to Otranto, one of the Apulian ports used by pilgrims heading for the Holy Land before Venice became popular, identifying it as "the way to Acre via Apulia." On the Cambridge manuscript (CCCC, MS 26) a special sign ⊖ at Otranto is similarly identified in two texts on the map.[125]

Various explanations of Paris's itinerary map have been offered. Still being recycled is the old explanation that this is a "pilgrim map," by which was meant a practical guide to the road to Rome and beyond. Other researchers focused more critically on the political context articulated in the chronicle. Susanne Lewis points to the coincidence of what may have been the first version of the itinerary (as early as the 1240s) with the marked increase in the number of English people travelling as litigants to the papal court in Rome after the fourth Lateran Council, held in 1215.[126] Like Konrad Miller at the end of the nineteenth century, Lewis and Richard Vaughan

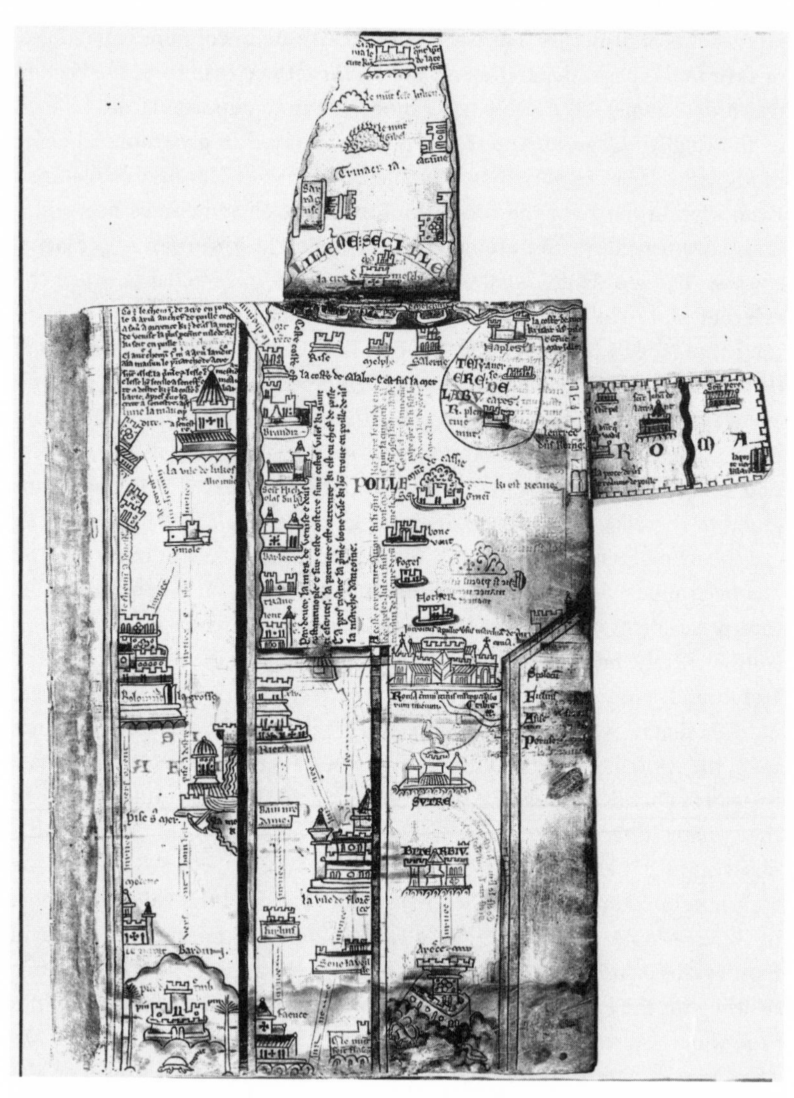

FIGURE 2.14 A map for a spiritual journey. The last page of the itinerary proper in Matthew Paris's *Chronica maiora.* Starting at the bottom left corner, the itinerary shows the route from Pontrémoli, north Italy (with its tortoise and pine trees) to Rome, after which the route opens out into a map of Apulia with (in the second column) the ports of Barletta, Trani, Brindisi, and Otranto—the latter marked as the port of Acre in Palestine by a special sign ⊖. The texts give details about the sea routes eastward, to Constantinople as well as Acre, and to the islands of Cyprus and Crete. The fold-out flaps show Rome and Sicily. B.L., Royal MS 14. C. VII. By permission of The British Library.

both note that since the last version of the itinerary appears to have been drafted in 1253–54 (judging from its reference to the pope's offer of the crown of Sicily to Richard of Cornwall), the map reflected a heightening of English diplomatic interest in southern Italy.[127] In support of a political reading, Lewis points out that in the part of the chronicle compiled by Paris, the focus is on events that can all be connected with places lying directly on the London-Rome axis or that Paris has added to the map irrespective of their relevance to the route depicted.[128] Besides the crusades, she points out, and from the time of Henry II onwards, England had interests in the Norman court of the Two Sicilies and in its Hohenstauffen successors. It would not be surprising, then, if Matthew Paris chose to illustrate the *Chronica maiora* with a record of contemporary history mapped out as an itinerary. Of one thing we may be clear, however: Matthew Paris's elaborate, enticingly detailed, and beautiful itineraries were not produced as practical travel aids. Not only did all four versions remain at St Albans, but Paris knew very well what route planning entailed. He had journeyed to London, 34 miles (54 km) away, possibly on more than one occasion, and he had been to Norway. An elaborate work of art is not needed for the journey or even as a library record of a specific route.

In a persuasive new look at the maps, Daniel Connolly turns instead to reflect on the more immediate context in which Paris lived—the monastery. Connolly points out that what was seen within Christian culture of thirteenth-century Europe as of the highest priority was not necessarily physical, but rather spiritual, pilgrimage to the Holy Land, the pilgrimage of the theologian, the contemplative, and the pious. Relatively few people in the Middle Ages were able to effect what many desired and the Church urged—a pilgrimage to Palestine. As Felix Fabri's account of the difficulty he had in persuading his superior to let him go even the first time confirms, monks were not usually permitted by their order or their superiors to travel whenever and wherever they might wish.[129] Most pilgrimages would have to be conducted "in the heart," not with the feet.[130] By turning the pages of the itinerary, Connolly notes, the viewer was already being led towards Jerusalem.[131] As he progressed, unfolding flaps appended by Paris to the pages depicting Italy, revealing first the city of Rome and then the island of Sicily, additional dimensions were added to the geographical space represented on the flat page. Where *mappaemundi* were available for study, contemplation of the earthly city of Jerusalem on the map would lead to focused meditation on the Heavenly Jerusalem. But there was no *mappamundi* at St Albans (as far as is known) other than Paris's own copy of one seen elsewhere, and Paris's cartographical surrogate of the pilgrimage route would have had to serve instead.

John Ogilby's circumstances could hardly be more different from those of Matthew Paris. Ogilby was an entrepreneur, full of ideas for money-making ventures, ready to exploit contacts, and doggedly starting afresh after each setback: "a typical Restoration figure of zestful and ebullient versatility."[132] Eventually he took up publishing. Lacking Paris's intellectual status and scholarly disposition, most of what Ogilby did was derivative; republishing the classics in fine editions, for instance. Even some of the most innovative aspects of his strip maps may have been the ideas of others, not his own.[133]

Today, Ogilby is celebrated for a single volume of text and maps entitled *Britannia* (1675). *Britannia* was a far cry from what Ogilby originally had in mind, namely an encyclopaedic volume of Britain in a five-volumed atlas of the world to be printed under the general title of *An English Atlas.*[134] The other volumes were to cover Africa, America, Asia, and Europe. Not one book contains more than a handful of maps. As Donald Hodson describes these geographical works, they "comprise translations of accounts of the travels of various figures, mainly ambassadors and Catholic missionaries, together with reports of strange customs and bizarre wonders with a little geography; as such they show little that is original."[135] The first three volumes appeared together with three additional Asian volumes—China (in two volumes, 1669, 1671), Japan (1670), Africa (1670), America (1671), and Asia (1673)—but with nothing on Britain. Ideas for the atlas of Britain changed radically, partly through delays but chiefly as Ogilby came into contact with Robert Hooke and others of the Royal Society. These "men of science," Hodson suggests, saw in Ogilby a talent for raising finance and experience in publishing books of a high technical quality, with the independence and freedom that ownership of a private press endows. They would provide the scientific expertise.[136] Thus by the middle of 1671, a quite different project from the original *English Atlas* had emerged. This was to be a three-volume original county-by-county survey of the whole country as a worthy successor to the by now hopelessly out of date Saxton atlas that would also incorporate maps and written descriptions. In 1671 a royal warrant appointed Ogilby as Royal Cosmographer to enable him to obtain the necessary information.[137] Interestingly, the warrant specified that Ogilby should "have help in the affixing of sufficient Marks for the better Direction of Travellers, and Ascertaining the Distances from Stage to Stage in Our said Kingdom," and by the end of 1672 it was reported that Ogilby "had made some beginning in ye measuring of Roads."[138] In the event, however, even this venture would not materialise. What was published the year before Ogilby's death was a single volume of road maps and text.

Britannia contains 100 pages of maps and 200 pages of written description in a leather binding and weighing over fifteen pounds (7.94 kg).[139] It portrays a single category of roads (designated post roads) in England and Wales. It is conspicuously pictorial in presentation (figure 2.15). Ogilby set out each road in a series of columns, to be read, like Matthew Paris's, from the bottom lefthand corner upwards, but Ogilby's conception applied to the page as a whole. Each column—there are seven or eight to a page—is disguised as part of a scroll that unrolls across the page as the road is "read" from left to right across the double-page spread. Where the idea of presenting the post roads as vertical columns or strips originated is a moot point. I have suggested elsewhere that it would have been perfectly possible for somebody in Ogilby's circle to have known about Matthew Paris's itinerary, having seen the copy of the *Chronica maiora* that was by then in Robert Cotton's collection in London.[140] The Cotton collection was in process of being catalogued and a number of copies of the catalogue soon circulated openly to Fellows of the Royal Society and other individuals. There is no record of Ogilby ever having set foot in the Cotton library, but it is scarcely conceivable that word about Paris's maps failed to reach him. It may even have been, as Hodson suggests, that he was simply instructed to lay out the road on the pages of *Britannia* vertically rather than horizontally by Robert Hooke or another of their scientist acquaintances.

The information Ogilby showed for each road came in part from questionnaires originally sent out for the three-volume project and in part from the survey of post roads he organized.[141] To draw the maps to a consistent scale throughout the volume, local linear measurements were converted to statute miles and furlong dots marked along the middle of roads to allow the computation of distance.[142] The roads are presented as more or less straight lines, with changes of direction indicated by compass roses, and are delineated in one of two ways. Solid lines indicate that the road was confined by hedges and walls; broken lines that it was surrounded by open-field cultivation or unenclosed moors and woods. Landmarks seen from the road, such as church spires or towers (whose bells would have been audible from afar with a fair wind) and large private houses are indicated. Other roadway details include bridges, fords and ferries, the direction of gradients, the destination of crossroads, and the location of blacksmiths. Ogilby rarely showed any alternative to the designated post road. One exception is found on the London to Chichester and Arundle road, where the Roman road continues in a straight line but the post road makes a large detour avoiding a valley bottom. Other features noted on the maps and described in the text, such as mineral workings and mills, served not only

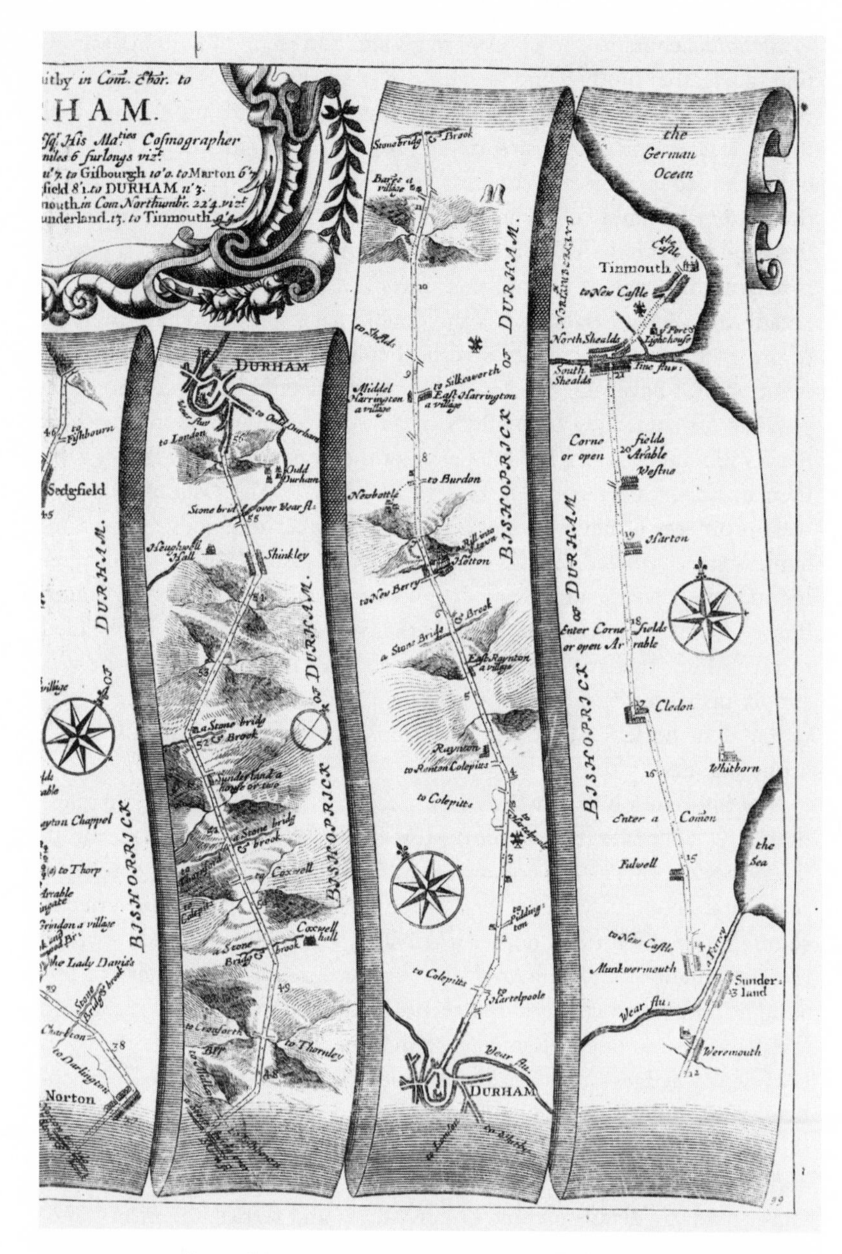

FIGURE 2.15 Road maps for the armchair geographer. John Ogilby's large and expensive folio volume *Britannia* (1675) was not a simple road book. An individual map sheet, if any were made available, could be taken in a carriage to study on the road, but the book as a whole was a portrait of a well-endowed and prosperous nation, serving the ends of nationalism rather than wayfinding. Detail from the road from London to Durham. Note the references to coal pits, cornfields, and open arable land. B.L., Maps C.6.d.8. By permission of The British Library.

to locate the traveler but also, Garrett Sullivan suggests, to emphasise the reassuring wealth of the nation.[143]

The maps in John Ogilby's *Britannia* have been hailed as "one of the half-a-dozen or so landmarks in British map-making" and "a key publication in understanding the developments of British cartography," but it is doubtful that they were properly understood when Brian Harley made this assessment.[144] Only now are they coming to be seen as much a functional enigma as Matthew Paris's graphic itinerary. Previously, assuming they served for wayfinding, Katherine Van Eerde asserted that "many owners did not hesitate to cut out strip maps of their local areas or of the places to which they were journeying, and presumably carried these strips with them."[145] As it happens, there seems to be scant evidence of maps having been cut out and used in this way, nor is the argument persuasive that owners of the volume would have encouraged such defacement. *Britannia* was far too expensive a book to have been used as a quarry for scissors-happy travelers, and there are no known examples of dissected maps having been used in this way, although separately printed pages of course could have been carried about if so desired.[146]

If Ogilby's maps of the road were not maps for the road, their explanation has to be sought in different, nonutilitarian, directions. As a starting point, both Sullivan and Donald Hodson emphasize the need to reposition *Britannia* in the context of Ogilby's geographical publications and to reassess it as a publishing venture.[147] Both Sullivan and Hodson pay particular attention to the "atypicalness" of *Britannia* when seen against Ogilby's other publications, notably his "English Atlas" project. Sullivan in particular underlines the contemporary popularity of the literary atlas genre. In his comment that "to read an atlas *is* to travel and traveling can take the form of reading an atlas," we may find an echo of Connolly's interpretation of Matthew Paris's itinerary.[148] To be sure, the circumstances in the second half of the seventeenth century were very different from those of Paris's day. Religious devotion was not in this case the motivating factor so much as nationalism and the politics of making a major map publishing venture pay. Sullivan points to the prospectuses issued by Ogilby for the projected three-volumed atlas and to the dedicatory letter addressed to King Charles II published at the start of *Britannia*. These express Ogliby's aim of producing "a True Prospect of This Your Flourishing Kingdom . . . and the Considerable Augmentation of It's Extent."[149] The key word is "augmentation," which, according to Sullivan, both was intended to be and would have been understood as referring not only to the extra acreage that the improved accuracy of the new surveys added to the dimensions of the kingdom, but to the ideas promoted by *Britannia*'s text and maps: that Britain was economically wealthy, its people prosperous, and its national security under no threat from any direction.

Ogilby's near contemporary, Thomas Gardner, appears to have understood better than most modern historians what *Britannia* was about when he classed it as "an Entertainment for the "Traveller *within* Doors" (my italics).[150] If the volume proved unwieldy as a wayfinding text, Sullivan concludes, this was not because Ogilby failed to achieve his aims but precisely because he was successful. Like Matthew Paris's improbably (for a wayfinding aid) artistic itinerary, Ogilby's *Britannia* was one of a series of texts "designed to gain him the approbation and support of monarchs, aristocrats and wealthy merchants" and its lavish form was "absolutely central to that task."[151] In the long run, while Ogilby's splendid volumes remain more or less intact on library shelves, as visual aids his strip map format came to be adopted in a new kind of road book designed for carrying on a journey. Although the pocketbook editions of Ogilby's *Britannia* may have found favor with the "carriage trade," they were much less likely to appeal to the professional man of the road, the trader.

Britannia was published in the autumn of 1675. Two further impressions surfaced within the first twelve months. By the middle of January 1676 there was an edition of the maps without the written descriptions, "Mr Ogilby's *Itinerarium Angliae,* or Book of Roads" and by April, individual maps were being sold separately.[152] It was not until 1719 that pocket-sized editions started to flood the market. First was Thomas Gardner's version and, in the same year, John Senex's *An Actual Survey of all the Principal Roads of England and Wales.* In 1720 came John Owen and Emanuel Bowen's *Britannia Depicta or Ogilby Improv'ed,* and throughout the eighteenth century, a whole train of further editions of each and their derivatives.[153] While both Gardner's and Senex's editions respected the clarity of Ogilby's maps, Owen and Bowen packed so much textual information around the maps as to render them virtually illegible (figure 2.16).

Distance Tables

In the planning of a journey, the distance between places can be a matter of great concern. Berlaimont's phrase book underlines the urgency of being somewhere safe by nightfall, and contemporary topographical maps, especially on the European continent, commonly distinguished between walled towns and villages and unwalled settlements, into which access could be more easily gained after dark. Calculating distances, and adding up the mileages between pairs of places, were arithmetical exercises at a time when education in mathematics had scarcely been introduced into

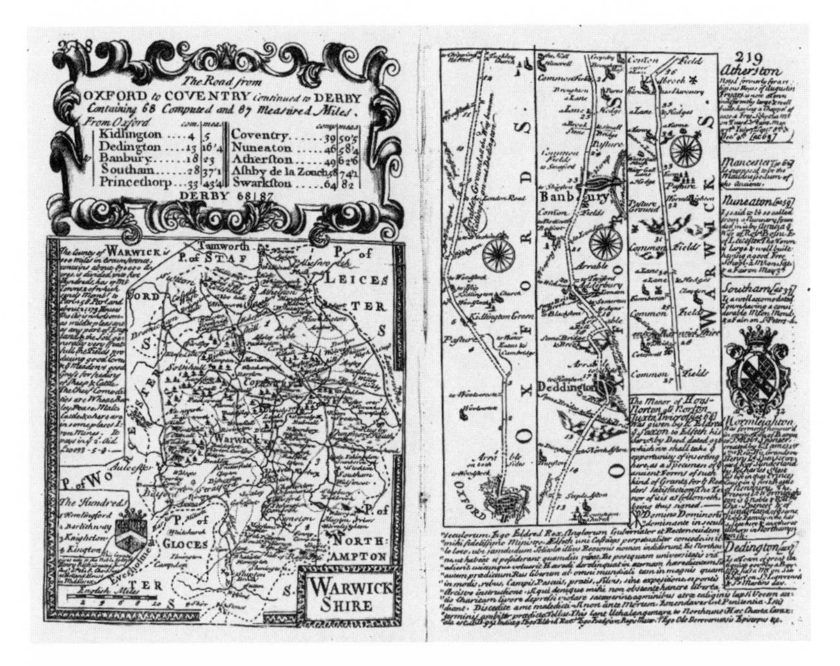

FIGURE 2.16 In Emanuel Bowen and John Owen's portable version of John Ogilby's road maps, *Britannia Depicta or Ogilby Improved* (1720), Ogilby's majestic presentation of the post-roads of England and Wales as a national icon has been reduced into a tourist guide. The text below the map gives an account of the origin of the manor of Hogsnorton, Oxfordshire, with a quotation in Latin from a deed of 951. B.L., Maps C. 27. a.13. By permission of The British Library.

universities, let alone schools.[154] Ready-made computations of the distance between major places would have been a welcome travel-planning tool.

When John Norden published a modest pocket-sized volume of *England: An Intended Guyde, for English Travailers, shewing . . . how far one Citie and many Shire-Townes . . . are from other* in London in 1625, he called it his "new invention." Like much else, however, the germ of Norden's idea was less than new. Triangular distance tables had been published over half a century earlier in Germany, and Norden's friend William Smith had lived in Nuremberg and introduced a number of continental practices to his own and to Norden's county maps.[155] Alternatively, it is not inconceivable, as Hodson suggests, that Norden took the idea from the triangular tables presented in the mathematician Leonard Digges's *Tectonicon* (1556) to demonstrate the calculation of area for land surveying.[156] Peter Meurer, however, points to Matthäus Nefe's textbook *Arithmetica*, published in Breslau in 1565, and a sheet with distances in Silesia printed in the same year, as the probable source of the idea to set out precalculated distances in a triangular table.[157]

Oxford shire with some confining Townes.	Oxford.	Woodstocke.	Chip.Norton.	Burford.	Witnye.	Banbery.	Tame.	Watlington.	Henlye.	Ew-elme.	Ricote.	Hocknorton.	Bicester.	Islip.	Dorchester.	Stoken-Church.	White-Church.	Lechelade.Glouc.	Faringdon. Bar.	Abbingdon. Bar.	Wallingford. Bar.	Reding. Bar.	Cleydon.	Dedington.	Cottesford.	VVendlebury.
Hieworth.Wilt.	17	17	17	9	11	27	27	24	29	24	24	22	25	19	19	27	23	4	4	15	20	28	32	24	28	23
Wendlebury.	7	7	13	17	12	12	10	14	22	15	11	12	3	4	14	16	22	21	19	12	16	26	16	8	6	
Cottesford.	13	11	15	21	17	9	13	19	25	20	14	12	4	9	19	19	26	26	24	18	21	30	12	7		
Dedington.	13	8	9	16	13	5	18	22	29	23	19	6	9	9	20	23	28	23	21	18	23	33	9			
Cleydon.	22	16	15	23	22	5	25	30	37	32	27	10	16	18	29	32	37	29	29	27	31	42				
Reding.Bark.	21	28	34	31	27	37	18	12	4	10	17	36	26	24	13	13	5	30	26	17	11					
Wallingford.Ba.	10	17	24	21	16	27	10	5	9	3	9	25	17	14	3	9	6	22	17	7						
Abbington. Bar.	5	11	17	14	10	22	13	10	15	9	12	20	15	9	5	14	12	15	11							
Faringdon.Bar.	13	14	15	8	8	25	23	20	25	20	23	20	22	16	15	25	21	5								
Lechelade.Glou	16	15	15	6	9	25	27	25	30	24	26	20	24	18	20	29	25									
White-Church.	16	22	29	26	22	33	15	8	6	6	13	32	22	20	9	11										
Stoken-Church.	14	20	27	27	22	28	6	5	8	6	5	27	15	15	10											
Dorchester.	7	14	21	19	14	25	10	6	11	5	9	23	15	11												
Islip.	4	5	12	14	9	14	11	13	20	14	11	12	6													
Bicester.	10	10	16	20	15	13	9	14	16	21	10	14														
Hocknorton.	16	9	5	13	13	6	22	25	33	26	23															
Ricote.	10	16	23	24	18	23	1	5	12	7																
Ew-elme.	11	18	25	23	19	28	9	3	7																	
Henlye.	18	24	32	30	25	34	13	8																		
Watlington.	10	7	24	23	18	26	7																			
Tame.	10	15	23	25	19	22																				
Banbery.	17	12	10	19	17																					
VVitnye.	8	6	8	5																						
Burford.	13	6	8																							
Chip Norton.	14	7																								
VVoodstocke.	7																									

The vse of this Table.

THe Townes or places betweene which you desire to know, the distance you may finde in the names of the Townes in the vper part and in the side, and bring them into a square as the lines will guide you; and in the square you shall finde the figures which declare the distance of the miles.

And if you finde any place in the side which will not extend to make a square with that aboue, then seeke that aboue which will not extend to make a square, and see that in the vpper, and the other in the side, and it will showe you the distance. It is familiar and easie.

Beare with defectes the vse is necessarie.

Inuented by IOHN NORDEN.

FIGURE 2.17 Distance tables in John Norden, *An Intended Guyde for English Travailers* (1625). Norden's "new invention," triangular tables providing ready-computed distances, had in fact long been used by German mapmakers. B.L., 577.h.27. By permission of The British Library.

Norden's book comprises forty pages of tables, one for each English county and one for Wales as a whole. Each table gives the distances between twenty-six selected places in each county cross-reference (figure 2.17).[158] The idea was evidently a commercial success, at least in England where Norden's book was reissued in various editions over the next fifty years, until outdated by Ogilby's road surveys. Nine years after Norden's death, the first new edition of his *Guyde* was published by Matthew Simmons, retitled *Direction for the English Travailler* (London, 1635). Simmons replaced the instructions that Norden had repeated on each page with a thumb-nail sized outline map of the county. Clearly indicating the four cardinal directions, the little map shows some of the main places listed in the matrix, allowing the user to see in which direction the next place on his itinerary lay. Further enhancements followed. In 1643 Thomas Jenner replaced Simmons's tiny sketch maps with more complex versions, adding compass dividers and a scale bar, and inserting a general map of England into the volume together with a matrix giving distances between all the shire towns. Jenner also explained how the three levels of the distance tables now operated. First, the reader was to refer to the triangular matrix in order to find the distance between two desired places. Then, "to know whether you are to travell, East, West, North or South," the reader was to look in the general map at the beginning of the book, and refer finally to the little map on the tables page to discover the names of any intervening places.

In one version or another, John Norden's distance-table book continued to serve travelers until *Mr Ogilby's Tables of his Measured Roads* appeared on the market in 1676.[159] According to Simmons, such books were a vital adjunct for the traveler, "a messenger which will go with thee, and direct thee from place to place . . . I dare say, had it beene found out the last great Snow, it had saved many a man much money, nay some their lives."[160] The distance tables were not itineraries, however, and the information presented was not geographically sequential but discrete, a scatter of isolated points. Moreover, the limited number of places in each county for which distances were supplied limited the user, in theory at least, to consulting the book only for travel involving places selected by the compiler of the tables. Travelers, in the minds of most seventeenth-century publishers, still largely followed the main national arteries and post roads, and their requirements were supposed to conform to the long-established pattern of major routes in England and Wales.

Maps for Travelers

Maps came to be used as wayfinding aids largely in the nineteenth century, when the depiction of roads on topographical maps could be relied on with some confidence and when all categories of roads were included. Until then, maps were normally used only in the course of planning a journey and for compiling an itinerary. Some of these maps were general topographical maps; others were specialist maps, showing only a single category of information—either routes (with distances) or roads. Such thematic maps were primarily useful to those responsible for the organisation of professional travel and travelers.

THEMATIC MAPS OF ROUTES AND ROADS

When, in the preface to *Britannia,* John Ogilby acknowledged his predecessors, it was the map once owned by Conrad Peutinger that he referred to, not Matthew Paris's itinerary. Yet while Ogilby's maps closely resemble Paris's, none of the maps he was responsible for has any connection, visual or otherwise, with the Roman map. The Peutinger map, so often held up as a typical Roman "map" (implying topographical map) or vaunted as the "earliest road map," must be one of the most misunderstood maps of all time. It is certainly an interesting map, not least for its remarkable dimensions: 6.82 metres in length but only 34 centimetres in height (about 20 ft × 14 in).[161] However, it shows not roads but a motley collection of routes, derived from a large number of itineraries of various dates, which interlink to form a network covering most of the Roman Empire, from Britain to Persia, and Germany to North Africa. The prototype is thought to have been made between 335 and 366 AD.[162] It is also generally asserted that it served as an official, (i.e., governmental) guide to the highways of the empire, whose use was restricted to certain classes of travelers, notably those on public business. In fact, the Peutinger map as we know it is unlikely to have been used in any such ways. It may not even have served any utilitarian function. It was far too large to carry about for a local journey; unlike the medieval Gough map, there is no hint that it existed in more than one exemplar, for use by different officials in different places; and, perhaps most crucially, many of the itineraries from which it was composed would have been several centuries out of date. Despite Benet Salway's and others' recent critical examination of the map, its context, and the use of itineraries in the Roman Empire, the origins of the Peutinger map remains a something of an enigma, although we may accept recent conclusions that it did not "belong to a venerable or a widely disseminated genre," that it was "the product of a private initiative rather than a state-sponsored project," and that it was from the start "a historical map" serving a primarily decorative or ornamental function."[163]

Possibly because itineraries rather than a geographical survey were the starting point of the Peutinger map's compilation, the outlines of Europe, North Africa, and the Middle East have been distorted to make the area shown fit the dimensions of a strip rather than being allowed to expand onto a vast rectangular wall map. The medieval Gough map, in contrast, is a compact map (115 × 55 cm, 45½ × 22 in) with a network of routes plotted on to a remarkably realistic outline of Britain. The compiler would have used itineraries to help locate places, but the structure of the map would have come from consulting other maps, possibly some of Roman origin or even an imported chart, and from such astronomically-determined locational data as was then available.[164]

Routes are clearly indicated on the Gough map by lines in red ink. Distances are noted in Roman numerals. Towns and villages are shown pictorially and are named.[165] Rivers are shown but there is no relief or vegetation. The routes do not interlock into a single network covering the whole country equally. They form one major integrated network and two isolated routes, one following the west coast of Wales and the other linking Southampton and Canterbury,[166] and they are absent in the north-east, apart from the coastal route from Beverly to Guisborough. No route to Berwick is shown, although the London to Carlisle route stands out clearly, and the concentration of routes in the York-Lincoln region led to the suggestion that while other versions of the map existed for use in other places, the extant copy could have been use in or near York or Lincoln. More convincingly than the Peutinger map, the Gough map gives the impression of recording operational networks, but whether the itineraries were for current use or whether, like the Peutinger map, they represent a historical collection is no easier to ascertain, and it too may well one day be shown to have been something of a propaganda or a vanity compilation.

It would certainly seem that both the Peutinger and the Gough maps were the outcome, like Matthew Paris's and John Ogilby's graphic itineraries, of unusual initiatives. For three centuries there was, as far as we know, nothing comparable to either the Peutinger or the Gough map. The vacuum is telling: there cannot have been much, if any at all, demand for so specialist a map when itineraries and experience provided the necessary information. By the time John Adams's distance line map was published in 1677, however, cultural, social, and economic conditions were quite different and the time was clearly ripe for the venture. Some eighteen versions or imitations of Adams's map were published in less than a century. Adams was a Shropshire barrister whose achievement lay in showing not so much practicable routes—which not all his lines indicate—as in producing a twelve-sheet map of the whole of England and Wales from which the distance between every place could be worked out (figure 2.18). Adams's map also allowed an itinerary to be compiled simply by picking out the relevant places, adding up the distances given by each line, and noting from the slant of the line the direction to be taken on leaving each place.[167] No roads are marked on Adams's map, and until friends pressed him to add them, there were no rivers either. The title of Adams's map—*Angliae Totius Tabula cum Distantiis notioribus in Itinerantium usum accommodata* [A map of the whole of England with the more important distances arranged for the use of travelers]—clearly identifies its route planning function.

Distance lines like those on Adams's map were not a seventeenth-century innovation, as we see from the Gough map and from a number of printed sixteenth-century maps, but it is always tempting to wonder again what led the author to produce the map in that particular format at that

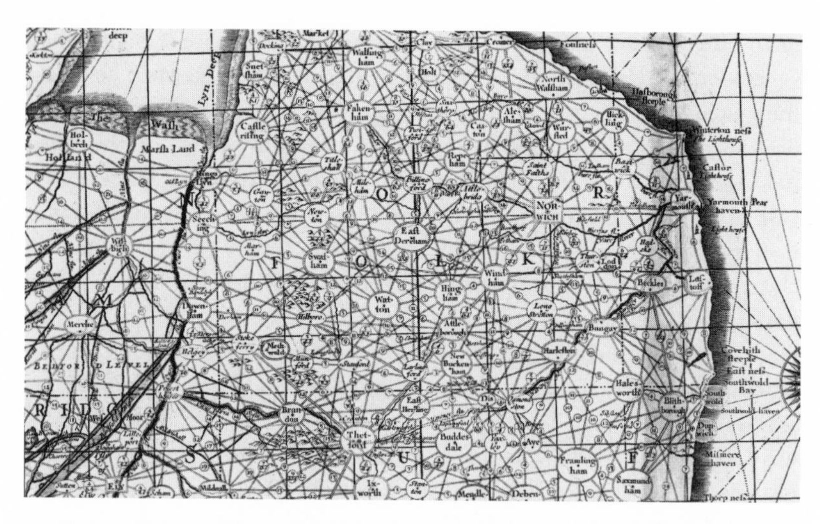

FIGURE 2.18 Maps with route networks. Detail from John Adams, *The whole of England with the more important distances arranged for the use of travellers* (1677), showing East Anglia. The urban hierarchy is indicated by the size and form of the frame around each place name. Rivers, county boundaries, and some major areas of heath or forest and higher land are discreetly indicated but the emphasis is on the distance lines and accompanying mileages. B.L., Maps 24. e. 22. By permission of The British Library.

particular time.[168] It is well known that John Adams was galvanized into thinking about the map from a friend's need to market fish in as many places as he could reach within a specified period.[169] It must have helped, too, that so much new information about road distances was becoming available.[170] Norden's distance tables were still in print—the latest (and last) edition, John Garrett's, was published in the same year as Adams's map—and, most significantly perhaps, Ogilby's *Tables of his Measured Roads* had appeared in 1676. Road travel in general had been on the increase since the end of the Civil War (1642–46), and a novel and successful travel aid was bound to be commercially viable. Adams's map was indeed a success. A cheaper and more convenient version in two-sheets was produced in 1679, and the map was widely imitated.[171] In the absence of subscriber's lists, it is hard to find out who the purchasers, not all of whom would necessarily have been engaged in regular travel, might have been. Some, it may be surmised, would have been happy simply to display the map at home, while others, such as roadside innkeepers, would have found it useful to have for clients to refer to before starting on the next stage of their journey.

Maps of road networks date mainly from the second half of the seventeenth century in England, somewhat later than on the continent. In France, Nicolas Sanson's *Carte geographique des postes qui traversent la France* (1632) was already being included in editions of Melchior Tavernier's atlas of France from 1637.[172] The word *géographique* in Sanson's title should act as a warning, though: Sanson's map is not strictly speaking a thematic map, despite its title. It shows the post road network clearly enough, with each road represented by a pecked line, but there are so many additional places between the roads, together with a prominent river network, that the visual impact of the road system is muted. Even more adventitious detail was added to later editions, on which roads are thus virtually lost.[173]

In England, a systematic classification of the road system was not initiated until 1633, the year following the first publication of Sanson's map, when Sir Brian Tuke was appointed as the first national postmaster.[174] Even so, Richard Carr's comprehensive network map, *A Description of al[l] the postroads in England*, was not published until 1668. Carr's single-sheet map was produced in Holland from material supplied by James Hicks, Carr's superior in the post office.[175] Like Sanson's, Carr's map is confused by extraneous matter: pictorial town signs, rivers, county boundaries, a mass of irrelevant place-names (especially around the coasts), and scrolls and ships in the seas (not to mention the outsize representation of the royal coat of arms) clutter the sheet such that the roads themselves, marked by parallel dotted lines, are scarcely noticeable. A few years later, Ogilby was more successful with a separate sheet map based on data from his post road surveys. His *New Map of the Kingdom of England & Dominion of Wales. Whereon are Projected all ye Principal Roads Actually Measured & Delineated*, published with *Britannia* in 1675, contains little other than the road network. Even more successful as a model of thematic clarity is George Willdey's *The Roads of England According to Mr Ogilby's Survey* (1713).[176] Willdey's map is dominated by the subject matter, the post roads themselves (figures 2.19 and 2.20). These are represented as broad double lines, beside which the names of places along the road are written and within which the distances between each pair of places are given. Major nodes ("Cityes") are marked by a rectangle, secondary nodes ("Shire towns" and "Market towns") by an oval. Single lines give cross-country distances between the main centres. No rivers are marked and no names clutter up the coasts. The fact that the coastal outline has been distorted so that England and Wales fit the circular format is of no consequence, since the topological relationship of the roads and the places along them is all that matters.

TOPOGRAPHICAL MAPS FOR ROUTE PLANNING

Unlike the thematic presentation of routes and roads, where only a single category of information is presented, topographical maps are by definition packed with heterogeneous information. Curiously, though, one of the

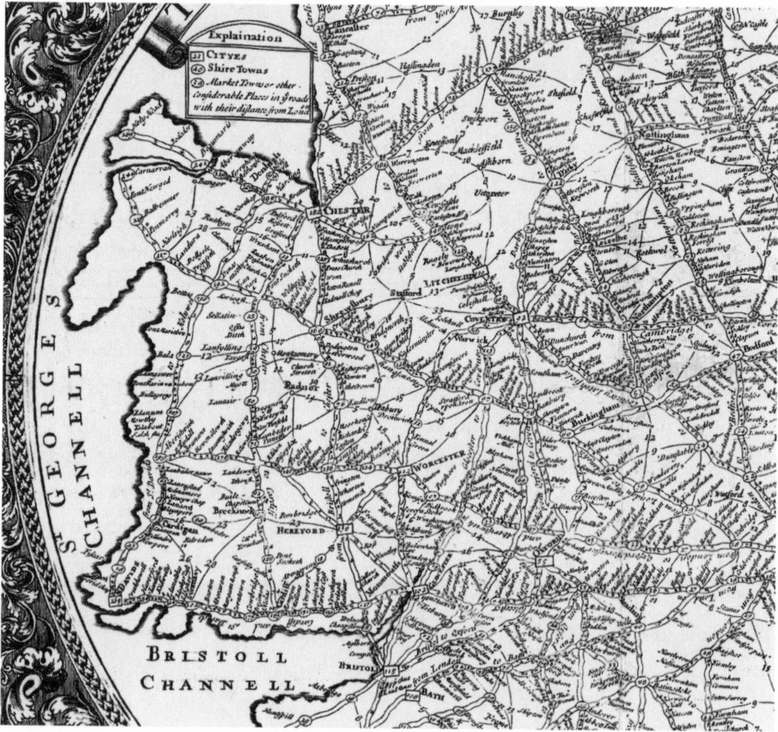

FIGURE 2.19 Roads and cross-roads are indicated on George Willdey's map of *The Roads of England According to Mr Ogilby's Survey* (1713). The detail shows the key to the status of major centres and the roads in Wales. B.L., Maps K. Top. V. 34. By permission of The British Library.

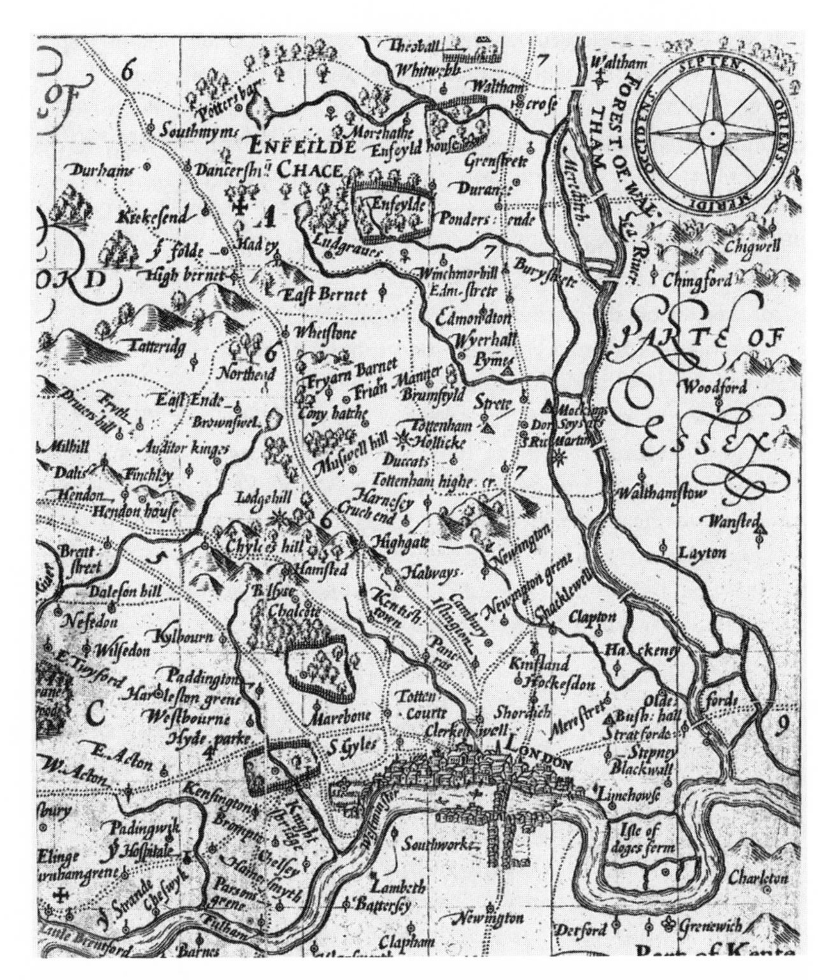

FIGURE 2.20 A topographical map with roads. Detail from John Norden's map of the county of Middlesex (1593), the first printed English map to include roads as well as relief, rivers, vegetation, six categories of settlement, castles, monasteries, hospitals, battle sites, "decayed places," forest lodges, and mills. B.L., Maps C.7.b.23. By permission of The British Library.

few landscape features absent from the vast majority of early printed topographical maps is the representation of routes and roads. A section of a particular road was sometimes singled out for its antiquarian interest, like the Roman *Via Emilia,* which was shown on sixteenth-century Italian printed maps of northern Italy and was still one of the major thoroughfares of Europe, carrying traffic between Venice and the Adriatic and countries north of the Alps. In England, even ancient roads are absent from Christopher Saxton's county maps, although many bridges are marked. Maps, as noted at the beginning of this essay, were simply not traditional wayfinding aids and adding a hierarchical network of tracks and roads to topographical maps would have cluttered the image. More importantly, horse riders and pedestrians hardly needed specially constructed roads and were both able and allowed to take cross-country routes of their own choosing across open fields, commons, and moors alike. They were expected to make their selection from the intricate maze of local paths and regional tracks that, like agricultural land itself, were so commonplace as to defy representation on topographical maps. Furthermore, inaccuracies in the depiction of roads were liable to prove troublesome.

A topographical map could be useful, however, to a ruler planning a major military operation by indicating important features such as extensive marsh or forest, wide rivers, and mountain barriers. In 1495 Charles VIII of France, preparing to invade Italy, asked Jacques Signot to reconnoitre suitable passes over the Alps. The result was a map of Italy with all the Alpine passes clearly marked—but no roads.[177] Not until the road network was included in full and reliably on maps, and there were people needing to use a map to find their way about the countryside, were topographical maps used for wayfinding, rather than just planning a route and compiling an itinerary.

The English public were late map users, but on the continent, by the turn into the sixteenth century, European mapmakers were marketing printed maps containing instructions for route-planning and for the measurement of distance. Erhard Etzlaub's map *Das ist der Rom Weg* [This is the road to Rome] was published in time for the Holy Year of 1500, when large numbers of pilgrims expected to make the journey to Rome. In part, Etzlaub's map embodied his attempts to produce a conformal projection, but, as a mathematician Etzlaub was also interested in the problem of how to represent both distance and direction accurately on a flat surface to meet the demands of a functional travel and road map.[178] Places are marked on his map by red circles and routes by a line of regularly spaced dots. A text at the bottom explains how distances were to be measured: a pair of dividers was opened until the points rested on each of the towns in question, then the open dividers were moved to the scale at the bottom of the map and the number of dots between the two divider points counted.[179] The distance was computed by multiplying the total number of spaces by the length of the local mile. Etzlaub also represented the hours of daylight at different latitudes by graduated markings along the side of the map for the traveler's benefit.

Etzlaub's example was soon followed by other continental mapmakers.[180] In 1515, Georg Erlinger simply copied Etzlaub's map, but in 1528 Lazarus Secretarius published a map of Hungary (*Tabula Hungariae*).[181] Lazarus also explained the potential usefulness of the map, pointing out that the map was a chorography and an itinerary (*Totius Hungariae Chorographia, itinerariaq*[*ue*]). Like Etzlaub, Lazarus clearly saw his map as an aid to compiling an itinerary. No routes or roads are indicated on the map, only the essential dividers over the scale bar reminded users of one way use for the map.[182] Etzlaub's, Erlinger's, and Lazarus's maps demonstrated to their contemporaries how the new cartographical genre—the printed topographical map—could be used in route planning.

Printed topographical maps came to be useful not only to the casual traveler, but also to the military. Again, this was not for information about roads but to see where there might be a reliable bridge across a major river, where there was a town or village from which supplies might be obtained and in which troops might be billeted, and, of course, where places to be guarded or attacked lay. Finding a way to all these places was the task of scouts in the field, not map readers at home or in the commander's tent. The attraction of a topographical map for use in a military context appears to have lain in its binding rather than its content. The English Civil War created something of a surge in the demand for printed maps of the regions involved. Skelton has suggested that separate maps from John Speed's *Theatre of the Empire of Great Britaine* were printed without the usual text on the back and that "these were sometimes bound to form a "campaigning atlas."[183] Skelton had in mind a particular bundle of maps stitched in a stout vellum roll, "presumably for attachment to a saddle," that had come up for sale at Sotheby's London auction house.[184] Other publishers took advantage of the demand for maps in a different way. In 1644, when Thomas Jenner published a much-reduced but still large-scale copy of Christopher Saxton's wall map of England and Wales (originally published in 1583), he explained that the map would be "useful for all Commanders—or quartering of soldiers, and all sorts of persons that would be informed where the Armies be."[185] The entry in Jenner's 1688 sales catalogue for his map gave the edition its nickname, "The Quarter-Master's Maps."[186] The originality of Jenner's publication lay in its modestly sized format, its binding, and the way the map (cut up into sheets of unequal size) was bound between stiff covers to form a compact volume measuring some 8 ½ × 3 ¾ inches (22 × 10 cm) that Jenner advertised as "Portable for Every Mans Pocket." Thomas Jenner also had quartermasters in mind when, updating Norden's distance tables in 1635, he described the new edition as a "work very necessary For Traveilers, Quartermasters, Gatherers of Breefs, Strangers, Carriers, and Messengers with Letters, and all others who know the name of the place, but can neither tell where it is, nor how to go unto it."[187] In America, a century later, the demand for topographical maps in handy format during the War of Independence (1775–81) led the London printers Robert Sayer and John Bennet to produce an American Military Pocket Atlas (1776), which stressed in the "Advertisement" placed at the beginning of the small, plump volume that "Surveys and Topographical Charts being fit only for a Library, such Maps as an Officer may take with him into the Field have been much wanted."[188]

TOPOGRAPHICAL MAPS FOR WAYFINDING

Finally—and almost last in the history of maps for travel—we can turn to the use of topographical maps on the road. The topographical map is arguably the least important cartographical aid for the vast majority of travelers, today as earlier. To be any use as a wayfinding tool, the scale of a topographical map should bear some relationship to the speed at which the traveler moves. For pedestrians and horse riders, a map on a scale of about an inch to one mile (1:63,000) should give adequate information and be manageable in the number of sheets needed. It can be expected to show the traveler which turning to take, what gradients lie ahead, where to find a bridge, ferry or ford, which hamlets, villages and towns will be passed through or seen from the road, what landmarks to look out for, and where to expect an inn. The map should also say something about the type of road selected by the traveler, its status (trunk road, cross road, local track), whether it is metaled or unmetaled, and whether it is fenced or open, level with the surrounding country side or deeply sunken. With such a map in hand, the independent traveler is liberated from prescriptive published itineraries and the leisured traveler is free to wander at will, seeking places to visit and vantage points for scenic views. The map-using traveler is also presumably an educated and proficient map reader.

John Norden and Philip Symonson in the 1590s, not Christopher Saxton in the 1570s, were the first to show roads on English printed topographical maps (figure 2.20).[189] In the first years of the seventeenth century, William Smith followed Norden's example, adding some roads and major highways between market towns. Overall, though, the inclusion of roads on early topographical maps was irregular and unsystematic. No traveler could have trusted without verification the few depicted, and it is difficult to discover the basis for the mapmaker's selection of those shown and those omitted on most seventeenth-century topographical maps.[190] Even in 1717, one writer on maps was driven to warn that "A MAP-maker should not take every MAP that comes out, upon Trust, or conclude that the newest is still the best."[191] He then went on to remark that "marking out the Roads is indispensably requisite in maps." He was as much concerned with the usefulness of showing roads as a way of structuring the map (without which "all Things appear in Confusion, and the Eye wanders as in the Dark") and of using the accuracy in which they were depicted as a guide to plagiarism ("Copiers who know they are careless in locating places omit the roads"), as with the traveler for whom it was "a Pleasure, as well as Advantage, to trace the Way, from one Place to another."

The situation began to change in the second half of the seventeenth century. The accelerating process of industrialisation led to new patterns of road use, and to ever-increasing road (and waterway) traffic, much of it commercial and bulky. In 1663 the first English turnpike trust was set up, permitting the levying of tolls to pay for local road improvements and a century later the General Turnpike Act of 1773 greatly accelerated the process of road improvement throughout the country. As the distinction between "good" and "bad" roads became sharper as a consequence of turnpiking, it became worth encouraging travelers to use the turnpiked roads by indicating them as "new" roads on maps. Ogilby's road surveys also promoted the compilation of new county maps. Two of these topographical maps are thought to have been Ogilby's own work (a map of Kent, engraved by Francis Lamb and described as a "new Map . . . actually Surveye'd and the Roads Deliniated"; and a map of Middlesex, engraved by Walter Binneman, both published in or about 1672). The third, a map of Essex eventually published in 1678, was the work of Ogilby's independent-minded assistant, Gregory King.[192] These maps all carry a comprehensive road network.[193]

The comprehensive depiction of roads on topographical maps was fully achieved only after the creation of national mapping agencies. In Britain, this was the Ordnance Survey, established in 1791. Even then, the process was slow, and the first sheet of the Ordnance Survey's one inch to one mile maps (Essex 1805) shows no more than four categories of road much as maps had throughout the eighteenth century: turnpikes (with distances), minor roads, unfenced roads, and roads fenced on one side only.[194] It was not until the survey had reached the north midlands, the northwest and the north, four decades later, that the number of categories more than doubled.[195] Finally, on the last of the one-inch maps issued before 1969 and the move to the metric scale of 1:50,000, no fewer than eighteen categories of roads and paths were differentiated.[196] Almost any one could use the topographical map, however they were travelling, for way finding as well as route planning.

From the start, the printed topographical map was "Everyman's" map. Its heterogeneous content was a result of the mapmaker's and the map printers' need to appeal to an unspecific and uncertain market. Even so, of all anticipated contemporary uses, wayfinding was not relevant at the beginning and not much thereafter until the eighteenth century. Only in the nineteenth century did wayfinding information come to be recognized as an essential component of a topographical map. Four centuries after the arrival of the printed map and the surge of social changes that created the early modern map user, the proliferation in the twentieth century of specialist (thematic) maps—from those for the regular professional driver to those for the occasional hiker, rambler, cyclist and car-based tourist—marked the coming of age of ubiquitous mobility and a map to match every travel need. Even so, the itinerary—a simple list of place-names in correct sequence—remains probably the commonest wayfinding aid; so commonplace, in fact, as to be generally overlooked. The salient difference between the majority of road travelers in the West today compared with those of medieval Europe is that they themselves have compiled their itinerary from maps to suit themselves, instead of being instructed to follow a prescribed route in the course of professional duty.

3

Surveying the Seas

ESTABLISHING THE SEA ROUTES TO THE EAST INDIES

ANDREW S. COOK

The sea chart differs from the land map not only in its form, but also in the way it is used. Humans are not at home on water: the sea is inherently hazardous, and the information on charts has always reflected this. The distinction is simple but fundamental to the different use of land maps and sea charts over centuries. Land is a stable environment: a land map shows the positions of places to which the observer can relate, at will, by processes of measurement. At sea, particularly out of sight of land, a mariner is placed beyond any observable relationship with known points, in a hostile medium in which the vessel is carried at rates and in directions that one can estimate but cannot accurately measure. Assuming the seaworthiness of the vessel, and fair weather, the "safest" part of any voyage is the period when the mariner is headed away from land to open ocean, and, with the problems in measuring progress at sea, the most hazardous time is when one tries to approach land again (or arrives unexpectedly at unforeseen land). Land maps enable travelers to define static relationships between their position with known points, or of known points with each other. The sea chart allows the recording of progress or movement only by a series of estimates of daily position—not in relation to known topography, but by astronomical observations for latitude, and by compass bearing, modified by informed guesses for distance covered, and the effects of wind and current. Aboard ship the chart, insofar as it provided a means of recording progress of the voyage, was another instrument, like the compass, the sextant (or backstaff), the sailing directions, the log and line, and the logbook (or account).

The history of the development of the practice of European navigation has evolved as a distinct discipline, based firmly on the study of treatises on navigation, such as Pedro de Medina's *Arte de Navegar* (1545) and Edward Wright's *Certaine Errors in Navigation* (1599), and on examination of surviving examples of the instruments with which mariners accomplished their voyages, from the history of the backstaff to the history of the chronometer.[1] Archaeological artifacts, fragments recovered from the seabed, play their part in this history, but the greater understanding of the technology of navigation instruments come from the study of museum pieces, the fine presentation examples of instruments that never went to sea. Similarly the study of the history of the sea chart has been developed as a separate genre of artifact history, away from the history of navigation practice, because of the disproportionate survival of presentation examples of portolans and chart atlases in royal and aristocratic collections.[2] Just as honorifically decorated land maps could instill and reinforce a sense of power and authority, so fine examples of charts of the oceans of the world could allow a sense of transoceanic power to develop in the recipient. Few enough examples of usable charts survive from the period before 1600, and fewer still show signs of surviving use at sea.[3]

Blank paper with graticule could, and often did, suffice to keep such a voyage record: geographical information regarding coastlines, rocks, shoals, dangers, and offshore islands cumulated from previous voyages could be presented separately either in books of textual directions or in cartographic form on charts. It is the customary presence of such contextual cartographic information that defines a chart (in *Cartographical Innovations*, for example) as "a map designed primarily for navigation."[4] After isolating the medieval portolan chart as a separate established genre, *Cartographical Innovations* distinguishes three types of charts by content: the nautical chart, the coastal chart, and the harbor chart. In the seventeenth and eighteenth centuries, these three types signified the oceanic navigation chart, the coastal approach chart, and the harbor plan, and embraced all charts in normal use among mariners. We need add only a fourth, the track chart—the cartographic record of a voyage done, with geographical features inserted according to the shipboard perception of them—to complete the cartographic resources available to a hydrographer (see figure 3.2).

The nautical or oceanic chart was used to suggest long courses to shape in open ocean to avoid landfall; the coastal approach chart to navigate straits or archipelagos, to approach coasts, or (less desirably in a sailing ship) to proceed along coasts; the harbor plan to identify entry paths (with leading and clearing lines and marks) and anchorages. With the coastal approach chart came coastal profiles or land recognition views, on the chart or separately, and with the harbor plan (usually on the plan itself) was commonly found the profile of leading or clearing marks to be brought into alignment for safe entry or anchorage. Each mariner's chart book, whether manuscript or printed, contained charts of all three types, with their associated profiles and often with text together.

As the navigational interests of European countries spread from the Mediterranean to Northern European, North Atlantic, and African waters, text "rutters" and portolan-type charts, including those of the Thames school in London, covered wider and increasingly disparate areas. But their manuscript dissemination was slow, and the mariner, with few charts to hand and no basis for the objective assessment of those he had, continued to rely, in Halley's terms, on "latitude, lead and lookout." The mariners who had most need of aids to oceanic navigation and of coastal approach charts were those of the trading companies, primarily of England, France, and the Netherlands, particularly the East India Company. The most complex navigation was that from Europe to the East Indies. The trip required first a passage of the Atlantic currents in both hemispheres, an eastward turn at the Cape of Good Hope, and either a passage through the Mozambique Channel and archipelagos to Bombay and Madras, or dead-reckoning runs in southerly latitudes, before turning northward to India or to the straits between the Indonesian islands to Batavia, China, or Bantam. In England, France, and the Netherlands, formal organization of chart production for ships' captains in the East Indies trade generally preceded and overshadowed the establishment of government hydrographic departments.

Unlike the French and Dutch East India Companies, the English East India Company had never established a hydrographic office. The Dutch East India Company had traditionally organized chart production as a secret central function, furnishing ships with charts only in manuscript until the eventual publication of the sixth part of Van Keulen's *De Nieuwe Groote Lichtende Zee-Fakkel* in 1753. The chart workshops in Amsterdam and later in Batavia produced coastal pilotage charts of the East Indies in large quantities, for issue to captains and officers who accounted for each item on inventories after each voyage, with money penalties for losses.[5] The French East India Company's hydrographic materials were in the hands of D'Après de Mannevillette, a retired captain, at Lorient in Brittany. As the East India Company's hydrographer he maintained an office independent of the Dépôt des Cartes, Plans, et Journaux de la Marine (founded in 1720) in Paris, until his death in 1780, when his collections reverted by agreement to the French crown.[6]

The East India Company in London had no formal mechanism for official chart publication, and generally exercised looser control over shipping and routes. Typically in the eighteenth century the East India Company controlled the provision of shipping by regulating the activities of consortia of private ship owners. As employee of the owners, the captain had responsibility for a ship's equipment, including navigation instruments and charts and inspection by company surveyors. The captain's qualifications and experience, too, were subject to ratification by the East India Company. His responsibility extended to chart supply: the company did not, at first, prescribe what charts were to be carried, only that a fair copy of the journal (embodying the daily log) was to be deposited at East India House at the conclusion of each voyage.[7] The accumulation of these journals—each with daily observations for position, hourly observations of conditions at sea, notes of land dangers observed, often hand-drawn coastal profiles, and sometimes sketch plans of harbors—was a potential resource for the analysis of voyages. Despite their wealth of information, no system of chart compilation, either through examination of these journals or independently, had ever developed in the East India Company in London. Ships' captains had to make do with published chart collections long out of date: John Seller's *The English Pilot* for the Oriental Navigation of 1675, revised and expanded by John Thornton in 1703, and continued in print by the chart sellers Mount and Page for most of the eighteenth century; or Cornwall's 1720 pilot book *Observations on Several Voyages to India.*[8] William Herbert's *New Directory for the East Indies,* in editions from 1758 onwards, was a popular English version of the coastal charts found in the 1745 edition of D'Après de Mannevillette's *Le Neptune Oriental.*[9] Each was a compendium of textual directions, oceanic charts, coastal charts, harbor plans, and views of land. Captains who had made harbor plans or amended coastal charts were free to sell their work to these and other publishers, and the resulting publications owed more to commercial opportunism than to systematic analysis.

The availability of printed chart collections from France and the Netherlands invited comparisons between *The English Pilot,* D'Après, and Van Keulen. Charles Noble, in a 1755 pamphlet comparing the French and English shipboard conditions, gave his opinion of the relative quality of charting from the two countries:

> The French have been at pains, to improve their Navigation and their Charts; Those of the Indian Seas, by Monsieur D'Après de Mannevillette,

a Captain in their Service, exceed every thing of the kind in Europe. It is a pity they are not translated into English, for the benefit of our Navigators. Those who understand the language, and have seen them, must have a despicable opinion, of our Indian Pilot, with which Messrs. Mount and Page have long imposed on, and picked the pockets of our Countrymen, and which are only fit for the Grocers and Chandlers Shops, or posterior uses.[10]

Company captains, particularly those experienced in navigating the "country" trade eastward from India, could see for themselves the deficiencies and inaccuracies of the European product in areas they knew well. When captain of a country ship working from Madras to Borneo, Canton and the China Sea in the early 1760s, Alexander Dalrymple kept a copy of the *Le Neptune Oriental* with him to use, and to improve his own observations. In a letter to his colleague Vansittart, Dalrymple explained: "I have also sent you a Chart to be joined to the first Sheet of M. D'Après of the China Seas as you will perceive by Condor & the Lines, in which I have laid down our Track with a Sketch of the Coast of Borneo which is extremely ill described in the French."[11] To be of general utility sea charts had to command general respect, and to command respect they had to be compiled to satisfactory standards.

The Atlantic Ocean, the routine passage from Europe for all fleets to the East Indies, offered many problems to navigation, few of which were recorded in charts. The Dutch East India Company and English East India Company each sent out four or five fleets each year to India, China, and Java. For the English East India Company alone this amounted to between twenty-five and thirty ships a year. Ships would try to keep company, but inevitably separations would occur, and it was always a relief to complete the Atlantic passage on the outward voyage, either by safe arrival at Table Bay or False Bay, or by sighting the Cape of Good Hope to take a new departure for the Indian Ocean. The Atlantic passage could well take four or more of the seven or eight months voyage to India or China. A letter written from the Cape of Good Hope in September 1775 by Dalrymple, then a passenger in ship *Grenville,* illustrates the nature of the passage:

I paid my respects to you from Madeira on the 15th May. We arrived at False Bay on the 21st August after a very tedious and uncomfortable passage. We had light winds after we passed the Cape de Verde Islands, none of which we saw sight of, leaving them to the westward. We had frequent soundings off the Coast of Guinea to the southward of Cape Verde, and saw the land to the westward of Cape Palmas, and again at that Cape, as we supposed. Notwithstanding which we met such violent currents, setting us for several days together above 70 miles a day to the westward, that we had sight of Trinidada, which we could not go to windward of. After passing the line, instead of meeting with a regular trade wind and smooth water, we had squally blowing weather and a great sea. And after we had passed Trinidada we had a violent gale of wind from the south-west.

We found the Marine Barometer of great use to us as we got to the southward, and is a great favourite of all the seamen, who never wish to go to sea without it.

some days after we had got into Table Bay the Gatton, which sailed two days after us from Madeira came into Table Bay. They passed to the westward of the Cape de Verde Islands, and fell in with the Easternmost part of Brazil. He was in 12 fathoms within a few miles of the land when the wind came round to south, with which they stood off ESE, and had soundings till out of sight of land. This wind continued a whole day, and came round again to the southward: they stood of ENE next day and then put about, and saw no more of the land as they stood to the southward.

The Coventry has not been here: I think it is very probable she has not weathered the coast of Brazil, as it not likely she would pass the Cape without touching here.

Our timekeeper has stopt three times, but Mudge's watch going very well, and, a regular register have been kept of all the watches, it was set a-going, and an account kept of the Times it had stopd. The Timekeeper gave us the land a little more than 1° to the westward of its true situation, which is very near considering the length of our passage from England. Our lunar observations were about ½ a degree too far East, but we had no sights for some days before we made the land.[12]

Such an account gives insights into the techniques of navigation in the Atlantic. Even for an experienced navigator like Dalrymple, there were many uncertainties. The importance of sighting land in passing is clear, but nowhere is a chart or pilot book mentioned. The winds are observed with considerable care, as is the force of the not unexpected current. The intention after leaving the Cape Verde Islands was to sight Trinidada, off Brazil, as a new point of departure, but *Grenville* was forced to the lee of the island. *Gatton* was less lucky, and got into soundings on a lee shore, before the wind changed to allow her to stand off. Each ship stood southward, or southeastward, to the latitude of the Cape of Good Hope, running that latitude to within sight of the Cape of Good Hope itself. In the case of

Grenville, a combination of lunar observation and chronometer brought it home, at least to within sixty miles. With the exception of the chronometer observations, this was navigation by latitude, lead, and lookout. Charts, if any, provided only a sense of the limits of the ocean, and approximate locations of islands.

The major problem of outbound Atlantic navigation was not recorded on charts: the ocean circulation pattern, the currents known to mariners only by experience. A modern diagram of current patterns gives clues (figure 3.1).[13] The prevailing currents in both north and south Atlantic drive

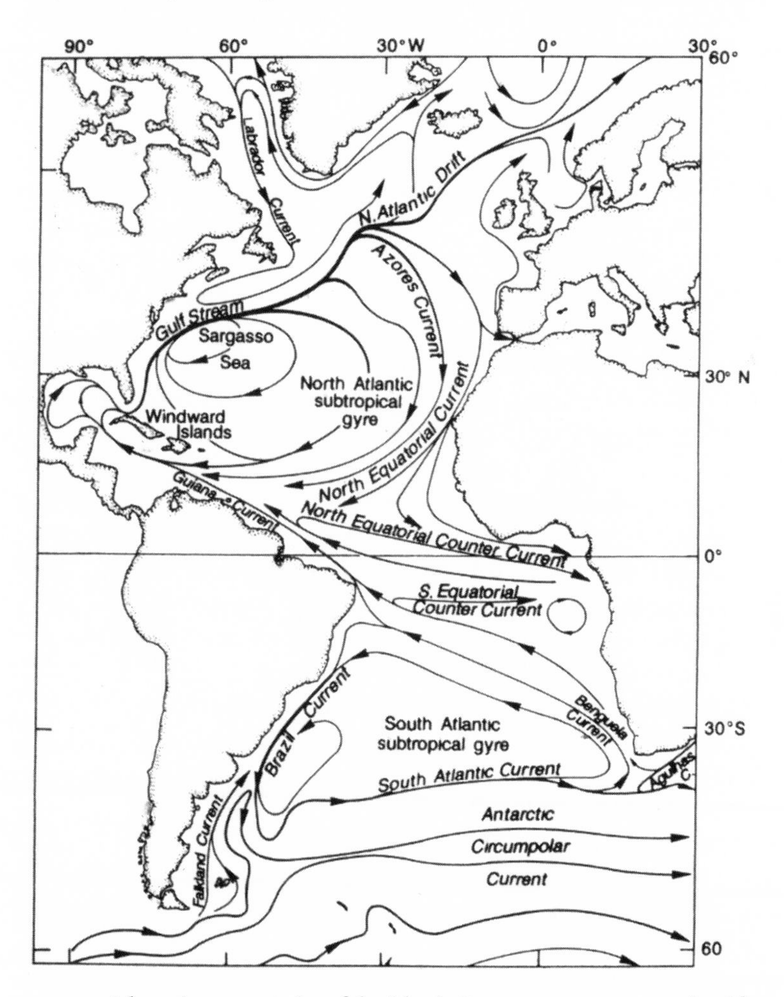

FIGURE 3.1 Schematic representation of the Atlantic Ocean current system, reprinted from John Gould, "James Rennell's View of the Atlantic Circulation: A Comparison with Our Present Knowledge," *Ocean Challenge,* 4, nos. 1 and 2 (1993), pp. 26–33, by kind permission of the Challenger Society for Marine Science.

toward the Caribbean and the Gulf of Mexico, the North Equatorial Current from the Azores and the west African coast, and the Benguela Current and the South Equatorial Current from the west coast of southern Africa. A ship sailing south-west from Europe encountered first the southward currents spinning off the Gulf Stream, and would try to sight the Canary Islands to take advantage of the Azores Current. Next sighting would be Cape Verde Islands, the sign to turn southward out of the main current as it swung toward the West Indies. A commander had to estimate how far to allow his ship to be carried toward the Gulf of Guinea by the countercurrent, before striking southward across the South Equatorial Current. Too far east and he would be in dead water in the Gulf of Guinea; not far enough and he could find himself on the Brazil coast without having turned Cape Recife (figure 3.2). The relative strengths of the countercurrents varied and were accentuated or mitigated by seasonal winds. The assessment of the westward set of the South Equatorial Current was one of the most important decisions a commander had to make, and on very little information. Determined to avoid a West African lee shore, his dead reckoning from the Cape Verde Islands was his chief input. His aim was for Trinidada or Martin Vaz, off the Brazil coast, as his next point of departure. He would then steer southward or southeastward, benefiting from the Brazil Current, to reach the latitude of the Cape of Good Hope, either to make Table Bay or to take a more southerly path into the Indian Ocean, avoiding the Lagullus Bank.

The navigation of the Indian Ocean offered different problems, but the critical part of any outward voyage was spent in the Atlantic. A ship that failed to make Cape Recife could easily lose a further four or six weeks in making a second attempt, and would usually require to water on the Brazil coast before continuing. It was simply impossible to sail direct for the Cape of Good Hope. Decisions at the three turning points all had to be taken in open ocean by dead reckoning and without reference to longitudes. Mariners preferred open ocean to lee shores. Isolated islands, such as Trinidada or Cape Verde Islands were ideal sighting points: they had a known relationship to the mainland, but were far enough out to allow ships to pass to either side before continuing. Latitude, lead, and lookout were the guides, as well as the measurement of daily distance by log, bearing, and the constant assessment of floating vegetation and of bird life. Before the capacity to make longitude observations these were the only guides a commander had. At any point he knew his daily latitude, he knew what land he had last seen (if he recognized it), he knew his bearing, he thought he knew how far he had sailed along that bearing, and he could estimate the lateral effect of wind. But he had to guess the effect of current.

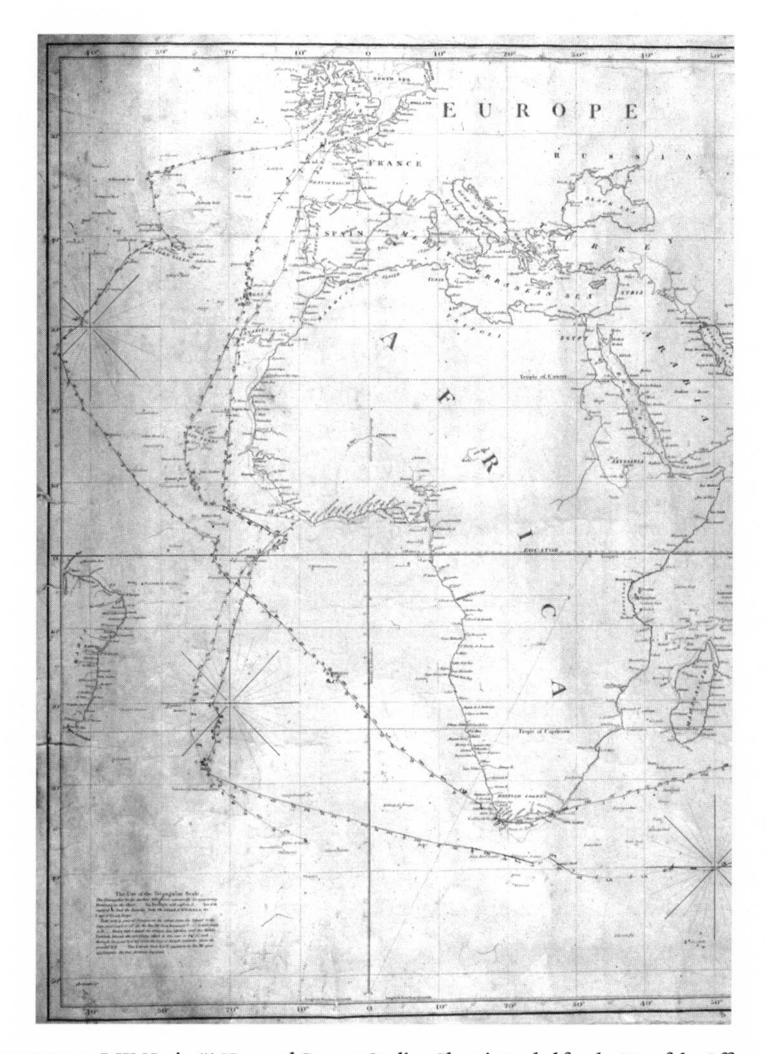

FIGURE 3.2 J. W. Norie, "A New and Correct Outline Chart intended for the Use of the Officers in the Royal Navy and Merchant's Service to Prick off a Ship's Track," 1833, additions to 1852. The full engraved chart extends from Brazil to New Zealand: the western or Atlantic part only is illustrated here. This example (London: The British Library, India Office Records X/10299) bears three manuscript tracks, two of outbound vessels approaching the Guinea coast before crossing the equatorial currents, and one of a homebound ship calling at St. Helena before crossing to the western North Atlantic Ocean to take advantage of the Gulf Stream.

Popular historians of the chronometer relate that John Harrison's invention solved navigation problems by enabling mariners to tell their position with accuracy. Manuals of seamanship gave much space to competing methods, from Mayer's *Tables* to Maskelyne's *Nautical Almanac,* and from Harrison's chronometer to Irwin's marine chair (suspending an observer in a cradle to observe the eclipses of Jupiter's satellites). A timekeeper was only a machine for carrying local time through a voyage: it was only as good as its maker, and it could stop. Pocket watches served on board ship from local noon to local noon, but a timekeeper had to be reliable over months, and consequently had to be checked by lunar observations. The reputation of the timekeeper had not yet been established: because single chronometers could not be relied on, three or more were later carried, but it was accepted that too firm a reliance "relaxed the vigilance which the known uncertainty of dead reckoning kept perpetually alive."

Secondly, in order to use knowledge of one's longitude effectively in navigation, one has to know the longitudes of the land one sights (and of the dangers to avoid) with the same accuracy. Anson's problem in clearing Cape Horn and running a latitude to Juan Fernandez would not have been solved by knowing his longitude, unless he knew the longitudes of Cape Horn and Juan Fernandez with equal precision. Knowledge of longitude is useful, then, only in relative terms. Without accurate charting of the coasts and islands of the Atlantic, knowledge of longitude was only another aid to navigation for the commander of the *Grenville*. And even with that knowledge, ships were still subject to winds and current.

To this system of experience, observation, and intuition the East India Company entrusted thirty ships a year, the generation of its wealth, and the safety of its employees. The oceanic navigation pattern of later years was not recorded on charts; only in the 1720s were a few limiting suggestions marked in *The English Pilot* (figure 3.3).[14] Though a wealth of track information was building up in East India House, there was no machinery for examining and connecting it with commercial chart production. Perhaps it was for this reason that Noble wrote scathingly about English charts.

The voyage of the *Grenville* was a watershed journey. It was the first East India Company voyage on which a chronometer had been carried and used daily in conjunction with other standard observations and calculations. The *Grenville* carried Lord Pigot to Madras to begin his second term as governor. Among his council members traveling with him was Alexander Dalrymple. Pigot had one of John Arnold's new chronometers: Dalrymple used it daily, and later published his journal of observations.[15] This voyage introduced Dalrymple to the uncertainties of oceanic navigation, particularly outbound from England; to the possibilities of the increased accuracy that chronometer longitudes could allow; and to the lack of precision with which longitudes of coasts and islands were marked on charts.

Dalrymple had gone out to Madras in 1752 as a young East India Company servant, and had spent some time in the early 1760s as commander

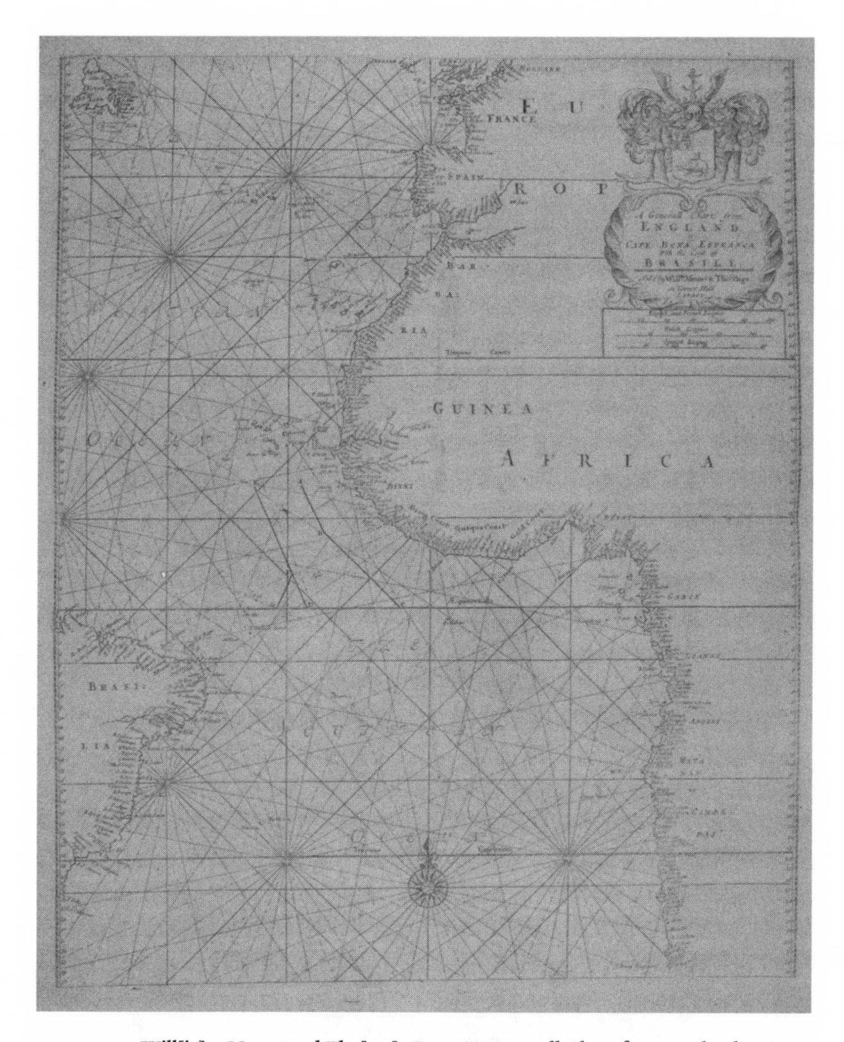

FIGURE 3.3 Will[ia]m Mount and Tho[ma]s Page, "A Generall Chart from England to Cape Bona Espranca with the Coast of Brasile," in *The English Pilot, describing the Sea-coasts [etc.] in the Oriental Navigation* (London, editions from 1723 onwards). This replaced a similar but smaller-scale chart in John Thornton's earlier editions of *The English Pilot, The Third Book,* which also incorporated the diagram (keyed to accompanying sailing directions by letters A to G) prescribing the best "zone" for crossing the equator southbound.

of a "country" ship in the China Sea, and Eastern Archipelago (figure 3.4). After his return to England in 1765 he spent time among the geographical materials and ships' journals in East India House, primarily to develop a series of charts of northern Borneo and the Philippines based on his own voyages. He produced a first chart of the South Atlantic to support the

FIGURE 3.4 Portrait of Alexander Dalrymple, by John Thomas Seton [1765]. Oil on canvas, 915 × 710 mm. *Private Collection, United Kingdom.* The only known portrait of Dalrymple as a young man, probably painted for his elder brother, Sir David Dalrymple, at Newhailes near Edinburgh after Dalrymple's return from India in 1765. In a pose characteristic of the Edinburgh portraitist Seton, the subject is portrayed with all the accoutrements of his position as an East India Company captain, and with globe, map, chart, and dividers for a navigator.

theory of a habitable southern continent (figure 3.5). He corresponded privately with D'Après de Mannevillette from 1767 about evidence of dangers on the route to India, using the ships' journals in East India House extensively, and also began in 1774 to publish harbor plans from his own manuscript collection.[16]

Dalrymple's second period in Madras, from 1775 to 1776, was in retrospect an interruption in a hydrographic career. He was sent back to London hurriedly in October 1776, with news of Governor Pigot's house arrest by dissident council members. Before leaving he recovered Pigot's Arnold chronometer from the *Grenville*'s captain (to whom Pigot had given it) and used it methodically on the return voyage via the Red Sea, Suez, and Marseilles. Back in London in April 1777 with the chronometer, Dalrymple soon made himself known to Arnold, commissioning a chronometer for himself and introducing other customers.[17] By January 1778 he had developed a scheme to assemble data from the deposited journals of oceanic voyages by East India Company ships; to predict the safest and quickest courses; and to explain the effects of currents, particularly in the Atlantic:

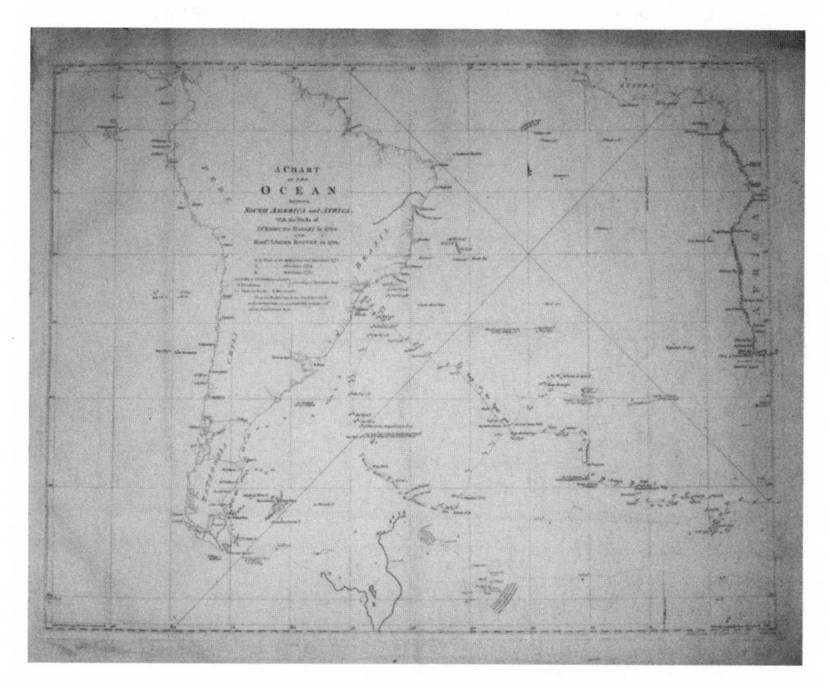

FIGURE 3.5 Alexander Dalrymple, "A Chart of the Ocean between South America and Africa," 1769. Dalrymple maintained this chart in print for at least twenty years, to show the known and suspected landmasses and hazards to navigation in the South Atlantic Ocean. This example is from the last-known published state of the plate c. 1786, and shows also the 1772–75 tracks of *Resolution* and *Adventure,* the ships of James Cook's second circumnavigation.

"A comparison of the several tracks of the Company's ships would, in a few years, determine with great precision what is, at every season, the most eligible course to pursue out and home. The improvement made lately in the art and practice of navigation, by the lunar observations for determining the longitude, and by the use of timekeepers, will be very conducive towards explaining the currents, that most curious and important phenomenon in nautical history."[18]

Dalrymple issued special journal forms to allow parallel recording of positions by account and by chronometer or lunar observation, and distributed with them blank charts calibrated at five degrees to one inch for marking tracks. The East India Company subscribed for 100 sets for the ships of the new season.[19] The following year Dalrymple added an equatorial chart specifically for the Atlantic, explaining: "As in low latitudes between Brazil and Africa the winds are generally faint and variable, and the daily runs consequently short and irregular, so as not to be easily expressed in so small a Scale as the general chart, I have thought it will be proper to give a plain Chart on a Scale of half an inch to 1°, from the Cape de Verde Islands to 21° S."[20] One example of this equatorial chart has recently been identified among the manuscripts in the Admiralty Library in London (figure 3.6). As issued, it had engraved frame and graticule, but the remainder was empty; Dalrymple used this copy himself, inserting the coastal outlines and tracks both "by account" (i.e., dead reckoning) and by a combination of latitude observations and chronometer, including that of the *Grenville* in 1775.[21]

For greatest accuracy, tracks had to show longitudes by chronometer based on established points of departure of known longitude. Dalrymple issued a list of useful places whose longitudes had been adequately observed: where a land observer had been stationed or longitudes established by short-run timekeeper voyages.[22] To achieve a useable series of chronometer tracks, Dalrymple needed to get commanders to buy chronometers, make series of observations, and complete not only track charts but extra journal forms as well. He had no authority to order the use of his system, and results did not materialize. But he had taken the first step toward the systematic collection and analysis of East India Company navigational information.

In 1772 Dalrymple wrote that he had "Nautical Remarks and Charts of various parts of India, collected, sufficient to make a more complete Set of Charts and Sailing Directions than any hitherto published. But to do this would be too great an Undertaking for any man, unless it was his particular occupation."[23] He had been in correspondence, not in competition, with D'Après de Mannevillette, and, though he had been disparaging about earlier chart publications, he was waiting with enthusiastic interest in 1774

FIGURE 3.6 Blank engraved chart for East India Company captains to enter tracks in the Atlantic Ocean from the latitude of Cape Verde Islands to 21° S, by Alexander Dalrymple [1779] (Portsmouth: Admiralty Library Vf 1/13). This example has been completed retrospectively by Dalrymple himself, with coastal outlines and tracks both "by account" and by a combination of latitude observations and chronometer, including the track of Dalrymple's own voyage as a passenger in *Grenville* in 1775.

and 1775 for the promised new edition of the latter's *Le Neptune Oriental.* Dalrymple was disappointed on his return from Madras in 1777 to find that D'Après had repeated in 1775, largely unchanged, the Indian Ocean coastal charts from his 1745 edition.

Dalrymple proposed himself for a new position "for examining the Ships' Journals from the earliest times that Notice may be given of every danger which has hitherto been discovered, and for publishing from time to time such Charts and Nautical Directions as a comparison of the various Journals and other Materials may enable him to do."[24] On April 1, 1779 the East India Company appointed him to this new responsibility. The immediate occasion on which the success of Dalrymple's proposal turned was the August 1778 loss of the East India ship *Colebrooke* on the Anvil Rock, in False Bay, Cape of Good Hope. He argued that the danger had been known since the *Caesar* struck in 1745, and was confirmed by the subsequent striking of the French ship *Mergé.* D'Après de Mannevillette had published the *Mergé*'s account in 1765, and Dalrymple was already corresponding with him privately in 1767 about the *Caesar*'s journal. A visual search for the Anvil Rock was one of Dalrymple's self-appointed tasks during his short stay in False Bay in 1775.[25]

In his memorial to the East India Company, Dalrymple argued that the proper coordination and publication of hydrographical information from fair copies of the journals routinely deposited in East India House by the commanders of returning ships would alert officers on subsequent voyages to particular dangers. He was supported by the chairman of the East India Company Council of Directors, Sir George Wombwell, whom he reported as remarking that "I had timed it very well," and that "the loss of the Colebrooke was more than the expence of such an Office to all eternity." It clearly did not work to leave to commercial chart publishers the job of publicizing promptly new dangers at sea. Though Dalrymple later called himself "Hydrographer to the East India Company," his position never appeared in establishment lists or salary books: he was paid a quarterly fee, always "For examining the Ships' Journals, &ca."[26] The provision for chart publication was that Dalrymple's expenses in engraving and printing 100 copies of any chart he chose to compile would be met by the East India Company, after which the plate became his property, and further impressions to use or sell as he pleased were at his own expense of paper and presswork. Editorial control was entirely in his hands, as was planning for publication. Market forces were conspicuously absent.

Dalrymple issued a notice on April 20, 1779, based on his memorial and chiefly for East India Company ships' commanders and officers. He explained his appointment and proposed compiling a new set of charts for navigation to the East Indies that would improve on D'Après de Mannevillette's *Le Neptune Oriental* by including the sightings of rocks and shoals from East India House ships' journals. His first task was to examine all the journals; his second to produce charts: "The Plan I propose to pursue is, with the utmost Expedition to get an Index made to all the Journals, containing the day of seeing every Land or shoal, or having soundings, so that as soon as possible the Company's Ships may have notice of every danger which has been discovered form the earliest times: and thus knowing every Ship which has seen any particular danger, the comparison of the different Journals will facilitate the laying down thereof."[27] If such an index was completed it has not survived. Dalrymple had no research assistant, nor any East India Company provision for a room to work in, and at his death almost twenty years later his executors had to ask the Court of Directors to take back over 700 ships' journals and others manuscript books found in his house.[28]

As his first year's output Dalrymple promised to compile and publish a small-scale chart (sixty nautical miles to one inch) of the Mozambique

Channel and Madagascar, and larger-scale charts of the Strait of Bangka, the Strait of Singapore, and the Parcelar Banks in the Strait of Malacca. The longer-term plan was for a series of twenty-seven coastal charts at a common scale of twenty nautical miles to one inch covering the most frequented coasts of the Indian Ocean and the Eastern Archipelago. The scheme included a group of seven charts to cover Madagascar, the Mozambique Channel, and the East Coast of Africa to Linde, and isolated charts for the Amiranté Islands and the Chagos Archipelago. A coastal chain of seven charts was to stretch from the Persian Gulf to the Bay of Bengal, itself already covered by Dalrymple's 1772 Bay of Bengal chart. A second chain of six charts was to cover the coasts of mainland South East Asia from Chittagong to Malaya, Indo-China and Macao, joining a group of four charts for Sumatra, Java, Borneo, and the Java Sea. One further isolated chart was envisaged for western Australia.[29] No charts were suggested for the Atlantic Ocean, as it lay outside the area of East India Company monopoly. Dalrymple tried unsuccessfully in July 1779 to obtain separate payment for two charts of the coast of Brazil he had prepared from Dutch sources before his appointment; he issued them later as East India Company charts.[30] The areas chosen included the passages most frequented by East India Company ships: many chose to water at Anjouan in the Mozambique Channel on the outward journey, then to proceed through known latitude channels in the Maldives to India.

To meet the court's requirement that he should demonstrate activity year by year, Dalrymple diverted time and energy to publishing single-source harbor plans, and plates of land views. Other, "special order" requirements came from above: the entry of the French and Dutch into the American War in 1781 placed the English East India Company's routes to China in danger, and Dalrymple was commissioned to produce twenty-eight extra plates between 1780 and 1782, chiefly harbor plans in the Celebes, Moluccas, and islands to Guinea and New Britain.[31] To complement these, he compiled in June 1782, at the request of the East India Company Secret Committee, a "Memoir concerning the Passages to and from China."[32] This analysis of the wide range of routes in the Eastern Archipelago, intended to help ships from England or India avoid French or Spanish interception, was the first synthesis of Dalrymple's ongoing examination of the ships' journals, but it detracted from his aim of producing charts. By 1783 chronometer observations were vital for Dalrymple to continue his chart work: "Chronometers will produce such a change in geography and navigation as must in a short period supersede any Hydrographical Work for the Indian Navigation, which can now be made, and render a new Set of Charts absolutely necessary, when we have the reciprocal longitudes of all places established. This consideration has influenced me to change my intention; instead of postponing the Publication of a Set of Charts, I shall, from the Charts at present in my collection, finish a Set forthwith, which set, though imperfect, will be less defective than any at present in use."[33]

After four years without publishing charts in the 1779 series, Dalrymple issued the first, "Chart of the Coasts of Guzurat and Scindy," in November 1783. In the accompanying *Memoir of a Chart from St. Johns on the Coast of India to Cape Arubah in the Gulf of Persia* he described its process of construction from fifteen manuscript source charts. In 1785 he had ready the "Chart of the Straits of Sunda and Banka," and published it the following year with a memoir detailing sources, which showed it to be a rapid composite of many known sources, among them a chart of Van Keulen "too dissonant to the Modern Observations to connect with them."[34]

But incoming information, often lacking a chronometer basis, was too disorganized to be used effectively, while at the same time conflicting with the topography in the older printed mapping. Dalrymple had no authority to order surveys to resolve these points. In 1784, he proposed, more as an academic exercise, a survey of the Coromandel Coast.[35] Only once, in 1786, was he asked by the Government of Bombay to advise on a proposed survey, a chronometer survey of the coast of Western India, which he said would provide data for the Malabar Coast charts of his own series. He sent instructions to Bombay for the use of Arnold chronometers:[36] "We would have the vessel proceed along the coast, from Bombay to Surat, determining carefully the Latitudes and Longitudes of the various points, as well as of the Peaks of Hills inland, with explanatory views of the Lands, taking altitudes for determining the time by Chronometer every hour; & taking the bearings and altitudes of the Lands &ca., by the Hadley at such time." John McCluer, of the Bombay Marine, was chosen to carry out the survey, on the basis of his reputation for similar surveys in the Persian Gulf in the early 1780s.[37]

Dalrymple continued to have a close association with Arnold, and he continued to recommend Arnold as supplier of chronometers for East India Company ships' captains who wished to take observations on their voyages.[38] The East India Company took a passing interest in the work Dalrymple encouraged captains to do, but all knew that the first concern of the captain had to be his ship, cargo, and orders. There was rarely time for surveying on the side.

So the scheme for systematic coastal chart compilation faltered. Dalrymple's table of sufficiently accurate longitudes was not comprehensive enough, and the firm longitudes not at close enough intervals, for charts to be compiled by filling in the intervening detail—at least not at his chosen scale. As a step toward remedying this deficiency, Dalrymple consistently advocated the collection of views of headlands as an aid to coastal navigation, and, in island groups, to aid recognition and position fixing: "A proper set of views should contain not only distant prospects to distinguish the land; . . . but also views still nearer the shoar which are useful to give the most competent description of the country."[39] Apart from his own collection of views, a single volume of which has survived unrecognized, Dalrymple rapidly located many views in ships' journals in 1779, and had others given to him. He wrote to D'Après: "I am induced to engrave a suite of views of lands for the whole course of the Indian Navigation."[40] By 1783, 30 of his 171 published plates were composites of land views, up to 30 to a plate, and 15 more plates were in proof. In that year Dalrymple proposed a voyage "for establishing the geographical situations of the remarkable headlands in the world," which he would take part in. "Three or four years of my remaining life would, I think, be very well employed, in ascertaining exactly the positions of all the remarkable lands from England to the remote parts of the East Indies: the expence of such voyage, though great, would be less than has been sometimes squandered by an individual in one Parliamentary election." The ship was to take five Arnolds to determine longitudes, but the expedition never materialized.[41]

The problems with using dead-reckoning longitudes irreconcilable with others are exemplified in Dalrymple's discussions with D'Après over the nonexistent islands Ady and Candy, thought to lie to the east of the Chagos Archipelago in the same latitude.[42] Circumstantial evidence from journals had encouraged mapmakers to place them on charts, but the islands' relationship to the Chagos could not be confirmed. Despite the increasing likelihood that the islands did not exist, they had to remain on Dalrymple's charts as a warning to shipping, even in the 1780s.

The same caution appears in the reciprocal longitude charts Dalrymple prepared in the 1780s. On one, showing part of Madagascar with the Indian Ocean islands, he could enter very little with certainty.[43] The published pilot books all showed different versions of the Mozambique Channel, and he had no basis for preferring one to the others. "Whoever knows anything of making Charts from a variety of materials, must know that it is impossible to reconcile those materials perfectly. Indeed it often happens that they are totally contradictory: where I find disagreement in particular charts, I have thought the best way was to engrave both, when I had nothing to enable me to decide on the merits of either."[44] In consequence he sent out five charts of the Mozambique Channel, each from a different source, while six representations of the Malabar Coast appeared on another plate.[45]

Dalrymple tried to collect and reconcile conflicting longitudes on the peninsula of India for a small-scale chart from Karachi to Sumatra. Five small areas only are shown: Gujarat and Sind from the 1783 chart, the Maldive Islands, Ceylon, the Madras coast, and the head of the Bay of Bengal—each area based on known longitudes. Another chart of the wider Indian Ocean has more detail, but also remained unpublished.[46] These small-scale essays are known in no more than two or three proof copies. Dalrymple's practice of publishing conflicting representations together without critical evaluation, and declining to issue compilation charts until he was sure of the longitudes, did not allow him to provide the standard set of charts he had promised. There were no market pressures on Dalrymple the researcher.

But by the early 1790s Dalrymple had systematized production and issue of the harbor plans and single-source coastal and archipelago charts that formed the bulk of his output. By 1789 he had produced twenty-eight charts, forty plates of views, and over 450 harbor plans and large-scale coastal charts. His compilation and draughtsman work was done at home; the engraving by a group of jobbing engravers in London (John Walker, William Harrison, Isaac Palmer, and Thomas Harmar); and the printing probably also on private premises.

The uniform style that identifies Dalrymple charts is a superficial uniformity of clear lines, lack of unnecessary ornamentation, and consistency of writing. Chart sizes were based on the 24 × 18 inch usable area on a plate 25 × 19 inches, accommodated to the size of the press. Within this regimen Dalrymple used full plates for coastal charts, cut half plates 25 × 9 inches for views of land, and cut quarter plates 9 × 12 inches for harbor plans. Larger plans or small charts fitted on half plates 12 × 19 inches. All except the full plate charts would bind up with the untrimmed quarto letterpress of his sailing directions or "nautical memoirs." Only in the late 1780s did Dalrymple gain the capacity to print from 28 × 23 inch plates. In 1783 he negotiated with James Whatman for special wove moulds, with "Dalrymple" wired in a lower corner, to make 13 × 27 inch paper for his views of land without a disfiguring central watermark.[47]

The corollary of adopting a standard plate size was a directly interrelated set of chart and plan scales. The combination of 24 × 18 charts and the 20-nautical-mile scale allowed Dalrymple in 1779 to specify plane charts (his preference for tropical waters) of 8° × 6° at 3 inches to 1 degree for his coastal chart series. Larger scale charts and plans were kept, as far as possible, to arithmetical multiples of the 20-nautical-mile scale: 10, 5, 2½, 1¼ nautical miles to 1 inch, 1.6, 3.2, 6.4, 12.8, 25.6 inches to 1 nautical mile. He enunciated this formally in the unpublished treatise "Practical

Navigation": "It is very desirable to have uniformity in the Scale of Charts and Plans, because it, insensibly, conveys great knowledge and precision of ideas, concerning the relative distances and magnitudes of places. The mind can readily conceive the proportion half, quarter, &c., but when it is distracted with fractional comparisons, men do not carry along with them that knowledge which uniformity would have produced. Hydrographers and Geographers seldom, however, attend to this uniformity."[48]

The system of production was matched by the system of cataloguing and sale. After the East India Company's requirement had been met for each plate, impressions were added to the stock for sale. One should imagine a series of stock shelves, with supplies of every chart, views plate or plan, each priced separately. Sets were available at discounts of up to 60 percent, as were the annual supplementary sets up to 1794. In 1789, twenty-eight whole plate charts were offered at prices from 2s.6d. to 5s., or 2 guineas the set; forty views plates at 2s. each or 2 guineas the set; 454 harbor plans and small charts, in eighteen geographical series, at 6d., 1s. or 2s. each (depending on the complexity of the engraving, or presence of views of land, or 6 guineas the set.[49] Despite this convenient marketing arrangement there was no editorial coherence to the collection: by no stretch of the imagination could Dalrymple's charts be called a pilot book. Samuel Dunn, who took over William Herbert's *New Directory for the East Indies* in the 1770s, issued editions in 1783 and 1788 incorporating impressions of Dalrymple's plates loaned to him, but only to supplement the coherent Herbert coastal chart series that still formed the basis of the book.[50] Laurie and Whittle's *East India Pilot* flourished after 1781, and when James Horsburgh, Dalrymple's eventual successor in the East India Company, proposed to publish privately his China Sea charts in 1804, Dalrymple encouraged him and recommended his own plan engraver, John Walker, to carry out the task.[51]

In 1779 Dalrymple had possessed a clear vision of the maritime community's need for a system of standard charts for navigation and reliable harbor plans in the East Indies. But the resources available only allowed him to examine the ships' journals and to publish single-source charts and plans. Many East India captains supplied him with their track charts or with harbor plans but, with responsibilities to their owners, none could go out of their way to conduct the coastal surveys he needed. Only the Bombay Marine had men and ships, and then only for locally authorized survey work in the Laccadives and Andamans.[52] The one occasion that Dalrymple persuaded the East India Company to direct a ship off-course to conduct a survey, it ended in disaster. The Court of Directors appointed Lestock Wilson, commander of the *Vansittart,* to spend some days surveying in the Gaspar Strait, east of Bangka, on his journey to China in 1790. Wilson wrecked his ship on the very bank he had been ordered to locate, though he brought his journal, observations, and charts home.[53] Francis Beaufort, later hydrographer to the Admiralty, was captain's boy on the voyage.

Yet out of this unsatisfactory East India Company situation Dalrymple was appointed hydrographer to the Admiralty in 1795, "to be intrusted with the custody and care of such plans and charts as now are, or may hereafter be, deposited in this office belonging to the public, and to be charged with the duty of selecting and compiling all such information as may appear to be requisite for the purposes of improving the navigation, and for the guidance and direction of the Commanders of Your Majesty's ships, in all cases wherever any knowledge in this respect may be found to be necessary."[54]

Dalrymple later claimed that he owed his appointment to Earl Spencer, first lord of the Admiralty from November 1794.[55] But the sequence of events owed much to the chain of Admiralty appointments in March 1795: Sir Philip Stephens rose from the secretaryship to a seat on the Board of the Admiralty, Evan Nepean became secretary to the Board of the Admiralty, and William Marsden was introduced as second secretary.[56] Stephens had been secretary for almost thirty-two years, and a friend of Dalrymple since at least the 1770s. Nepean had been undersecretary to Henry Dundas since 1782, and Marsden was an East India Company servant and orientalist. All three had long connections with Dalrymple, either through the Royal Society and the Royal Society Club, or (in Nepean's case) in the conduct of colonial business for Henry Dundas as home secretary and secretary for war. Dalrymple had been called to advise government officials on geographical matters from Nootka Sound to the Red Sea and China with increasing frequency in the late 1780s and early 1790s, partly because of his association with Sir Joseph Banks. Dalrymple was already undertaking an investigation for Nepean and Dundas in February 1795, using East India Company ships' journals to compare journey times between England and the Cape of Good Hope by the Guinea Coast route, via St. Helena, and by the open westerly route.[57]

The demands of Dalrymple's new appointment were quite unlike those of his East India Company retainer. His first responsibility was the organization of an official hydrographic collection that had accumulated, until his arrival, in the care of a clerk.[58] Dalrymple was immediately faced, at the age of fifty-eight, with running an office with subordinate staff, a position quite different from the personal appointment by the East India Company "for examining the ships' journals" that he had executed practically alone, building up his private collection of manuscript and printed charts and atlases as sources for his publications. He secured the appointment of

Aaron Arrowsmith as his assistant from September 7, 1795 at a salary of 100 pounds a year.[59] The work of the new office was first to sort, classify, and arrange the materials collected, and then to evaluate them to select and compile information for the use of ships. No specific mention was made of printing or publishing charts, though when Dalrymple had approached the East India Company for permission to accept the appointment, he mentioned "forming and engraving charts."[60]

By convention, naval officers treated surveys made in the course of duty as their private copyright, and many had their charts published through the map trade in London. This practice was foreign to Dalrymple's East India Company experience, and he later assembled a list of such publications. Surveys commissioned directly by the Board of Admiralty from appointed surveyors formed the nucleus of the collection in the care of the admiralty clerk before 1795. It was the lack of a system for coordinating and compiling this information, and for disseminating it securely to ships of the fleet, which Dalrymple had first to address. He assessed charts for their continuing utility, arranged them geographically in bundles, and consigned the outdated to storage attics. At the same time new private surveys were arriving that needed to be copied for the Hydrographic Office before being released for commercial publication.

Investigative work without tangible product was satisfying to Dalrymple, but less so for his assistant. Arrowsmith was already an established map and chart publisher by the 1790s, and he lasted just over a year as Dalrymple's office assistant. Dalrymple had been sending out draughtsman work, reduction, and copying of charts to John Walker (the engraver whom he had long used for East India Company charts), and Walker was formally appointed Dalrymple's assistant late in 1796 or early in 1797.[61] The work continued to range widely: Dalrymple took in the charts and journals of the D'Entrecasteaux voyage, brought to St. Helena by Rossel when the expedition broke up in Batavia, and he employed Rossel, then in exile, to compute the results of observations on the voyage.[62] Dalrymple responded to a variety of questions referred to him by Nepean, advising him which of a collection of Danish charts and sailing directions might be copied for the Admiralty; recommending maps, charts, and globes for the Portsmouth Naval Academy; and assessing deviations in the homeward course of *Sphynx* from St. Helena in 1799 for a possible "sinister motive" on the part of her commander.[63] Dalrymple put his private collection and East India Company experience to Admiralty use in June 1798, advising Marsden on the navigation of the Red Sea and on the passage time to the Strait of Babelmandel, as well as furnishing a set of his East India Company charts, plans, and memoirs for Commodore Blankett's precautionary expedition against a possible French move from Egypt toward India. He invoiced the Admiralty for published items, and for copying manuscript accounts and sets of views.[64]

The first phase of work in the Admiralty Hydrographic Office, the organization of the existing charts and plans, was complete early in 1800. Dalrymple later referred to it as the time "when the Hydrographical Office was made efficient in 1800."[65] Dalrymple and Walker had organized the materials sufficiently to be able to answer a request from the Admiralty Board by promising Nepean on March 22, a list of charts and plans in the Office "fit to be engraved." Nothing more advanced than using the London map trade for engraving and publication seems to have been envisaged by the Board of Admiralty, although in 1808 Dalrymple claimed, in a heated moment, that from its creation, "an avowed Purpose of the [Hydrographic Office] was to publish accurate Charts for The Use of the Royal Navy, which purpose could not ever be carried into effect if MS Charts &ca. given into the Hydrographic Office were delivered to private Chart Sellers to be mixed with other Materials of unknown or doubtful authority."[66]

Dalrymple accompanied his 1800 list of charts with an application to install a press in the Admiralty building. The board's response was immediate and favorable, and a press was in place later that year. The establishment of a rolling press necessitated changes in the complement of the Hydrographical Office: two plan engravers were employed, John Cooke and Isaac Palmer, each of whom had a private business outside the Admiralty, as did Thomas Harmar, employed on piece-work rates as writing engraver. The Hydrographic Office press was seen as a security press, not subject to commercial constraints. The capital costs, salaries, and materials were borne by the Admiralty, and there was no provision for selling charts. Dalrymple's budget for expenses allowed only for proofing plates, not for extended print runs, and the press was often at a standstill.[67] Late in 1801 Dalrymple demonstrated the capacity of the Hydrographic Office to function as a security press when, as part of the Admiralty's involvement with surveys being carried out for Philip Gidley King as governor of New South Wales, he had engraved a sketch from King of Bass Strait with the discoveries made in *Lady Nelson,* sending proofs to Nepean in January 1802.[68]

Dalrymple had no authority to commission surveys, only to engrave charts from materials supplied by ships' officers, supplemented by manuscripts in the Hydrographic Office and by foreign printed charts. To fill extensive gaps in the Hydrographic Office collections for the East Indies, Dalrymple negotiated for the Admiralty to buy 100 copies each of over 800 of his East India Company plates. The Hydrographic Office press was kept busy producing these 81,700 impressions in 1804 and 1805.[69]

Dalrymple designed coherent series of charts for the south coast of England, following his East India Company practice of standard related scales. The lack of original surveys in the Admiralty was a recurring theme, and a reflection that Dalrymple had no authority to order surveys, and little control over the results. The habit of surveying officers to retain their surveys, a habit left over from the time when no effective Hydrographic Office existed, proved difficult to break. Keeping track of privately published surveys by naval officers was a continual problem: in October 1807 Dalrymple produced a list of more than fifty such charts and plans, doubting whether the Admiralty could reengrave them without infringing private copyright.[70]

To encourage the transmission of nautical information from ships of the fleet, Dalrymple designed (or revised) in February 1804 a "Form of Remark Book" with a cover letter that he proposed sending with orders to ships for officers to collect information about ports visited.[71] Few responded, but of the officers serving in home waters in the early years of the office, William Bligh was the most sympathetic to Dalrymple's aims. Bligh was sufficiently conversant with Hydrographic Office procedures to assume temporary responsibility for the office in the spring of 1804, when Dalrymple was ill for six weeks.[72]

Dalrymple did not compete with commercial publishers in providing small-scale coasting charts, and he unwittingly provoked the disagreement that led to his dismissal in 1808 by openly disclaiming any capacity to judge the merits of charts of waters he did not know. The Admiralty of 1807 had changed since 1795, and the demands made of Dalrymple had changed as well. When Marsden retired on grounds of ill-health in June 1807, to be replaced by William Wellesley Pole, the entire "office memory" of the circumstances of Dalrymple's appointment and his achievement since 1795 was erased. The Board of Admiralty, through Pole, developed a different concept of a Hydrographic Office's function, seeing it as an office for assessing the products of the London map and chart trade, and as an agency for supplying bought-in charts in wholesale quantities for fleet use.

In 1807 Dalrymple had been asked to purchase and arrange "a compleat Set of all Charts published in England"[73] and to make a selection of "the best and most necessary Charts and Plans of Ports." He declined the latter task on the grounds that very few privately published charts had accompanying memoirs by authorities, and suggested "a Committee of Officers who have the necessary experience."[74] The Admiralty straightaway appointed a committee of three serving officers, Home Popham, Edward Columbine, and Thomas Hurd, who proposed first to examine commercially published charts of the English Channel, North Sea, and Baltic Sea. By mid-December they had set up a parallel office in the "Chart Committee Room," but were not immediately in conflict with Dalrymple, who made no claims for charts he had not produced. By May 1808 they had formulated a severe critique of Dalrymple's unsystematic chart publication: "There are many important parts of the World, of which we have no tolerable Charts Published, the materials to supply the deficiencies, may in a considerable degree be obtained from Manuscripts, in the Hydrographic Office, and they should be engraved in the Common course of business in that Office, but we fear that little benefit can be expected from that Quarter, unless Their Lordships should be pleased to command Mr. Dalrymple to employ the engravers on such Charts only, and in such a progressive Order, as they might direct."[75]

This, with derogatory comments on Dalrymple's choice of scales and projections for charts, seems to have been calculated to provoke an outburst justifying his dismissal. But it was overshadowed on May 28, when Dalrymple refused to hand over his security copies of the D'Entrecasteaux charts of New Britain, surrendered by Rossel in 1795. Dismissed instead for his celebrated refusal, Dalrymple died three weeks later.[76]

Dalrymple's legacy was to transform an office housing inert accumulation of manuscript material in the care of a clerk to a working Hydrographic Office, with draughtsmen, engravers, and proofing press, and an initial body of charts, sixty-five completed and a further twenty or thirty in preparation, done with a cautiousness of compilation and restrained style of presentation and printing, which set the style for Admiralty charts produced from then on. Still, the funding and staffing the Admiralty Board accorded his office did not yet allow it to respond to fleet requirements.

During Dalrymple's tenure as hydrographer to the Admiralty he continued to publish charts for the East India Company, encouraging the retired "country trade" captain James Horsburgh in his own publication of charts and his compilation of *Directions for Sailing to and from the East Indies, China, New Holland, Cape of Good Hope, and the interjacent Ports,* which appeared in two volumes in 1809 and 1811. Dalrymple's death opened the way for Horsburgh to be appointed hydrographer to the East India Company in December 1810.[77] (While on Dalrymple's dismissal by the Admiralty in June 1808, Thomas Hurd was immediately appointed hydrographer, to carry into effect the plans of the Chart Committee.)[78] From 1810 both the East India Company and the Admiralty had active Hydrographic Offices, operating in effective cooperation.

In addition to purchasing commercially produced charts of the Atlantic Ocean, Mediterranean, and North Sea for the Admiralty, Hurd had to bring Dalrymple's plates to a state fit for publication. By November 1808 many charts in the English Channel sequence were ready, and the chart series inherited from Dalrymple was sufficiently advanced by 1811 for Hurd to publish the two atlases *Charts of the English Channel* and *Charts of the Coasts of France, Spain and Portugal.*[79] Hurd replaced Dalrymple's imprint with his own on many of these charts. Increasingly Hurd's new publications replaced or supplemented Dalrymple's charts, but the Mediterranean Atlas of 1813, and the 1815 atlas of North Sea and Baltic charts, contained impressions of many Dalrymple plates. The first issues of small-format atlases that Hurd prepared for other ocean areas relied heavily on the stocks of impressions of East India Company plates that Dalrymple had allowed the Admiralty to print in 1804.

Dalrymple's will had bequeathed all his atlases and manuscript charts to the Admiralty, including a group of 160 manuscripts from Byron's circumnavigation, as well as a collection for the Oriental Navigation which included over 100 old Dutch charts on vellum.[80] Hurd acquired a large number of Dalrymple's East Indies copper plates as scrap, and reprinted from about 400 of them, changing Dalrymple's imprint for his own, until uncertainty about the risk of being held responsible for outdated content outweighed the credit of appearing as the republisher.

Like Dalrymple, Hurd was able to publish new charts only sporadically, reacting to requirements for office compilations, or responding to the provision of particular manuscript surveys. In 1814 he reviewed the deficiencies in British hydrography of the world's coasts, pointing particularly to China and the eastern seas, to the coasts of Africa, the east coast of South America from the River Plate to the West Indies, and the southern shore of the Mediterranean, as well as the coasts of Scotland and Ireland. This marked the first statement of intent to provide chart coverage worldwide. With the decision in 1821 to price and sell Admiralty charts to the public through a group of agents, Hurd turned the Hydrographic Office into the chart depot envisaged in 1808. The first Admiralty chart sales catalogue was issued in 1825. But even in Francis Beaufort's long tenure as hydrographer, from 1829 to 1855, catalogue pages showed impressions from old Dalrymple plates, rescued as scrap in 1810, alongside coastal charts from systematic series such as Owen's.[81] It is to Beaufort's tenure that one can date the modern concept of the Admiralty chart, which relied for its authority on the Hydrographic Office as publisher rather than on users' assessments of the merits of individual surveyors or compilers. And yet Beaufort had direct connections in hydrography back to Dalrymple. The hydrographer in post in the Admiralty during the Baltic War of 1854–56 had served in 1789 as captain's boy to Lestock Wilson on the ill-fated surveying voyage of the *Vansittart*. The same man, as the young commander of the store ship *Woolwich* in 1805 and 1806, was then in regular correspondence with Dalrymple as hydrographer. As well as publishing Beaufort's survey of Montevideo, Dalrymple had given him a complete set of his East India Company charts and sailing directions.[82] Among these was Dalrymple's 1779 proposal for the recording of ships' tracks by chronometer and by dead reckoning on blank charts to aid in chart compilation. This found a ready audience in Beaufort in 1805: even the wind scale that Beaufort introduced in the 1830s, on which commanders could record their weather observations, he had copied into his 1806 journal from Dalrymple's 1779 pamphlet.

Dalrymple may have been fifty years ahead of his time in his advocacy in 1779 of systematizing and standardizing the collection and dissemination of hydrographic information in charts and sailing directions, and in developing means of assessing, analyzing, reconciling, amending, compiling, and sometimes commissioning, charts and sailing directions for mariners. But he highlighted the problems of oceanic, coastal, and archipelago navigation, particularly the developing commercial navigation of European powers to the East Indies with its hazards and its requirements for reliability. The discovery of the longitude (or more precisely John Harrison's invention of a reliable means to carry the local time of one's point of departure as a datum of measurement on one's journey) happened at the time when Dalrymple was searching for means of establishing reciprocal longitudes for chart construction, and he consequently encouraged the use of chronometers from John Arnold's workshop on board East Indiamen as an effective means of reducing observations of rocks, shoals, and currents at sea to the common datum necessary for their reconciliation and promulgation in standard form. The problems he encountered in compiling systematic series of charts detracted from the expectations he raised, both in the Admiralty and the East India Company, with overambitious schemes of chart series, and were to contribute to his eventual dismissal. Then, following a period of expediency in chart publication after Dalrymple's death, the Admiralty and the East India Company separately ordered and provided for the type of systematic coastal survey that Dalrymple had proposed. During Sir Francis Beaufort's hydrographership from 1829 to 1855, the Admiralty chart and printed sailing directions increasingly became the international authority for navigation. With the eventual transfer of East India Company chart publishing to the Admiralty Hydrographic Office in the 1850s, the fruit of those fifty years—from Dalrymple in 1779 to Beaufort in 1829, and for the East Indies as for other parts of the world—is the creation of the international system of hydrographic charts and sailing directions we know and use today.

4

Mapping a Transcontinental Nation

NINETEENTH- AND EARLY TWENTIETH-CENTURY AMERICAN RAIL TRAVEL CARTOGRAPHY

JERRY MUSICH

More than any other economic or technological factor of the nineteenth century, American railroads helped build a continental nation. Most dramatically, the first transcontinental railroad tied together the Atlantic and the Pacific coasts in 1869. But, as symbolic as the link-up between the Union Pacific and Central Pacific surely was, of greater importance from 1830 onward was the fact that railroads freed American settlement and economic development from their ties to waterways—the ocean seacoast, the coasts of the Great Lakes, and the banks of major riverways. Railroads made all areas of the nation accessible, transforming the United States from a coastal nation to a continental one.

Railroad cartography both documented and facilitated this transformation. As Andrew Modelski notes, "Railroads could not have developed in North America without maps. The builders of the railroads depended on surveys to help determine the best routes, and when subsequent lines were in operation, railroad companies used maps to promote them."[1] As railroad development changed the face of the nation, travelers, shippers, investors, employees, government officials, and the general public sought out current maps to understand the transformation as well as the geographical details of America's emerging rail network. Railroads (and the map publishers and printers who served them) strove to provide this information in abundance—not just to satisfy the demand for information, but also to help create demand for railroads themselves.

Tens of thousands of different maps and map-enriched railroad publications were produced in the United States between 1830 and the early twentieth century. These ranged from highly detailed surveys of individual routes produced by and for the men who planned, constructed, and operated them, to small and rudimentary diagrams appended to handbills and advertisements. Indeed, railroads produced so many maps themselves and fostered the production of so many by others that we are challenged simply to get our arms around the subject of nineteenth- and early twentieth-century American railroad travel maps.

In a very broad sense, railroad maps of this period divide into five general categories on the basis of the audiences they were designed to serve: investor maps, land promotion maps, general reference maps, industry maps, and passenger travel maps. This essay will begin with a brief survey of the first four categories before looking in depth at the most public of them all: railroad travel maps. By riding trains, Americans came to understand how thoroughly railroads were transforming travel and to see firsthand the extraordinary growth of the country fostered by the iron horse. Travel maps interested potential travelers in riding the trains and then interpreted for them where they were going. Of all the categories of railroad maps, the travel map most invites study.

A Historical Typology of American Railroad Maps

1. INVESTOR MAPS

Railroads produced maps both for prospective investors and, after the railroads were operating, for current investors. Prospective maps provided a broad overview of a possible route (or routes) to initial or potential investors, usually within published prospectuses or annual reports. The scale of such maps was quite small, and, because precise detail was not the point, the maps were often schematic. Prospective maps were intended to demonstrate the approximate route of proposed and existing lines and to give some general information about communities served. On occasion, they also depicted connecting lines.

The importance of prospective maps is clear when we consider the directors and investors for whom they were intended. Many of the early examples were commissioned by the boards of directors of newly forming railroad companies as part of their effort to determine the feasibility and

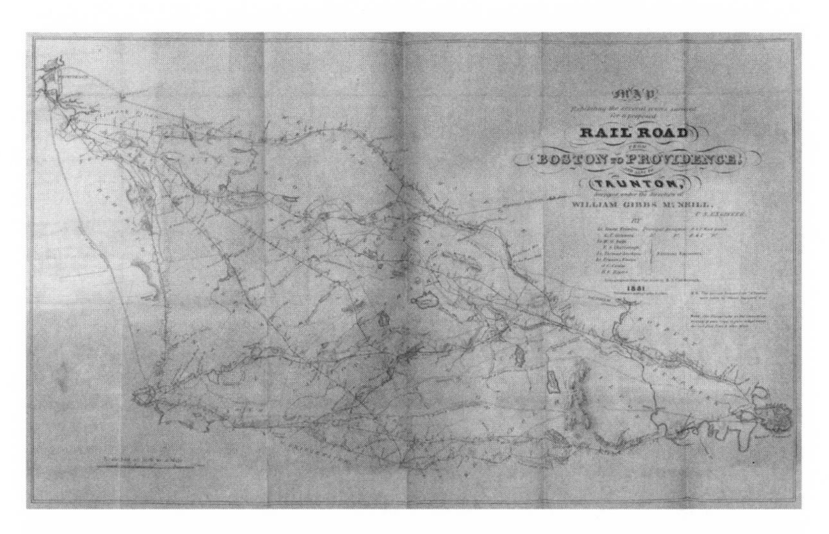

FIGURE 4.1 "Map Exhibiting the Several Routes Surveyed for a Proposed Rail Road from Boston to Providence and also to Taunton, Surveyed under the direction of William Gibbs McNeill, U.S. Engineer" (Pittsburgh: Wm. Schuchman, 1831). Personal collection of the author.

cost of constructing lines. The board of a railroad would commission an engineer to survey a number of possible routes, provide a survey map, and estimate the cost of building. An early example is the 1831 *Map Exhibiting the several routes surveyed for a proposed Rail Road from Boston to Providence and also to Tauton* (figure 4.1). Prepared by William Gibbs McNeill, the engineer in charge of the initial survey of the line, the map was bound into his report to the railroad's board of directors.[2] The report included commentary on each of five proposed routes; geographical hurdles to be overcome; and a fold-out sheet twice the size of the map it accompanied called *Profiles of Routes for the Proposed Boston & Providence Rail Road.*[3]

The investor audience was as important as the boards of directors. During the first decades of railroad building in America, most investment capital was found either in eastern cities, such as New York and Boston, or in London. While such investors might know the location of major communities on the East Coast, they remained unfamiliar with the geography of inland communities. In the mid-1840s and the 1850s, when railroad building took hold in the Midwest, prospective maps helped investors understand the relationship of a proposed railroad to the waterway system of the area, to other existing or proposed rail lines, and to those towns and agricultural areas that the railroad might serve.

Early midwestern railroads especially had to develop creative financing schemes. The mid-1830s had been a time of wild railroad speculation.

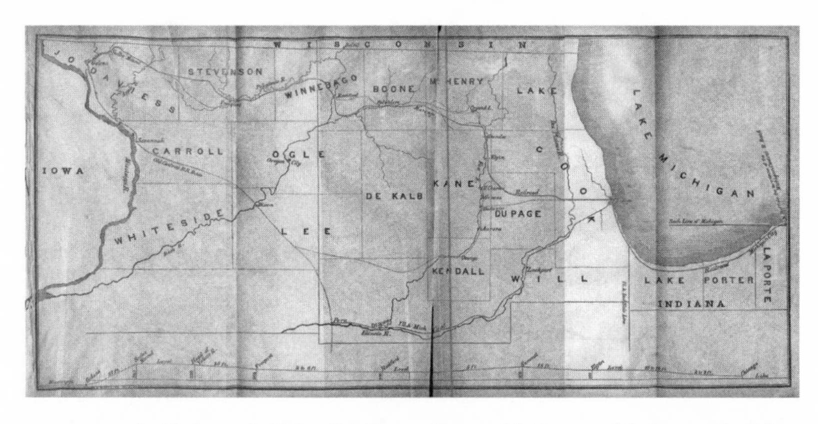

FIGURE 4.2 Untitled map, in Richard P. Morgan, *Report of the Surveys of the Route of the Galena and Chicago Union Rail Road together with the Original Charter of the Company, and Amendments thereto* (Chicago: Daily Tribune Print, 1847). Courtesy of the Newberry Library.

In 1836, the state of Illinois, for example, proposed building an ambitious network of six railroads. When a bank panic devastated the nation's economy in 1837, the Illinois plans collapsed and the state repudiated its bonds. When the idea of building midwestern railroads revived in the mid-1840s, eastern investors well remembered having been burned in the 1837 experience and turned a deaf ear to requests for capital. As a result, early midwestern railroads were partially financed by selling stock subscriptions to farmers, townspeople, and even local governments along the proposed route of each railroad. Prospective maps were critically important for local investors needing to see if the railroad would run close to their property or community.

An instructive example of an early midwestern prospective map is the 1847 survey for the Galena & Chicago Union Rail Road (G&CU), one of the first to operate on a sustained basis in Illinois (figure 4.2). On this map, a possible route was drawn to show which towns lay near the rail line. Because of the railroad's shortage of capital, the proposed route was inexpensive to build, following river valleys wherever possible to make use of favorable, gently graded terrain. For example, while a straight route between Belvidere and Rockford would have reduced the trackage by half, Richard Morgan, the surveying engineer, feared that the grading to accomplish this would impose too high an initial cost for a railroad with so little financing. Thus, he suggested that the railroad follow river valleys between the two towns. This approach gave some railroads a circuitous quality, making them much longer than more direct routes would have been. In the 1850s, as financing and technology improved, a number of new railroads built what were known as "air lines"—routes laid out on straight lines, as though drawn on air—that corrected earlier route inefficiencies.[4]

In 1847, the success of any midwestern railroad venture was considered highly speculative. In addition, the G&CU's organizers did not have the money to produce a more lavish map. For both reasons, the published survey map was a relatively small, modest production. But the G&CU was an instant success, turning a profit within its first year, and it set off a near frenzy of railroad organizing throughout the Midwest. The Aurora Branch in Illinois, progenitor of the Chicago, Burlington & Quincy, was organized in 1850, as was the Rock Island & La Salle Rail-Road, a predecessor of the Chicago, Rock Island & Pacific.

The difference between the earliest eastern and midwestern prospective maps and those from just a few years later is striking. The pent-up demand for affordable shipping and travel was so great the second wave of railroad promoters were confident about their prospects. Knowing that investors were now interested in new railroad ventures and realizing that each new company was competing with several others for investor attention, some newer railroads published large, handsome surveys to suggest how promising their prospects were. The title page of the Rock Island & La Salle's 1850 *Report of the Chief Engineer . . . together with the Act of Incorporation* prominently advertised its inclusion of "a map showing the line of the road, and also the line of a rail-road from Rock Island to Council Bluff [*sic*]." The accompanying fold-out sheet measures twenty-five by forty-five inches and includes three maps. The first shows the rail connections from Rock Island to Boston and from Rock Island to Council Bluffs, though almost none of these lines existed at the time. The second is a map of the proposed route of the Rock Island & La Salle, and a third shows the city of Rock Island.[5]

After railroads were operational, they needed to inform their investors of their progress and of any plans for new construction. Maps were an effective way of doing this and thus often appeared in annual reports. Many of these maps were handsome, quite detailed, and large enough to require folding. They depicted both the existing lines and any extensions or alterations that the company proposed. The map included by the Galena & Chicago Union in its 1854 annual report (figure 4.3) gives a clearer, more precise picture of the actual route than that provided by the survey map of 1847. In addition, the newer map depicts several proposed routes or extensions, supporting explanations in the annual report's text about the company's intentions. For example, a diagonal line from Cottage Hill to Elgin depicts a route the G&CU proposed as early as 1853—although the company never built it. The newer G&CU map further hints at the company's plans through its typography and artwork. The bold lettering

FIGURE 4.3 “Galena and Chicago Union Rail Road, the only Rail-Road Route from Chicago through Northern Illinois to Wisconsin, Iowa and Minnesota” (Chicago: Henry Acheson [1854]), in *Galena and Chicago Union Railroad, Seventh Annual Report* (Chicago: Daily Democrat Book and Job Printers, 1854). Courtesy of the Newberry Library.

“from Chicago through Northern Illinois to Wisconsin” that dominates the upper portions of the map describes the existing service area, while Iowa and Minnesota are identified in outline type at the two lower corners as regions for future expansion.[6]

Another example is the *Report of the Directors of the Michigan Central Railroad Company to the Stockholders* for the year ending June 1875, which includes two fold-out maps. The first, titled “No. 1,” shows the Michigan Central as it was twenty-five years earlier, in 1850 when it made the decision to go around the southern tip of Lake Michigan to Chicago. Map “No. 2” shows the railroad in 1875. In addition to the main line from Detroit to Chicago, we are shown the Michigan Central Air Line that roughly paralleled the mainline but ran on a straight line; the Grand River Valley Division; the Jackson, Lansing & Saginaw Division; and the Detroit and Bay City Division.[7] While the text does not refer directly to the maps, the purpose for including them was to celebrate the growth of the railroad and to remind investors of how successful their company had been.

While the usual venues for investor maps were prospectuses and annual reports, a few appeared on railroad bonds and stock certificates. The earliest example of such use is an 1853 New York Central bond (plate 1). The map occupies the right edge of the bond and features small panoramic views of Albany and Buffalo, with the railroad connecting the two cities. The New York Central had been formed that very year through a merger of seven shorter and older lines that roughly paralleled the Erie Canal, so the map on this bond likely was intended to acquaint the investor community with the new company. Other railroads incorporating route maps on their stocks or bonds included the New York, Ontario & Western and the Illinois Central.

From the early 1850s to the Bank Panic of 1857, and then again after the Civil War, railroads were some of the most profitable investments available to individuals in the United States and abroad. As the railroad economy grew, independent publishers began to produce maps for the investor community, which was anxious to learn as much as it could about the promising industry. Henry Poor began publishing his annual *Manual of the Railroads of the United States* in the late 1860s; by the 1880s the *Manual* included railroad maps. In 1909, John Moody began publishing *Moody's Analyses of Investments: Steam Railroads,* and within three or four years he was including fold-out maps of several of the railroads he covered. While informative, many of Poor's and Moody's maps lacked the size and aesthetic appeal of the much earlier maps produced by the railroads themselves.

In the face of these independent publications, and because investors had become more familiar with railroads and their routes, railroads curtailed publishing more lavish maps for their investors and replaced them with smaller, simpler productions. Investor maps produced by the railroads became little different than maps designed for other audiences, such as traffic agents. Unhappily, except for such notable exceptions as the Southern Pacific annual report maps of the first years of the twentieth century and the maps published by the Chicago, Burlington & Quincy into the 1920s, investor maps became simple, prosaic affairs.[8]

2. LAND PROMOTION MAPS

Railroads recognized that their long-term growth strategy had to include promoting the idea of locating along the railroad. Thus, many nineteenth- and early twentieth-century railroad maps were aimed at another category of investors—potential settlers and businesses. The most important publishers of such maps were the various land grant railroads. Starting in the early 1850s with the Illinois Central, Congress awarded substantial grants of land to midwestern and western railroads to foster the building of lines.

The grants saved the railroads the expense of having to buy the land for their routes, but, more importantly, the sale of land to settlers also became a major source of capital. Moreover, the availability of such inexpensive land helped attract the settlers and business investors who would in time provide the freight and passenger traffic for the company. The more settlers a railroad could entice to buy land—especially land owned by the railroad—the more traffic the railroad guaranteed itself for future years.

It was therefore incumbent on land grant railroads to produce maps that attracted purchasers for their land. A good example of a land grant promotional map is an 1872 Burlington and Missouri River Railroad Company piece advertising land for sale in Iowa and Nebraska. The B&MR produced it shortly after the completion of the transcontinental railroad from Omaha to Sacramento, at a time when Nebraska was much on people's minds. The document describes the railroad, how to get to the areas under sale, the availability of a Land Buyers' Exploring Ticket, the cost of credit, the terrain and natural resources of the area, and the history of recent population growth. It also repeats popular promotional phrases of the day, such as Bishop George Berkeley's "Westward the Course of Empire Takes Its Way." Another fine example is an 1882 folded brochure produced by the land department of the Texas & Pacific Railway (plate 2). The brochure's map, produced by Woodward & Tiernan of St. Louis, was intended to educate potential settlers about Texas and encourage them to buy land from the Texas & Pacific. Emigrants could buy "land explorers' round-trip tickets" to Texas and travel on special "Emigrant Sleeping Cars." Emigrants could then use a lodging house in Baird, Texas, while they determined where to settle. An illustration on the right-hand column of the brochure depicts the Immigrant House, a building where "families and baggage can be comfortably left, *free of charge,* while the heads of such families are looking out for permanent houses."[9]

Numerous such maps were distributed, sometimes as single folded sheets, sometimes as parts of promotional booklets. An example of the latter is the forty-eight-page *Guide to the Northern Pacific Railroad Lands in Minnesota,* published in 1872. The inside front cover provides a map of northern Minnesota (figure 4.4).[10] The railroad's projected route is in bold, while the territory for sale along the right-of-way is shaded. The back cover of this booklet provides a map of the United States, with the projected line of the Northern Pacific once again in bold. The intent of this map apparently was to demonstrate how the lands of northern Minnesota were readily accessible through the nation's railroad network.

Some railroads even printed their standard system map in brochures advertising special excursions for prospective settlers. One such brochure, the

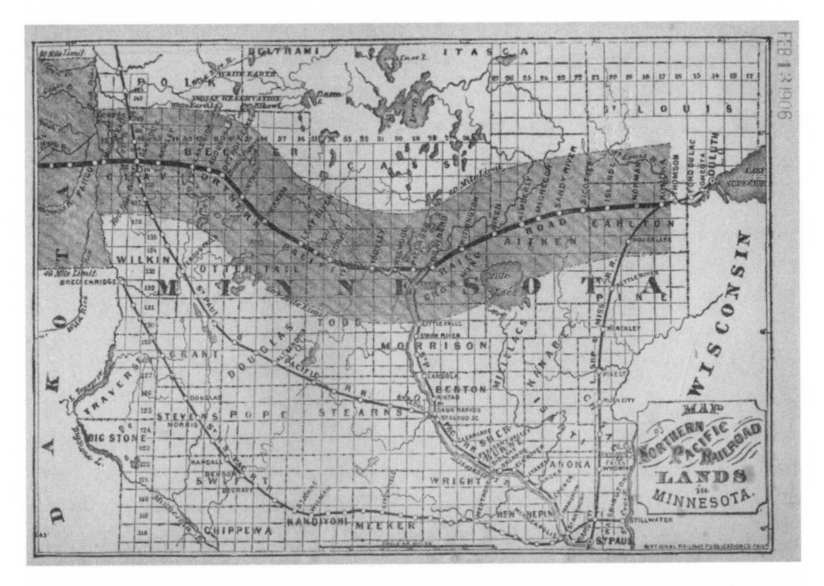

FIGURE 4.4 "Map of Northern Pacific Railroad Lands in Minnesota" (Philadelphia: National Railway Publication Press, n.d.), in *Guide to the Northern Pacific Railroad Lands in Minnesota* (New York [1872]). Personal collection of the author.

four-panel *Harvest Excursions via the Chicago Milwaukee & St. Paul Railway on August 29, September 10, September 24, 1895, to see A Golden Harvest,* explains that its excursions were planned "In order that the people of the more Eastern States may see and realize the magnificent crop conditions which prevail along the [railroad's] lines." To further encourage easterners to settle in areas served by the CM&StP, the brochure urges interested parties to contact its "General Immigration Agent" for tickets. The brochure's map emphasizes the "Iowa, Dakota, and Minnesota Lines" of the railroad, showing each station in great detail. Wisconsin and Illinois, the other two states through which the CM&StP operated, receive a much freer and less detailed treatment. Only major stations in those states are indicated, and Wisconsin is drawn with considerable liberties: Milwaukee, which is due north of Chicago on Lake Michigan, is shown about thirty miles further west than it is, distorting the coastline of Lake Michigan accordingly.[11]

Such promotional techniques succeeded. The Illinois Central, the first land grant railroad and recipient of more than 2.5 million acres in Illinois, reported average sales in the summer of 1855 of nearly 40,000 acres a month. The average price of those acres was about $13—a reasonable price at a time when farmers were reporting annual profits of about $15 an acre in good years.[12]

Railroads were not the only entities interested in promoting settlement and business investment. The emerging cities, towns, and agricultural districts of the Midwest and West realized they could best attract new business and new residents by highlighting the rail connections that made them easy to reach and easy to ship from. When business leaders in Milwaukee published *A Map exhibiting the System of Railroads centering at Milwaukee, with their connections with the East by the great Northern Route* in 1862, Milwaukee had already grown into a sizable city. But as recently as 1848, when Wisconsin entered the Union, the city had only 14,000 residents. The rapidly growing community accordingly published the map to announce its new status and demonstrate that it was primed for further growth.

A document serving a similar purpose is an untitled 1871 map printed in *The Great Resources and Superior Advantages of the City of Joliet, Illinois.*[13] At the time Joliet was a community of 10,200 residents, but the map (figure 4.5) suggests the city rivaled Chicago as a transportation hub. The Illinois and Michigan Canal and seven existing or projected railroads are shown reaching Joliet, versus only six running out of Chicago. A second map in the pamphlet is a street map of Joliet showing the routes of the various railroads that served the town.

Maps that trumpeted the promise of various communities, no matter how exaggerated, were especially important in the mid-nineteenth century, when many cities west of the Appalachians were relatively new and competing for settlers and investment. Minnesota boosters published maps showing the state's favorable geographical relationship to Lake Michigan, Milwaukee, and Chicago. "Bailey & Wolfe's Minnesota Railroad Map, Showing Connections with Lake Michigan"[14] appeared in *Minnesota Railroad and River Guide for 1867–68.* The 486-page publication by A. Bailey included "historical sketches of railroads in operation, and of principal towns in the state; an alphabetical directory of business firms in each town . . . , [and] railroad and river distances."

In the 1870s and 1880s, the desire to highlight the modernity of one's community, including its rail connections, expressed itself in maps and pictures in county atlases. Whenever possible, each township map in such atlases emphasized the location of rail lines, and numerous images included locomotives passing by community landmarks to portray even rural counties as modern members of the wider world. The map of Wayne Township, which encompasses the city of Fort Wayne, Indiana, from an 1880 *History of Allen County, Indiana,* shows three railroads: the Wabash; the Pittsburgh, Fort Wayne & Chicago; and the Fort Wayne, Muncie & Cincinnati, as well as the Fort Wayne & Little River Turnpike.[15] County atlases

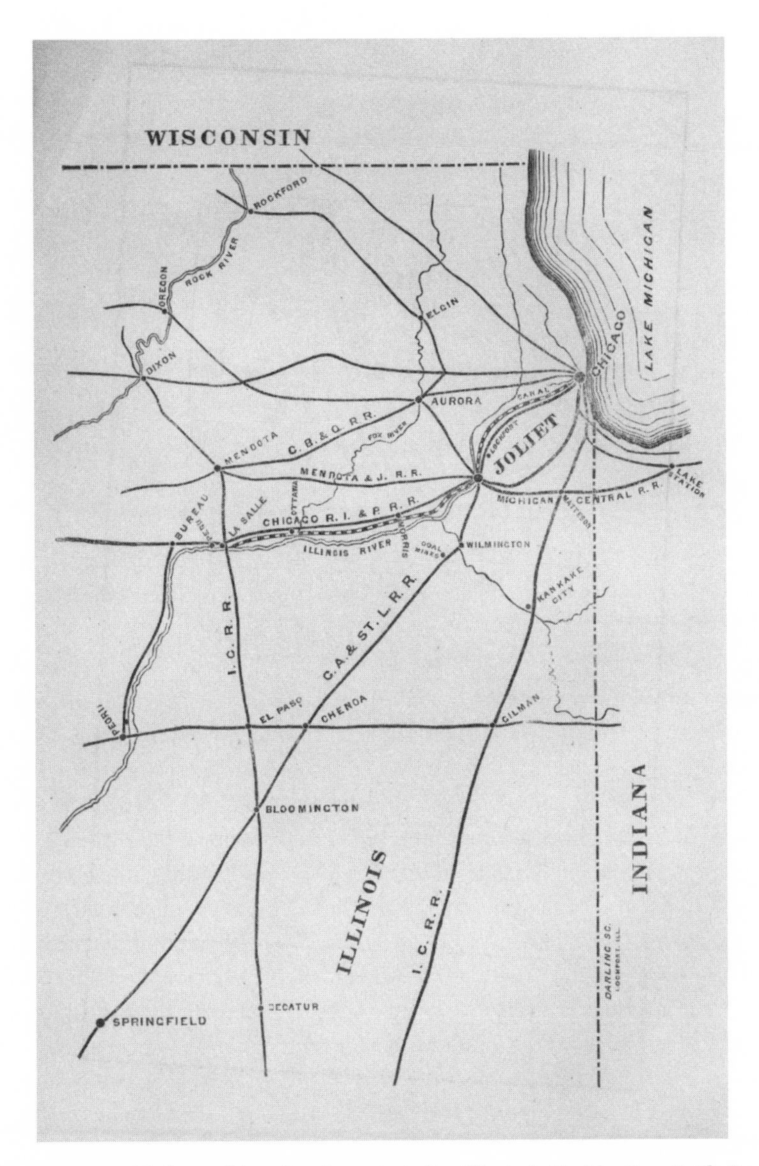

FIGURE 4.5 Untitled map of the railroads serving Joliet, Illinois, in Hopkins Rowell, *The Great Resources and Superior Advantages of the City of Joliet, Illinois* (Joliet: Republican Steam Press, 1871). Personal collection of the author.

also featured illustrations of city scenes, factories, and farms, and where appropriate included the image of a train to suggest the area's modernity (figure 4.6).[16]

Railroads, states, and communities seeking to lure European immigrants to North America prepared some promotional maps in the language of

FIGURE 4.6 Farm Residence of Frederick S. Druckamiller, Esq., in *Illustrated Historical Atlas of Elkhart County, Indiana* (Chicago: Higgins, Belden & Co., 1874), p. 58. Courtesy of the Newberry Library.

the country targeted and designed others expressly for travel agents. The March 1877 "Pennsylvania Railroad Co. Special Information for Agents Forwarding Emigrants to the United States and Canada, North America" touted itself as "a complete and convenient Guide for Agents." Several panels of the brochure list fares for travel from Liverpool to Philadelphia or Antwerp to Philadelphia. Half of this side of the unfolded sheet is then given over to a list of the emigrant fares from Philadelphia to approximately 800 destinations. The verso side includes a small world map entitled "Round the World Across the American Continent, via the Pennsylvania Railroad." The map includes the major oceanic routes; the route of the Pennsylvania Railroad to Chicago; the route of the Chicago & North Western; and the Union Pacific. In a larger, more detailed "General Map of the Pennsylvania Railroad and Its Connections," the railroad route is drawn in a heavy black line, with major stops, while the boundaries of the various states are set off with hand coloring.[17]

3. GENERAL REFERENCE MAPS

The growth of railroads was one of the nation's most dramatic, even breathtaking, events of the mid-nineteenth century. At the same time the burgeoning rail system opened up unsettled portions of the Midwest and West, large numbers of Europeans sought emigration to America. Sections of the nation simply exploded in population. Chicago had 20,023 residents in 1848 when its first railroad began operation. Five years later, in 1853, its population had mushroomed to 60,692. Many older eastern cities such as Buffalo, N.Y., saw a similar rate of increase due to the growth of inland railroads. Much initial midwestern rail traffic terminated at one of the Great Lakes, with its freight and passengers then taking lake ships to Buffalo and transferring either to the Erie Canal or the New York Central (or its predecessor lines) for the trip to eastern ports. Buffalo's population grew from 33,600 in 1845 to 49,740 in 1850 to 74,261 in 1855. Smaller inland communities also developed almost overnight as a result of the coming of the railroad. As George Douglas writes: "Tiny hamlets were not 'reached' by the railroad; the hamlet was literally dropped down out of the railway cars. Where the railroad ran, there was something, or at least a pretense of something. Where there was no railroad, there was no civilization at all."[18]

Spurring on the explosive expansion of American settlement, the network of American railroads thickened dramatically in the East and Midwest, began to reach across the Great Plains, and in the 1860s crossed the Rocky Mountains to the Pacific. This dramatic growth and the quickening pace of migration profoundly affected the nation's economy and politics and were clearly on Americans' minds, whether or not they were travelers or shippers. Americans needed current maps, and general reference maps published by independent commercial publishers paid increasing attention to the nation's transportation network—including roads, canals, and railroads—and by the late 1850s they highlighted railroads above almost any other cultural feature.

The individual maps in Samuel Augustus Mitchell's 1839 *School and Family Geography* emphasize existing and projected railroads, as well as the other elements of the contemporary transportation network. Canals are depicted, and rivers are given special emphasis as transportation routes. The overall lengths of major rivers such as the Mississippi and the Missouri are given, while tributaries such as the Wisconsin and Illinois Rivers are marked as navigable and their distance from the sea noted. Interestingly, the maps in Mitchell's 1858 edition of the *School and Family Geography* omit the interpretation of river travel, and, while canals are still shown, the emphasis is on railroads. The maps of American states in Henry Schenk Tanner's *A New Universal Atlas* for 1845 delineate canals and railroads in equally prominent blue and brown lines, respectively. Major roads and turnpikes also appear on the maps but are not colored or mentioned in the map legends.[19] Similar maps from the 1860s also show roads, canals, and railroads, but in works such as the 1862 *Johnson's New Illustrated Family Atlas,* railroads appear as the boldest of the three lines representing transportation systems, becoming the most prominent transportation features indicated.[20]

General reference maps were usually more expensively produced than other maps from the American railroad era—as commercial publications, which required considerable investment both on the part of the publisher and the consumer, they are among the most enduring and attractive records of the American rail network during the nineteenth century. Their steel-engraved lines were finer, clearer, and more elegant than the lithographs or wax-engraved maps preferred by railroads, and they were often hand-colored. These more expensive techniques were possible because the publishers sold their atlases and maps, thus recouping their expenses. And because several publishers were competing in this market, each publisher sought to make his maps as handsome as possible. In contrast, as we have seen, the maps produced by the railroads themselves were often crude, small in scale, and schematic. Maps made for travelers were intended to be used on an actual trip and thrown away when the trip ended. Indeed, as we shall see, they were often printed inside timetables or travel brochures. Investor maps, too, were designed to be disposed after a year's use. For example, the SooLine Railroad printed a current map in each of its annual reports from 1905 through the 1930s, thus making the previous year's map obsolete. The general reference maps published by Mitchell or Colton were bound into atlases, folded inside cloth covers, or published in other substantial books. They were designed to be used in the home or office and were not thought of as disposable.

At the same time, the growth of railroads and thus of communities was so rapid that general map publishers struggled to keep up. Though the atlases they produced might come out annually, or nearly so, they did not entirely remake their maps with each edition. Instead, a map bearing the same copyright date would be updated periodically to account for changes in railroad lines and other geographical details.

4. INDUSTRY MAPS

Our first three categories of maps were documents intended for people interested in the railroads as an idea, an abstraction. Investors, settlers, and entrepreneurs wanted to learn about the location of railroads and to envision how that railroad presence might affect their interests. The general public wanted to see how railroad activities were transforming their nation. Our final two categories involve maps for people who actually used and operated the railroads. The first of these is a broad range of maps intended for railroad employees, government, and freight traffic agents and shippers.

Maps for Employees Railroads needed maps for use by their train crews, engineering departments, maintenance-of-way departments, and contractors. A variety of small-scale divisional maps, larger scale engineering maps, and track charts filled this need. Train crews relied on maps showing the layout of the division on which they operated. Often these maps were printed as stand-alone documents; a few were included in the railroad's employee timetables. The New York Central, for example, routinely printed divisional maps with its in-house timetables (figure 4.7).[21]

While employee divisional maps tended to be somewhat stylized and small-scale, engineering drawings were large-scale and highly detailed. They showed line gradients and curvatures, helpful to train operating crews who needed to anticipate where to slow down for curves and stations. They also showed bridges and grade crossings for easy reference by maintenance-of-way crews. They even showed when rails and ties were installed and referenced the weight of rail and frequently the size of ties to provide engineering departments with a snapshot of the physical plant, useful in planning capital renewal projects.[22]

As a result of their detail and scale, engineering maps tended to be quite large. For example, a Chicago, Burlington & Quincy map of its trackage in Aurora, Illinois, dating from 1919, measures a remarkable 28 feet in length (figure 4.8). These were working maps, and surviving examples bear evidence not only of use but also of the life history of the railroad itself. The CB&Q Aurora map, for instance, has fifteen years of additions and corrections written onto its face.

Engineering drawings encompassed almost any feature of a railroad's right-of-way, including the track itself, buildings, and other structures. Track charts, a third, more specialized type of map used in railroad operations, focused on the company's trackage, including the location of switches and sidings.

Land Maps Railroads acquired land adjacent to their trackage to add sidings or build new trackside facilities. For this purpose they produced maps that showed the parcel of land owned by a private party and detailed where the proposed siding or facility would cut across the property in question. Since railroads enjoyed the right of eminent domain, they had to clearly delineate property they hoped to acquire. Such maps were then used in property negotiations and in courthouse filings. Maps of particular parcels were often relatively small, often on an 8 ½ × 11 inch sheet of paper, though their scale is quite large. An example is "'Exhibit A': 3 Miles East of Kokomo Ind., Proposed Purchase of Land from Mary A. Langley for Right of Way." This map was produced by the Office of the Engineer of Maintenance of

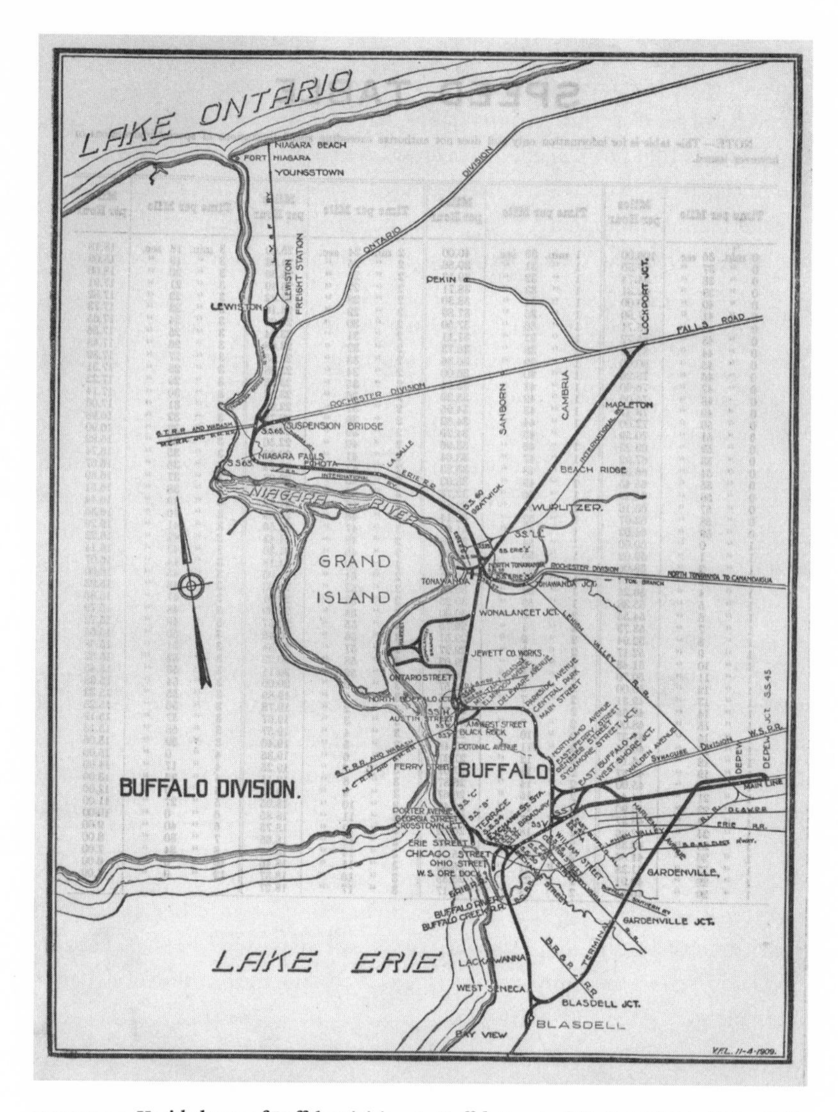

FIGURE 4.7 Untitled map of Buffalo Division. In *Buffalo Terminal (Buffalo Division) Time Table No. 10A for Employees Only* (Buffalo: Mathews-Northrup for the New York Central Railroad, 1919). Personal collection of the author.

Way for the Pennsylvania Railroad, Richmond Division of the Lines West of Pittsburgh.[23] The map is a simplified diagram of the section lines, for purposes of orientation, and the 40 × 413.5 foot parcel of land the railroad intended to acquire.

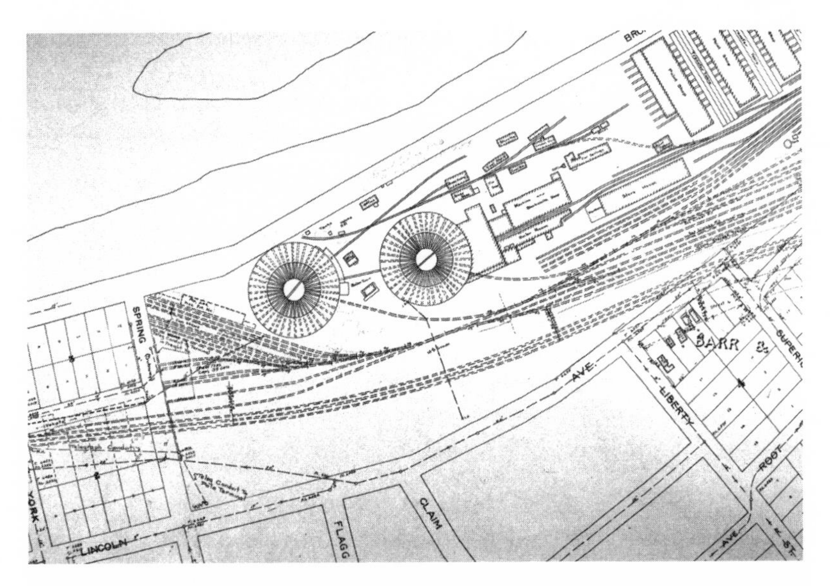

FIGURE 4.8 Detail from *C.B. & Q.R.R. Proposed Track Elevation, Aurora, Ill.* (Chicago: Chicago, Burlington, and Quincy Railroad, 1913–19). Personal collection of the author.

Traffic Maps for Shippers and Businesses Companies that relied on railroads to ship their freight also needed specialized maps to help them understand how to get their merchandise, produce, equipment, or raw materials from one locale to another. The railroads' traffic agents, who served shippers, relied on maps as well to help route freight and calculate shipping charges. Accurate maps describing the nation's rail system in its entirety, without bias or distortion, were essential to these users. No single railroad served all regions of the nation, and maps produced by the railroads themselves were inadequate because they showed the routes of connecting and competing lines incompletely or with prejudice. A number of independent publishers produced traffic maps to fill the void. These were published either as separate maps of individual states or regions or collected in sturdy atlases.

One particularly useful set of maps is the Asher & Adams maps of the 1870s. Published as part of the *Commercial and Statistical Atlas and Gazetteer*, these 16 × 22 inch maps emphasize the routes of each state's railroads, clearly labeling them all. The only other significant details included are cities and counties, while rivers and roads are deemphasized. The most widely used of these shippers' atlases during the late nineteenth century may have been Rand McNally's annual *Business Atlas* (later *Commercial Atlas*), which appeared in 1876. Soon thereafter, Rand McNally began publishing the individual state maps from the atlas as separate *Pocket Maps &*

Shippers' Guides, a practice the company continued well into the twentieth century.

A later example of shippers' maps was William Arthur Shelton's *Atlas of Railway Traffic Maps*, published first in 1913 by LaSalle Extension University for use by traffic agents and by those studying to be traffic agents. The atlas claimed to be

> the first attempt to present comprehensive and accurate information concerning the freight rate territories throughout the country on maps prepared for that purpose. . . . These maps were prepared primarily to be used in connection with the LaSalle course in Interstate Commerce and Railway Traffic. . . . Independent of this course, however, the maps will furnish to traffic men and students an invaluable reference work. The information contained on these maps is not to be found elsewhere in one place.[24]

The atlas consists of fold-out maps with such titles as "Freight Routes in the Southern Territory" and "Established Grouping of Territory for Rates to St. Paul, Minn."

Government Maps Government's involvement with railroads shifted dramatically during the nineteenth and early twentieth centuries. Initially, governmental units at all levels avidly promoted railroads. They provided land grants totaling 132 million acres, backed the bonds of some railroads, produced promotional literature, and in other ways lent enthusiastic support. This congenial attitude began to sour as early as the mid-1850s. Railroads had become remarkably profitable, and their stock had passed into the hands of eastern and London investors. As a result, little of the railroads' profits passed to the locations the railroads served. Farmers, merchants, and the general public came to resent the high tariffs and fares they paid when they knew those high prices were passed on to out-of-state investors in the form of extravagant dividends—in extreme cases reaching a return of 25 percent. In addition, as Harold Dunham notes, railroads that were recipients of land grants "had almost uniformly evaded the obligations and abused the privileges which were the object of this generosity."[25] Thus, the earlier governmental impulse to support railroads became one to regulate them.

Many states found means of overseeing their railroads, including the creation of commissions that needed maps of their state's rail system.[26] In 1868 the state of New York published an *Annual Report of the State Engineer & Surveyor* that included 774 pages of reports on individual lines, along with a large fold-out "Map of the Railroads of the State of New York."[27] Many of these maps, especially around the turn of the twentieth century, were quite expansive; some were designed to hang on a wall or even on rollers so they could be pulled down for use. Others were folding maps, some of the sturdiest having a fabric backing for extra support.

5. PASSENGER TRAVEL MAPS

Railroad maps for passenger travel were the most widely used of all railroad maps and the most important. The remainder of this paper is devoted to them. Often distributed directly to travelers, they were important instruments in the promotion of rail pleasure travel, helping to create the modern American tourism industry. Their ubiquity and variety make them valuable aids to the study of railroad history.

The chief purpose of rail travel maps, of course, was to demonstrate to travelers and to travel agents how rail passengers could travel from one point to another. While freight agents needed to know the route of every branch line, travelers only needed to know where passenger trains ran. Furthermore, travelers needed to carry their maps with them, so travel maps were portable while maps for the general public might be bound into a sizable atlas or geography text. Because clarity and portability were important, passenger travel maps generally avoided the full detail of a traffic or freight map, showing instead only the lines that served passenger trains.

Passenger maps also promoted the very idea of travel, and as promotional pieces they appeared in a wide variety of contexts. Though passenger maps sometimes were published as stand-alone documents, they often formed part of a larger promotional publication that included pictures and word pictures of scenes along the way. In such works, the maps were of no greater importance than the promotional text, pictures, and graphics. Many of these maps were therefore quite small, both because they needed only to provide a small bit of information and because space was needed to accommodate the other material. All of these factors encouraged the development of a schematic and stylized map design that showed rail routes as straight lines between points and that sacrificed detail and accuracy for simplicity and promotional impact. As Modelski observes, these "distorted schematic railroad map[s] . . . were made to show the importance of a particular railroad by the intensity of the line symbol, distortion of scale, and manipulation of particular lines to portray them favorably in relation to competing lines."[28]

By 1930, railroads had peaked and, for all practical purposes, had introduced all the variations of railroad travel maps they were ever to produce.

Studying the changes in maps over the first 100 years of railroad operations provides a unique perspective on the changes that occurred with railroads themselves.

Four Stages of Railroad Passenger Maps, 1830–1930

Maps for railroad passengers underwent significant changes over the period from 1830, when American railroads began construction, through the first decades of the twentieth century, when railroads were at their height. Four distinct phases of development in these hundred years reflect the development of American railroads themselves. The initial phase saw the construction of the first, rather modest railroads. This phase began in the East and Southeast in the early 1830s and in the Midwest in the late 1840s and 1850s. In the second phase, starting in the 1850s, existing short rail lines began to consolidate into much longer lines, and the main trunk lines began aggressively to build a web of branch lines. The third phase, covering the final decades of the nineteenth century, began with the construction of the new transcontinental railroads, along with the first active promotion of rail travel for sightseeing. The final phase culminated with the building of interurban railroads in the 1890s and early part of the twentieth century, a new order of railroads that fostered the further growth of suburbs and the development of new exurban communities.

1. "LOOK OUT FOR THE CARS!" BUILDING THE INITIAL ROADS

The American railroad era began in May 1830 with the opening to traffic of the Baltimore and Ohio Rail Road. Over the course of the 1830s American supporters of internal improvements enthusiastically embraced railroad construction. While a serious bank panic in 1837 brought a great number of the more speculative railroad schemes to a halt, by the end of the decade some 2,800 miles of railroads were in operation. The 1840s saw a tripling of that mileage to roughly 8,500 miles. Approximately 22,000 additional miles of railroads went into operation in the 1850s, a decade of truly breathtaking growth in which midwestern states threw their energy and capital into railroad building.[29]

Railroads as Components of a Larger Transportation System Supporters of the first American railroads did not see themselves creating a rail network to replace other modes of transportation, except perhaps in their wildest dreams. Railroads of the 1830s and 1840s served one of two purposes. As George Taylor and Irene Neu note, most early railroads served

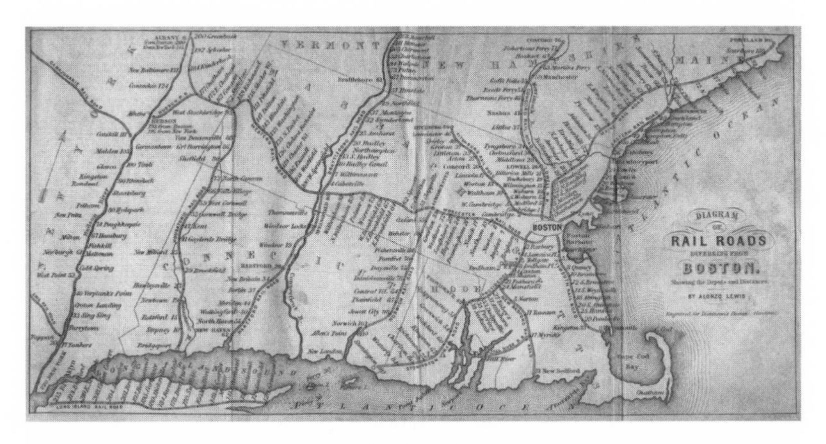

FIGURE 4.9 Alonzo Lewis, "Diagram of Rail Roads Diverging From Boston, Showing the Depots and Distances," from S. N. Dickinson, *Boston Almanac for the Year 1846* (Boston: Thomas Groom, 1846). Personal collection of the author.

nearby and local needs—such as controlling trade of the region immediately surrounding their headquarters community.[30] A contemporary map illustrating this point is the "Diagram of Rail Roads Diverging from Boston, Showing the Depots and Distances," which was included in S. N. Dickinson's *Boston Almanac for the Year 1846* (figure 4.9).[31] Some seven railroads are shown converging on Boston, all of them connecting nearby towns with the area's central seaport.

Other, more ambitious railroads were built to complement the existing transportation network, especially as a means of connecting two or more waterways. The first major eastern railroads were designed to connect an Atlantic seaport with an interior waterway, such as a major river or a major lake, while early midwestern railroads were intended to connect two lakes, such as Lake Erie and Lake Michigan, or Lake Michigan with a major river such as the Mississippi.

The prospectus map bound into an 1850 booklet published by the Bellefontaine and Indiana Railroad shows this geographical logic at work (figure 4.10).[32] The Baltimore & Ohio, still under construction, strives to connect the Baltimore seaport with the Ohio River at Wheeling, Virginia, a point it finally reached in 1853. The Hudson River Rail Road connects New York City, emerging as the preeminent eastern port, with Albany and an amalgam of ten very short lines running west from Albany to Buffalo on Lake Erie. Similarly, the New York & Erie has reached Owego, New York, on its way to Dunkirk, New York, further west on Lake Erie than Buffalo. Not to be outdone, Philadelphia supported the Pennsylvania Rail Road in its efforts to connect its port with Pittsburgh at the confluence of three rivers,

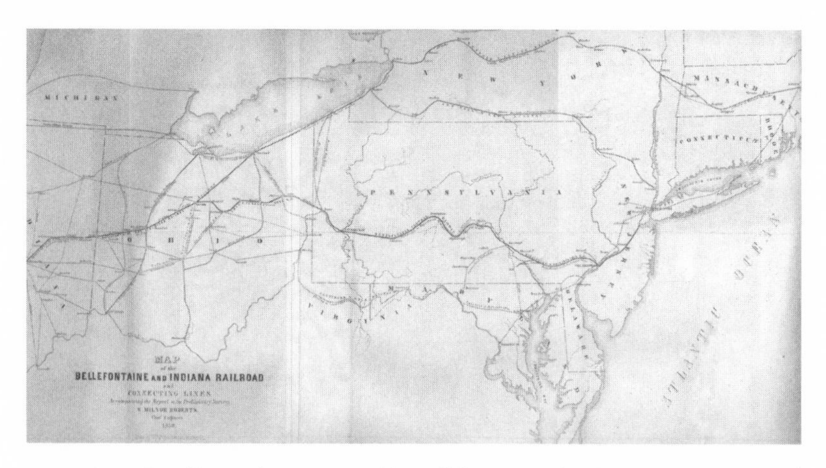

FIGURE 4.10 W. Milnor Roberts, "Map of the Bellefontaine and Indiana Railroad and Connecting Lines" (Pittsburgh: Wm. Schuchman, 1850), in *Report on the Preliminary Surveys for the Bellefontaine & Indiana Rail Road Company* (Pittsburgh: Johnston and Stockton for the Bellefontaine & Indiana Rail Road Co., 1850). Personal collection of the author.

while the seaport of Boston is connected with the Hudson River at Albany by several smaller railroads. These lines transported travelers and freight by rail between the East Coast and various inland waterways, where they were transferred to an appropriate lake ship or riverboat. The Baltimore & Ohio momentarily offered a variation on this theme, for it reached Cumberland, Maryland, in 1842 and thus tapped the National Road traffic from the West.

In the Midwest, new railroads were initially constructed to carry waterborne traffic further west. The Michigan Central's first plan was to build west from Detroit to its intended terminus at New Buffalo, Michigan, on Lake Michigan. From there, freight and passengers were to be transported across Lake Michigan to Chicago or Milwaukee. Likewise, the rival Michigan Southern and Northern Indiana initially intended to carry traffic from Toledo on Lake Erie to Michigan City, essentially duplicating the intentions of its competitor.

As the Michigan railroads progressed, the Galena & Chicago Union Rail Road started building west from Lake Michigan at Chicago in 1848 in anticipation of this new lake traffic. Its initial intention was to be the vehicle for connecting Lake Michigan and the Mississippi. Its plan was to reach Galena and then Dunleith, Illinois, on the Mississippi River, where its traffic would transfer to river vessels heading north to the western reaches of the new state of Wisconsin and to the Minnesota Territory or south to Iowa's river ports and St. Louis.

As the Bellefontaine and Indiana map suggests, an extended trip of this period could involve traveling by river vessels, by lake ships, by canal boats, by stage, and by railroad. In his 1838 *Valley of the Upper Wabash, Indiana,* Henry William Ellsworth contemplated the various modes by which travelers could travel from St. Louis to New York if only a railroad was built from Alton, Illinois, on the Mississippi opposite St. Louis, to Lafayette, Indiana, on the Erie and Wabash Canal:

> If the whole of this channel of communication is once finished, a journey may be performed from St. Louis to New York within a week. . . .
>
> Suppose the traveler leave the bank of the Mississippi on Monday morning at 4 o'clock, by railroad, for Lafayette—210 miles the distance at fifteen miles per hour, would be run in fourteen hours time, and he would arrive there at six o'clock the same evening. He here takes his supper, and starts again at 10 o'clock, P.M., on board a canal-boat, for Maumee Bay—distance 215 miles, at the rate of four miles per hour, which will require 54 hours, and he will, therefore, arrive at Lake Erie at 4 o'clock on Thursday morning. He leaves here by steamboat at 8 o'clock A.M., for Portland—distance 225 miles, at the rate of ten miles an hour, which will bring him to Portland on Friday morning at about 6 o'clock. Leaves here again at 8 o'clock, A.M., by railroad for New York, distance 450 miles, which, at the rate of only thirteen miles an hour, will bring him to the city of New York on Saturday evening before 8 o'clock; that is, a journey of 1,100 miles will be performed with only one night's travel by land, in five days and sixteen hours.

Ellsworth went on to note that, with such railroads in place, this 1,100-mile journey will be some 550 miles shorter than the alternative route by way of the Ohio and Mississippi Rivers and the Miami Canal.[33] In a similar vein, the 1852 report on the completion of a rail line from Boston to the St. Lawrence River notes that Boston was then only 1,006 miles from Chicago via rail to Ogdensburgh, New York, then by the St. Lawrence and lower Great Lakes to Detroit, and then by rail across lower Michigan to Chicago.[34]

Limitations of Early Railroads and Early Rail Travel Early railroads, designed to be components in a mixed transportation network, were relatively short lines reaching only 30 to 200 miles in length. When the Charleston & Hamburg Rail Road was completed in 1833, it was, at 136 miles, the nation's longest railroad. Even into the early 1850s, trunk line railroads of more than 150 or so miles were regarded as ambitious undertakings, while the vast majority were branch lines of some 30 or 40 miles or less.

The modest length of these early railroads resulted largely from four factors. First, railroads were capital intensive. A railroad of 100 miles might cost in the neighborhood of $2 million or more to build and furnish with depots and equipment—a substantial investment at the time. Therefore, early railroad builders set modest goals, intending to see if their investment paid off before committing to longer routes. Second, bridging major rivers was a serious financial and engineering challenge. As a result, many early railroad companies simply proposed building to a town on a major river and relying on ferry companies to carry passengers and traffic to the other shore, where another railroad might operate.

Third, railroad corporations presented new challenges to the nation's legislative thinking. Individual railroad companies operated by state charter and were accorded powers of eminent domain by individual states. State legislators cautiously felt their way through the challenges of deciding how to empower a corporation of one state to operate within another state. After reaching Cumberland, Maryland, western progress of the Baltimore & Ohio stalled from 1843 to 1847 because the Virginia legislature refused to give the line approval for building in Virginia.[35] As a result, many railroads simply ran to the border of the state in which they were chartered, where they proposed connecting with a railroad chartered in the neighboring state. The 1852 charter from the State of Wisconsin for the Green Bay, Milwaukee & Chicago Rail Road stated that the "company is authorized and empowered to connect its road with the road of any rail road company or companies in the state of Illinois, or to become part owner or lessee of any rail road in said state, and any rail road company of said state of Illinois, duly organized under the laws of said state of Illinois, may connect its road with the road of said company."[36] While the charter permitted connection with an Illinois chartered company, it did not and could not authorize the Wisconsin company to construct track in Illinois.

The final impediment to establishing rail lines of great length was technological. The simple locomotives of the 1830s and even 1840s were lightweight affairs that generally averaged a speed of fifteen miles an hour. They required stops for water and fuel every thirty or forty miles, making numerous trackside facilities necessary. Track was primitive, militating against high speed. Much early trackage was strap rail—two parallel wooden rails laid on ties and capped with iron straps spiked down atop the timbers. Early passenger equipment was equally primitive. The very earliest cars were constructed on patterns drawn from stagecoaches, though more substantial passenger cars quickly replaced them. But the second generation cars were uncomfortable—low, with little head room, furnished with benches for seats, lacking diner facilities, and ventilated only by windows. Furthermore, most early railroads did not operate at night, so travelers expected to lie over for the evening. A railroad journey of 150–200 miles or so was thus the furthest a traveler might anticipate for one day's journey.

Maps of the 1830s and 1840s Their short length and dependence on other established modes of travel discouraged the early railroads from producing travel maps of their own. Operating as they did over relatively short distances, railroads saw little need to help passengers understand their route. If a particular line ran no further than from Boston to Lowell, or Syracuse to Oswego, New York, or Indianapolis to Lafayette, Indiana, then travelers had no compelling need for a map of the route. The early railroads' shaky finances may also have played a part. The weak economy of the late 1830s and 1840s made railroad companies especially cost-conscious. The early midwestern railroads built in the late 1840s and early 1850s in particular struggled to raise capital even for construction. The production of maps for travelers was thus a luxury not to be contemplated, especially given the cost of map production. Materials were expensive; cheap wood pulp paper had not yet had an impact. And, map production itself was labor-intensive, as mapmakers relied on engraving and hand coloring rather than on less expensive lithographic techniques.[37]

As an inexpensive alternative to travel maps, American railroads printed the sequence of their stops on the back of their seat checks—the small, ticket-sized pieces of cardstock that passengers received upon boarding to give proof of having paid their fare. Examples from the early 1850s include seat checks from the Toledo, Wabash & Western, from the Galena & Chicago Union, and from the Peru & Indianapolis. The Peru & Indianapolis check records a sequence of sixteen stops between its two termini and lists the mileage from Indianapolis of each stop. The Toledo, Wabash & Western seat check (figure 4.11), issued for travel from Fort Wayne to Lafayette, Indiana, lists fifteen stops in the 109 mile trip between the two cities.

A second reason that early American railroads did not produce travel maps is that the industry had not yet embraced the idea of marketing, other than to potential investors. At most they might pay to have their timetable printed in a local newspaper, although some of the larger lines also had their timetables printed in such national papers as the Boston-based *American Railway Times*. Even so, the editor of the *Railway Times* shared the view of the *Rochester American* by quoting the latter's 1852 lament: "Railroad companies lose by not advertising their arrangements in newspapers at a distance. . . . All Railroads advertise we presume, in the papers of their own immediate vicinity. This however, is not enough. It is not there alone that

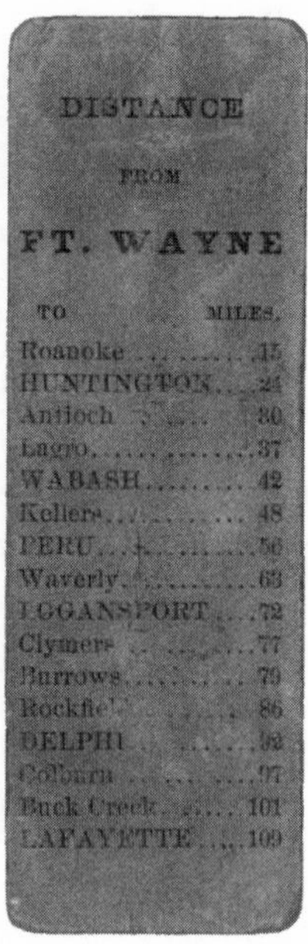

FIGURE 4.11 Seat check. Toledo, Wabash & Western Railway, n.d. Personal collection of the author.

information is needed. For example, a man here wishes to go to Watertown. He very easily finds out when he can leave Rochester, but how shall he learn when the cars reach Rome or depart from that place for his destination? And so of many other routes."[38] An industry that did little advertising of even its timetable, other than by posting handbills in its own locales or running ads in its local newspapers, was not an industry prepared to produce and distribute more expensive maps.

Finally, railroads chose not to produce maps for travelers because maps issued by commercial publishers already filled the existing need. As we have seen, as early as the 1830s established publishers, chiefly in New York and Philadelphia, produced maps and atlases for the general public that noted the progress of rail construction while also depicting the routes of other modes of travel. They also published similar maps specifically earmarked for travelers. Samuel Augustus Mitchell was the most successful publisher to address this niche at the early stages. In 1832 he published the first edition of his *Travellers Guide through the United States,* containing a map of the United States showing roads and steamboat and canal routes. He continued this and similar publications through the 1840s.

Both Mitchell and his competitor J. H. Colton published a series of "pocket" traveler's maps and guides of individual states and regions.[39] An example is Colton's 1840 "Guide Through Ohio, Michigan, Indiana, Illinois, Missouri, Wisconsin & Iowa." As noted by this map's subtitle—"Showing the Township lines of the United States Surveys, Location of Cities, Towns, Villages, Post Hamlets, Canals, Rail and Stage Roads"—all major forms of transportation were provided the reader or prospective traveler (plate 3).[40] Since railroads still complemented the prevailing water-based system of travel and shipping, the Colton map features an inset table that explains "Steam Boat Routes" and provides the distance from major city to major city by ship.[41]

2. *"RAILROADS AS THICK AS GRASSHOPPERS"*: MAPPING THE CONSOLIDATION OF RAIL ROADS INTO RAILWAY SYSTEMS

Even as early eastern railroads achieved their goal of completing lines to inland waterways, their focus began to shift. The success of their railroads had spurred the development of new lines—branch lines that radiated off established main or trunk railroads and new inland railroads that promised to connect with the eastern lines.

Railroads as a Transportation System of Their Own Some existing lines gave up on their initial plans to serve as feeders to lake or river transport and worked to become the sole link between two termini. For example, Michigan Southern decided in 1849 to abandon its initial plan of terminating at Michigan City and transferring its westbound traffic to Lake Michigan ships. Instead, it announced its decision to build around the southern tip of Lake Michigan and reach Chicago, bypassing the need for any water transportation between Toledo and Chicago. The Michigan Central, which had intended to have its terminus at New Buffalo on Lake Michigan, also altered its route and by 1852 completed a line from Detroit to Chicago. Having achieved their separate goals of reaching Chicago, both Michigan railroads struck deals with Chicago railroads building to the Mississippi to forge westward links.

New railroads were also created that served as branch lines for existing mainlines. A vast number of railroads were chartered in the Midwest in the 1850s—125 in Wisconsin alone, although only a fraction of these were ever built. While some of these new lines intended merely to connect two communities lacking any rail service, a substantial number were designed to serve as branch lines to the new mainlines in their area. The Aurora Branch (eventually the Chicago, Burlington & Quincy) was chartered in Illinois in 1850 to connect the city of Aurora with the Galena & Chicago Union fourteen miles northeast. Similarly, the Mineral Point Rail Road was chartered in 1852 to connect the lead mining region of southwest Wisconsin with the Illinois Central.

The 1850s also saw the first railroads to be conceived as systems from the very start. This was true of the Illinois Central, the beneficiary in 1850 of a very large land grant from the federal government, making possible a major north-south rail route roughly paralleling the Mississippi River.[42] Still other railroads merged with one another, forming larger entities. In 1853, various short lines in New York consolidated into the New York Central.[43]

Others worked out collaborative arrangements that effectively extended their geographical reach while retaining their own separate identities. The Baltimore & Ohio, upon reaching the Ohio River at Wheeling in 1853, began a second line west from Grafton, Virginia, to Parkersburg, also on the Ohio. There, in 1857, the B&O connected with the Marietta & Cincinnati, which in turn connected with the Ohio & Mississippi, a line being built west to St. Louis. Travelers could soon go entirely by rail from Baltimore to the Mississippi on this American Central Route of three connected lines.

By the early 1850s a movement toward creating a rail network had caught fire—a network of consolidated short lines, as in the case of the New York Central; interconnected mainlines; the federal government's support of major enterprises; and the development of an intersecting web of trunk lines and branch lines. As the *Chicago Democrat* noted with only moderate hyperbole, railroads had become "as thick as grasshoppers."[44] Indeed, by 1859 a mile of railroad operated for every 1,000 Americans—two and a half times the ratio existing in England.[45]

As discussed above, a related development was the construction of air lines. Recall that many early railroads closely followed river valleys to reduce construction costs, and so they often took somewhat circuitous routes between two major points. Sometimes they laid out a route between two cities that served a series of smaller towns along the way, expecting residents of each town to contribute to construction costs. Since few towns were laid out in a straight line, the resulting railroad itself did not follow a straight line. However, railroads with such circuitous routes were at a competitive disadvantage. By custom, passengers and freight were charged a set amount per mile—for passengers, often 3¢ a mile. Thus, air lines, which were straight and more direct, were more attractive to travelers and to shippers, in terms of both time and cost.

Maps of the 1850s are filled with air line routes. An air line proposed between New Haven and Boston in 1852 shaved some twenty-eight miles off the existing, competing route.[46] Similarly, the Galena & Chicago Union, the nation's most financially successful railroad by the mid-1850s, changed its entire strategy in 1853 when faced with the threat of a proposed competitor, the Chicago, St. Charles & Mississippi Air Line. The G&CU's initial course was to follow a circuitous route toward Galena, near the Mississippi.

When the backers of the St. Charles Air Line proposed a straight-line railroad to the Mississippi, thus standing to capture the Chicago-Iowa traffic, the G&CU reacted quickly. The company stopped construction at Freeport, Illinois, of its originally proposed line to Galena and began building its own air line straight west from Junction to Fulton, Illinois, on the Mississippi. Not to have done so would have seriously undercut its profits and future prospects.

This air line phenomenon signified the emerging dominance of railroads in the American transportation system. No longer laid out primarily as connections to other modes of transportation, they now served as the sole or primary links between two termini. Indeed, railroads and their backers demonstrated how much railroads had become the dominant transportation system by constructing some 22,000 miles of new trackage in the 1850s, much more than tripling the nation's total. By the end of the decade, the United States had some individual railroads of substantial dimensions. *Ashcroft's Railway Directory* of 1862 describes 421 U.S. and Canadian railroads, including some under construction. Of this number, seven American companies and one Canadian line now each encompassed more than 400 miles of trackage: the Illinois Central, 704 miles; the New York Central, 654 miles; Michigan Southern & Northern Indiana, 528 miles; Baltimore & Ohio, 518 miles; Mobile & Ohio, 488 miles; Pittsburgh, Fort Wayne & Chicago, 467 miles; Erie Railway, 461 miles; and the Grand Trunk Railway of Canada, 1,096 miles.[47] Admittedly, 289 of the railroads listed were shorter than 100 miles, but a trunk line railroad now controlled a substantial number of them, converting them into part of a larger system. Taking note of how much railroads now dominated passenger travel in the areas they operated, the *Chicago Democrat* stated this in 1859: "When railroads began to be extensively introduced into our country, . . . we expressed the opinion that, so far as the [existing] passenger traffic was concerned, it would be almost annihilated, and that vessels would have to depend almost entirely upon freight. This has literally proved true."[48]

The State of Rail Travel at Midcentury Dramatic improvements in railroad technology contributed greatly to this expansion. American railroads settled on a large, powerful locomotive for the time, the 4-4-0 American. Fitted with larger diameter driving wheels, this style engine could achieve substantial speeds—sixty miles an hour or more, though few ran consistently at that speed. The locomotives also featured larger tenders, which carried more ample supplies of water and wood, and later coal, thus needing fewer stops for refueling. Track and roadbeds improved dramatically. While some midwestern railroads were still built with primitive strap rail up to 1850, safer and more substantial T-rail had become the rail of choice. The development of an American iron industry freed the nation from a British monopoly and made this more substantial and safer rail affordable.[49]

Passenger equipment improved in response to a demand for greater passenger comfort. The clerestory roof was introduced to passenger cars in the early 1860s, providing an additional eighteen inches of headroom and a means of indirect ventilation. Springs were added to cars to provide a smoother ride. And in the second half of the 1860s sleeper cars were introduced, allowing passengers to travel greater distances with less fatigue. George Pullman unveiled his first sleeper in 1869.[50]

Finally, while rail travel in the 1850s involved making several transfers between different railroads and following circuitous routes in some places, it also offered speedier travel than the alternatives. The 538-mile trip from Buffalo to Chicago took only twenty and one half hours by way of the Lake Shore & Michigan Southern. Commenting on the completion of the railroad from Milwaukee to Janesville, Wisconsin, in 1853, the *Weekly Chicago Democrat* noted that the new route would save time even for travelers from Milwaukee to Chicago. Milwaukee is ninety miles directly north of Chicago, but the roads were poor and lake shipping was closed for four or more months in the winter. The fastest route, especially in winter, in 1853 was to take the evening train seventy miles west from Milwaukee to Janesville and stay overnight. In the morning one could take the stage to Rockford, thirty miles south of Janesville, and catch the evening train there to travel ninety-two miles east to Chicago. Given the poor quality of the roads, one-hundred and ninety-two miles going west, south, and east by trains and stage was a day faster than a direct ninety-mile trip south by stage alone.[51]

Maps of the 1850s and Beyond Expanding rail networks went hand in hand with expanding markets. The first railroad to operate out of Chicago—the Galena & Chicago Union—ran its first train in the fall of 1848. By 1856, experts estimated that Chicago's railroads moved as many as 3.5 million people a year.[52] With vastly improved travel times, more comfortable equipment, some 30,000 miles of track in place by 1860, and millions of travelers a year, railroads had reached a point of needing maps focused on rail travel. The time had passed for relying on general maps that also highlighted canals, stage roads, navigable rivers, and lake steamboat routes.

In the 1850s railroad companies could not yet produce individual maps on their own. But passengers' growing demand for comprehensive rail information in the decade generated a proliferation of traveler's guides by a number of publishing firms. As Brad Lomazzi notes, "The earliest, mid-nineteenth century railway travel guides usually contained nothing but timetables, solicited from railroads by independent publishers. The publisher would then reproduce these with a minimum amount of editing. Sometimes maps of railroad lines were included. This system was a practical alternative for those railroads that could not afford to print and distribute timetables to ticket agents on all railroads across the nation."[53] Among the names of contemporary travel guides Lomazzi lists are *Dinsmore's American Railway Guide,* begun in 1850 and later published as *Batterman's American Railway Guide; Appletons' Railway and Steam Navigation Guide,* begun in 1857 and later called *Appletons' Illustrated Railway Guide; The Travelers' Official Railway Guide,* first published in 1868; and *Rand McNally's Travelers' Hand Book to All Railway & Steamboat Lines of North America,* a monthly begun in 1870.

A look at an 1857 edition of the *Locomotive Courant: A Monthly Record of Material Progress* demonstrates how such guides helped travelers understand the number of transfers they might make. This New York monthly provided timetables, information on fares, and a guide to routes by which a traveler could get from one city to a distant one in the mid-1850s. The section titled "Combined Railroad Routes" lists two ways to travel between Baltimore and Cincinnati. One is by way of the B&O to Wheeling and then by a series of three interconnected railroads to Cincinnati. The other is by way of the Northern Central R.R. to Harrisburg, Pennsylvania; then the Pennsylvania R.R. to Pittsburgh; next the Pittsburgh, Fort Wayne & Chicago R.R.; and finally the Cleveland, Columbus, & Cincinnati R.R. to Cincinnati. The *Record* lists four routes for the trip between Boston and New York. While three of them involved some travel via steamboat, one was completely by rail and the other three involved travel by at least two railroads each.[54]

Taylor and Neu have argued that such guides published the best contemporary railroad maps.[55] The *Locomotive Courant,* for example, included small maps of principal cities, showing the lines that served each region. The map of the "Vicinity of Philadelphia" from the April 1857 edition shows the routes of ten railroads serving Philadelphia and the Delaware valley. Other maps, all measuring 3½ × 4–4½ inches, feature Albany, Baltimore, Boston, Buffalo and Niagara Falls, Chicago, Cincinnati, Indianapolis, Montreal, New York, Pittsburgh, and Portland, Maine.

The state of Ohio saw the publication in 1854 of the *Ohio Railroad Guide, Illustrated: Cincinnati to Erie, via Columbus and Cleveland.*[56] This 135-page hardbound book provided information about the route between Cincinnati and Erie, Pennsylvania, complete with the railroads to be traveled and illustrations of many depots and bridges along the way. It also included a

$7\frac{1}{2} \times 14$ inch fold-out panoramic "View of Station Building and Lake Erie at Cleveland" by Shipley-Stillman. This fold-out shows the train stations, lake docks, loading area for stages, and railroad tracks.

While many of these guides emphasized the transfers that were necessary between rail lines, an example of a guide focused on a single railroad is H. F. Walling's *The Erie Railway and Its Branches . . . , Illustrated with Maps.* Published in 1867 by Taintor Brothers of New York, this pocket-sized book includes sixty pages of text and hand-colored maps, as well as numerous timetables and advertisements. Each of the twenty $3\frac{1}{4} \times 5\frac{1}{4}$ inch maps depicts a portion of the railroad's route.

An 1858 advertisement for *Dinsmore's Railroad and Steam Navigation and Gazetteer* provides a further sense of what such travel guides offered:

> Guide and Route Book
>
> Contains Official TIME TABLES, with connections by Steamboat and Stage, also with other roads; *COMBINED Railroad Routes,* with Names of principal places; Roads over which you go, distances, & c.; STEAM NAVIGATION Gazetteer; Maps of Railroad Centres; also *a splendid Railroad Map, corrected monthly,* setting all the new Roads—corrected by the Superintendents themselves, making it RELIABLE. The whole work is in TYPE 256 pages, price 25 cts.; Cheap edition, with Time Tables only, 12 cts.
>
> DINSMORE & CO., Publishers
> New-York.[57]

While the Dinsmore company refers to steamboat and stagecoach travel, the emphasis was now clearly on rail travel. The small maps depict railroad facilities and routes within a major city, while the large map emphasizes the relatively new combined routes—steamboats and stagecoaches merely provide connections to the trains. The advertisement also notes that the main railroad map is corrected monthly, indicating both the breathtaking pace of railroad expansion and the relative ease with which cartographers could now respond to such change. New map printing technologies emerged, such as lithographic transfer and wax engraving, that lowered the cost of printing maps and facilitated updating existing ones.[58] Finally, Dinsmore & Co. notes that it draws its information from railroad personnel themselves. Railroad superintendents recognized that helping publishers with their maps spared railroads the task of producing their own.

The most important development in the realm of railway guides and their production of maps was the establishment in 1868 of the *Travelers' Official Railway Guide* in 1868. Originally published by H. H. Wheeler and

Edward Vernon, the *Official Guide* was acquired by the National Railway Publication Company in June 1870, with Vernon continuing as editor. The monthly publication included the timetables of American railroads and of some that served Canada. The term "official" in the title derived from the fact that the National General Ticket Agents' Association considered it to be "the recognized organ of this Association."[59] Publication of the monthly *Official Guide* continued into the 1970s.

In addition to timetables, early editions of the *Official Guide* included descriptions of various railroad developments and dozens of railroad system maps. Some of the maps occupied a portion of a page, others were full-page, and a few might be spread across two pages. Because the *Official Guide* was a monthly publication, the publishers did not invest substantial funds into printing the individual maps included with the various railroad timetables. However, by 1870, the *Guide* included a large fold-out *Railway Map of the United States and Canada Showing All the Railroads Completed and in Progress.*[60] This was a more ambitious undertaking than were the maps contained on the pages of text, and as a way of underwriting some of the production costs, the publisher included paid advertising on the map's verso side.

While ticket agents comprised the principal audience for the *Official Guide,* the title page notes that it was "For Sale by all Periodical Dealers, News Agents and Booksellers, also on Trains, and at the several Railroad Depots and Ticket Offices in the United States and Canada." Thus, frequent train travelers had access to authoritative travel information in the form of timetables and a substantial selection of maps. An example, drawn from the October 1882 issue, is the *Condensed Outline Map of the Bee Line* (figure 4.12). The Bee Line Route was the interconnected route of three railroads: the Cleveland, Columbus, Cincinnati & Indianapolis; the Indianapolis & St. Louis; and the Dayton & Union Railroads. This small map complements two pages of timetables for the three railroads in question.[61]

It appears that the maps produced for the *Travelers' Official Guide* were available for use in other publications as well. The 1873 book *The Pine and the Palm Greeting; or, The Trip of the Northern Editors to the South in 1871, and the Return Visit of the Southern Editors in 1872* includes two maps.[62] The first is a hand-colored foldout map entitled "Map of Six Hundred Miles around the Jordan Rockbridge Alum Springs; Rockbridge County, Virginia." The line of the Chesapeake & Ohio Railroad from Richmond, Virginia to Cincinnati is drawn in a heavy red line, while the main line railroads of the East are shown in very light black lines.[63] Of greater interest, though, is a smaller, black-and-white foldout "Map of the Chesapeake & Ohio Railroad and Its Connections." Fisk & Russell of New York engraved the map for the National Railway Publication Company, the publisher of the *Travelers'*

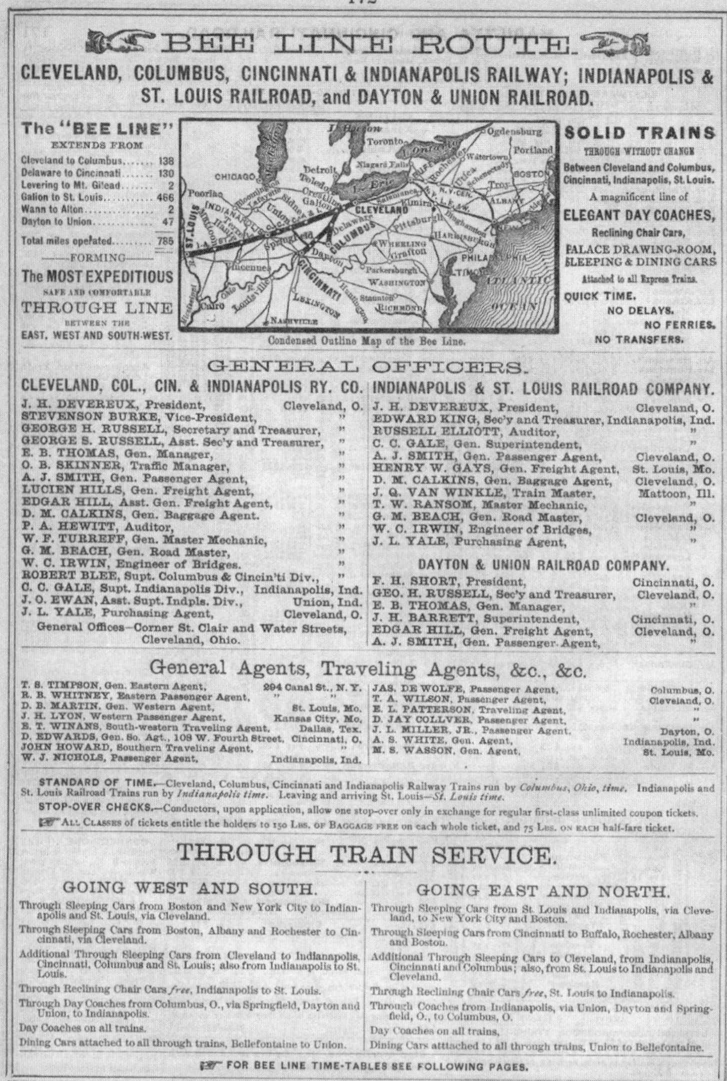

172

BEE LINE ROUTE.

CLEVELAND, COLUMBUS, CINCINNATI & INDIANAPOLIS RAILWAY; INDIANAPOLIS & ST. LOUIS RAILROAD, and DAYTON & UNION RAILROAD.

The "BEE LINE"
EXTENDS FROM

Cleveland to Columbus	138
Delaware to Cincinnati	130
Levering to Mt. Gilead	2
Galion to St. Louis	466
Wann to Alton	2
Dayton to Union	47
Total miles operated	785

—FORMING—
The MOST EXPEDITIOUS
SAFE AND COMFORTABLE
THROUGH LINE
BETWEEN THE
EAST, WEST AND SOUTH-WEST.

Condensed Outline Map of the Bee Line.

SOLID TRAINS
THROUGH WITHOUT CHANGE
Between Cleveland and Columbus, Cincinnati, Indianapolis, St. Louis.
A magnificent line of
ELEGANT DAY COACHES,
Reclining Chair Cars,
PALACE DRAWING-ROOM, SLEEPING & DINING CARS
Attached to all Express Trains.
QUICK TIME.
NO DELAYS.
NO FERRIES.
NO TRANSFERS.

GENERAL OFFICERS.

CLEVELAND, COL., CIN. & INDIANAPOLIS RY. CO.

J. H. DEVEREUX, President, Cleveland, O.
STEVENSON BURKE, Vice-President, "
GEORGE H. RUSSELL, Secretary and Treasurer, "
GEORGE S. RUSSELL, Asst. Sec'y and Treasurer, "
E. B. THOMAS, Gen. Manager, "
O. B. SKINNER, Traffic Manager, "
A. J. SMITH, Gen. Passenger Agent, "
LUCIEN HILLS, Gen. Freight Agent, "
EDGAR HILL, Asst. Gen. Freight Agent, "
D. M. CALKINS, Gen. Baggage Agent. "
P. A. HEWITT, Auditor, "
W. F. TURREFF, Gen. Master Mechanic, "
G. M. BEACH, Gen. Road Master, "
W. C. IRWIN, Engineer of Bridges. "
ROBERT BLEE, Supt. Columbus & Cincin'ti Div., "
C. C. GALE, Supt. Indianapolis Div., Indianapolis, Ind.
J. O. EWAN, Asst. Supt. Indpls. Div., Union, Ind.
J. L. YALE, Purchasing Agent, Cleveland, O.
General Offices—Corner St. Clair and Water Streets, Cleveland, Ohio.

INDIANAPOLIS & ST. LOUIS RAILROAD COMPANY.

J. H. DEVEREUX, President, Cleveland, O.
EDWARD KING, Sec'y and Treasurer, Indianapolis, Ind.
RUSSELL ELLIOTT, Auditor, "
C. C. GALE, Gen. Superintendent, "
A. J. SMITH, Gen. Passenger Agent, Cleveland, O.
HENRY W. GAYS, Gen. Freight Agent, St. Louis, Mo.
D. M. CALKINS, Gen. Baggage Agent, Cleveland, O.
J. Q. VAN WINKLE, Train Master, Mattoon, Ill.
T. W. RANSOM, Master Mechanic, "
G. M. BEACH, Gen. Road Master, Cleveland, O.
W. C. IRWIN, Engineer of Bridges, "
J. L. YALE, Purchasing Agent, "

DAYTON & UNION RAILROAD COMPANY.

F. H. SHORT, President, Cincinnati, O.
GEO. H. RUSSELL, Sec'y and Treasurer, Cleveland, O.
E. B. THOMAS, Gen. Manager, "
J. H. BARRETT, Superintendent, Cincinnati, O.
EDGAR HILL, Gen. Freight Agent, Cleveland, O.
A. J. SMITH, Gen. Passenger Agent, "

General Agents, Traveling Agents, &c., &c.

T. S. TIMPSON, Gen. Eastern Agent, 294 Canal St., N. Y.
R. B. WHITNEY, Eastern Passenger Agent, " "
D. B. MARTIN, Gen. Western Agent, St. Louis, Mo.
J. H. LYON, Western Passenger Agent, Kansas City, Mo.
E. T. WINANS, South-western Traveling Agent, Dallas, Tex.
D. EDWARDS, Gen. So. Agt., 108 W. Fourth Street, Cincinnati, O.
JOHN HOWARD, Southern Traveling Agent, "
W. J. NICHOLS, Passenger Agent, Indianapolis, Ind.

JAS. DE WOLFE, Passenger Agent, Columbus, O.
T. A. WILSON, Passenger Agent, Cleveland, O.
E. L. PATTERSON, Traveling Agent, "
D. JAY COLLVER, Passenger Agent, "
J. L. MILLER, JR., Passenger Agent, Dayton, O.
A. S. WHITE, Gen. Agent, Indianapolis, Ind.
M. S. WASSON, Gen. Agent, St. Louis, Mo.

STANDARD OF TIME.—Cleveland, Columbus, Cincinnati and Indianapolis Railway Trains run by *Columbus, Ohio, time.* Indianapolis and St. Louis Railroad Trains run by *Indianapolis time.* Leaving and arriving St. Louis—*St. Louis time.*

STOP-OVER CHECKS.—Conductors, upon application, allow one stop-over only in exchange for regular first-class unlimited coupon tickets.

☞ ALL CLASSES of tickets entitle the holders to 150 LBS. OF BAGGAGE FREE on each whole ticket, and 75 LBS. ON EACH half-fare ticket.

THROUGH TRAIN SERVICE.

GOING WEST AND SOUTH.

Through Sleeping Cars from Boston and New York City to Indianapolis and St. Louis, via Cleveland.

Through Sleeping Cars from Boston, Albany and Rochester to Cincinnati, via Cleveland.

Additional Through Sleeping Cars from Cleveland to Indianapolis, Cincinnati, Columbus and St. Louis; also from Indianapolis to St. Louis.

Through Reclining Chair Cars *free*, Indianapolis to St. Louis.

Through Day Coaches from Columbus, O., via Springfield, Dayton and Union, to Indianapolis.

Day Coaches on all trains.

Dining Cars attached to all through trains, Bellefontaine to Union.

GOING EAST AND NORTH.

Through Sleeping Cars from St. Louis and Indianapolis, via Cleveland, to New York City and Boston.

Through Sleeping Cars from Cincinnati to Buffalo, Rochester, Albany and Boston.

Additional Through Sleeping Cars to Cleveland, from Indianapolis, Cincinnati and Columbus; also, from St. Louis to Indianapolis and Cleveland.

Through Reclining Chair Cars *free*, St. Louis to Indianapolis.

Through Coaches from Indianapolis, via Union, Dayton and Springfield, O., to Columbus, O.

Day Coaches on all trains,

Dining Cars atttached to all through trains, Union to Bellefontaine.

☞ FOR BEE LINE TIME-TABLES SEE FOLLOWING PAGES.

FIGURE 4.12 "Condensed Outline Map of the Bee Line," in *Travelers' Official Railway Guide for the United States and Canada* (New York: National Railway Publication Co., October 1882). Personal collection of the author.

Official Guide. The text surrounding the map and the advertisement on the reverse side strongly resemble what likely was a pair of two-page spreads in the *Travelers' Official Guide*.

A new development in railroad passenger maps occurred as the 1860s dawned. In the 1850s, railroads had begun to extend their routes and acquire other lines, but expansion especially gathered steam after the Civil War. As a result, travelers now had two or three or more choices of routes between major cities. For example, John S. Wright could report in 1870 that "five rival railroads" now served Chicago from the East alone:

> the Michigan Central, by which travelers reached New York via Detroit, Canada's Great Western Railway to Niagara, and the New York Central;
>
> the Michigan Southern, running to Toledo and then east by way of the American Lake Shore Railroad Line (comprising the Cleveland & Toledo, Cleveland & Erie, and Buffalo & Erie), and then connecting with the New York Central;
>
> the Pittsburgh, Fort Wayne & Chicago, connecting with the Pennsylvania Central to Philadelphia, New York, and Baltimore;
>
> the Chicago & Great Eastern, previously known as the Columbus, Chicago & Indiana Central, which connected east with the Baltimore & Ohio;
>
> and the Lafayette, Indianapolis & Central route through Ohio to connections with Baltimore.[64]

In this competitive situation, railroads finally had an incentive to produce travel maps on their own. For the first time, consumers traveling to some destinations might be able to choose from among several possible train routes. To maintain or expand their share of the market, railroad companies now needed to demonstrate that their route was the shortest or most scenic or that their trains were the most comfortable. This called for maps presenting information in ways that gave apparent advantages to particular rail companies and their lines. Though similar in appearance these differed in intent from maps published in independent railway guides, which, by their very nature, gave individual lines no particular advantage.

In addition, as their miles of trackage grew, railroads became sizable corporations, a fact that attracted major capitalists—often the same ones for several lines. John Wright noted:

> One of the strongest points of this argument in favor of the certain continuation of the railway system as now instituted, and its spread indefinitely into the Great Interior, is the direct interest which capitalists of eastern roads, mainly residents in New York and New England, have in the construction of extending lines with ramifying branches. An offshoot of the Michigan Central, running into its depot, a large part of the stock held by the same parties, with the same capable President [James F. Joy] over both, is the Chicago, Burlington & Quincy.[65]

Shareholders and management of one line that served a major city such as Chicago had a vested interest in making arrangements with another line feeding that same city from another direction. If the Michigan Central could guarantee that eastbound traffic arriving in Chicago over the CB&Q would transfer to their railroad, for example, they stole a march on their eastern competitors. Companies thus built depots that served both railroads and struck arrangements to be "connected roads." And they had reason to produce maps that emphasized their connections with their partners, such as the Kankakee and Seneca Route map reproduced later in this chapter as figure 4.15.

Railroads from the 1860s on had a newly realized need, as well as the financial resources, to publish their own passenger maps. Moreover, as we have seen, the cost of map production dropped with the advent of new wood pulp paper, new wax-engraving techniques, inexpensive lithographic processes, and new high-speed printing presses.[66] In the 1860s and 1870s railroads also developed the perfect vehicle for including their maps—passenger timetables. In the earliest years, railroads depended on handbills, posters, and newspaper notices to inform the public about train schedules. As Lomazzi points out, this was done in simple list form, recording a town and all arrival times and then another town and all its times and so on. By the 1850s and 1860s, however, railroads changed to a tabular form, a "time table" that traced a specific train and its times city by city. This format was easier to read although it took more space. According to Lomazzi:

> It was necessary to devise new ways to present, to as many potential passengers as possible, schedule information for many trains serving a growing number of towns. That information had to be in a practical format that was easy to read and sized to fit in one's pocket. Schedules also needed to include information on: fares, regions, land availability, towns, and some advertising. In the 1860s, two types had evolved: the folder or folding brochure—a single sheet, usually folded into many

separate panels to resemble a brochure, and the stapled pocket booklet or pamphlet.[67]

An obvious addition to this newer, more elaborate timetable was a map with which individual railroads could control their message to customers in a way they could not in railway guides. Railroads could publish times and fares alongside whatever claims they wanted to make about them, as the reader could not turn a page as he could in a railway guide to make comparisons with competing lines. Pictures could be included as well.

The earliest example known to me of a timetable containing a map is the 1861 "Condensed Time Table of the Lake Shore and Mich'n Southern R. R. Line between Buffalo and Cleveland, Toledo, Chicago, Milwaukee, Cincinnati, St. Louis, and the West and South-West."[68] This sixteen-panel folding timetable includes three panels of timetables for LS&MS "thro' trains" and local trains. It also includes four panels of timetables for express trains on connecting roads, including the Chicago & Rock Island; the St. Louis, Alton & Chicago; the Chicago & Milwaukee; the Galena & Chicago Union; and the LaCrosse & Milwaukee. Half of the space of the eight inside panels is given over to a map of the LS&MS (figure 4.13), which emphasizes the connections at both the eastern and western ends of the LS&MS. Also included is a list of fares from Buffalo to the main cities on connecting roads. The timetable notes that the LS&MS and its connected roads offered "'Union Depots' throughout the entire Line, Thus avoiding the inconvenience of Ferries and Omnibusses."[69]

Another early example is the 1874 Chicago, Milwaukee, and St. Paul Railway map reprinted in Modelski's *Railroad Maps of North America*. This map was part of a ten-panel brochure that included a list of stations, mileage between major cities, timetables, a small map of Chicago's rebuilt business district, and pictures depicting scenery along the Milwaukee Road.[70] And, of course, an enticing route map could be a major feature, one that presented journeys in the best possible light by implying that the railroad operated the shortest and therefore quickest possible route. One way to make a route look enticing was to use map design tricks, including distortion. Modelski addresses what he calls the "geographic inaccuracy in railroad maps" by quoting an 1879 Rand McNally booklet: "Map 'designing' to other than a railroad official, might seem a peculiar phrase, but the majority of railroad maps have some 'peculiar designs' hidden under the careful pencil of the draughtsman. . . . The various friendly interests must be shown to best advantage, and rival interests disposed of in a manner that 'no fellow can find out.'"[71]

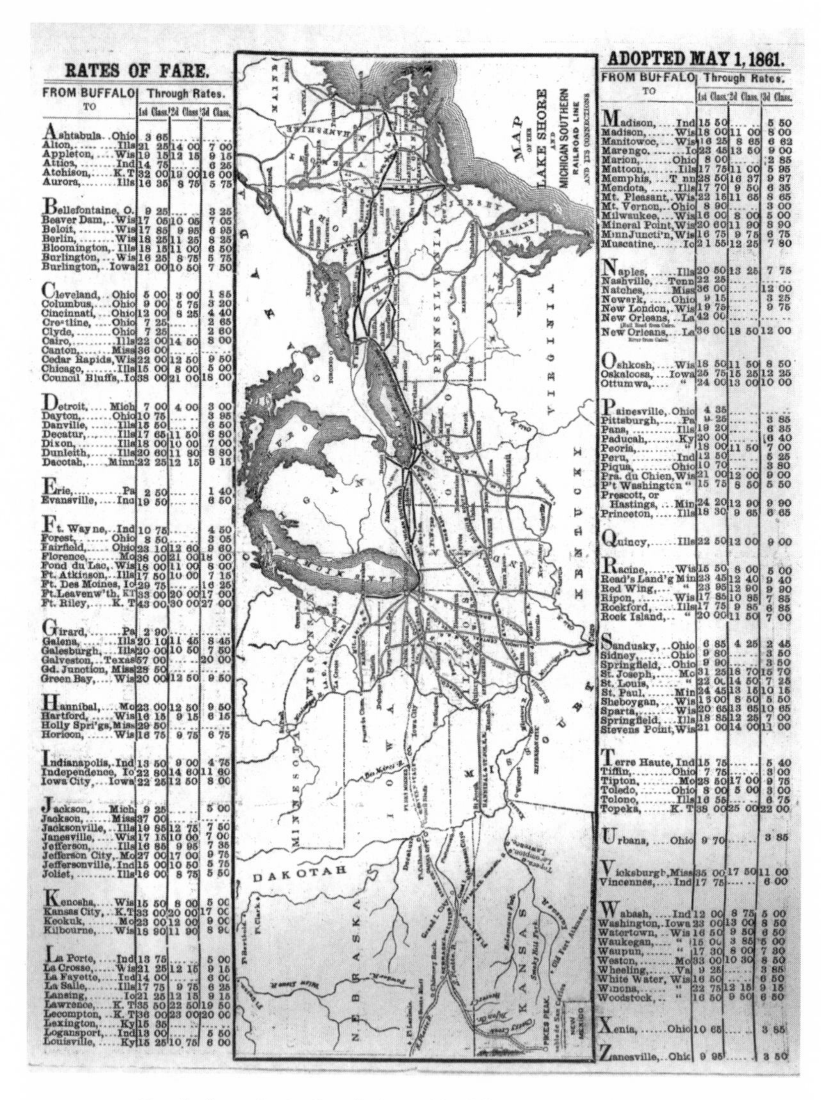

RATES OF FARE. ADOPTED MAY 1, 1861.

FROM BUFFALO TO	Through Rates. 1st Class.	2d Class.	3d Class.
Ashtabula. . Ohio	3 65		
Alton, ... Ills	21 25	14 00	7 00
Appleton, ... Wis	19 15	12 15	9 15
Attica, ... Ind	14 75		6 25
Atchison, ... K. T	32 00	19 00	16 00
Aurora, ... Ills	16 35	8 75	5 75
Bellefontaine, O.	9 25		3 25
Beaver Dam, . . Wis	17 05	10 05	7 05
Beloit, ... Wis	17 85	9 95	6 95
Berlin, ... Wis	18 25	11 25	8 25
Bloomington, . Ills	18 15	11 00	6 50
Burlington, . . . Wis	16 25	8 75	5 75
Burlington, . . Iowa	21 00	10 50	7 50
Cleveland, . . Ohio	5 00	3 00	1 85
Columbus, ... Ohio	9 00	5 75	3 20
Cincinnati, . . . Ohio	12 00	8 25	4 40
Cre*tline, ... Ohio	7 25		2 65
Clyde, ... Ohio	7 25		2 60
Cairo, ... Ills	22 00	14 50	8 00
Canton, ... Miss	36 00		
Cedar Rapids, Wis	22 00	12 50	9 50
Chicago, ... Ills	15 00	8 00	5 00
Council Bluffs, . Io	38 00	21 00	18 00
Detroit, ... Mich	7 00	4 00	3 00
Dayton, ... Ohio	10 75		3 95
Danville, ... Ills	15 50		6 50
Decatur, ... Ills	17 65	11 50	6 80
Dixon, ... Ills	18 00	10 00	7 00
Dunleith, ... Ills	20 60	11 80	8 80
Dacotah, ... Minn	22 25	12 15	9 15
Erie, ... Pa	2 50		1 40
Evansville, ... Ind	19 50		6 50
Ft. Wayne, . . Ind	10 75		4 50
Forest, ... Ohio	8 50		3 05
Fairfield, ... Ohio	23 10	12 60	9 60
Florence, ... Mo	38 00	21 00	18 00
Fond du Lac, . Wis	18 00	11 00	8 00
Ft. Atkinson, . . Ills	17 50	10 00	7 15
Ft. Des Moines, Io	29 75		16 25
Ft. Leavenw'th, KT	33 00	20 00	17 00
Ft. Riley, ... K. T	43 00	30 00	27 00
Girard, ... Pa	2 90		
Galena, ... Ills	20 10	11 45	8 45
Galesburgh, . . . Ills	20 00	10 50	7 50
Galveston, . . Texas	57 00		20 00
Gd. Junction, Miss	28 50		
Green Bay, ... Wis	20 00	12 50	9 50
Hannibal, ... Mo	23 00	12 50	9 50
Hartford, ... Wis	16 15	9 15	6 15
Holly Spri'gs, Miss	29 50		
Horicon, ... Wis	16 75	9 75	6 75
Indianapolis, . Ind	13 50	9 00	4 75
Independence, Io	22 80	14 60	11 60
Iowa City, . . . Iowa	22 25	12 50	8 00
Jackson, ... Mich	9 25		5 00
Jackson, ... Miss	37 00		
Jacksonville, . . Ills	19 85	12 75	7 50
Janesville, ... Wis	17 15	10 00	7 00
Jefferson, ... Ills	16 85	9 95	7 35
Jefferson City, . Mo	27 00	17 00	9 75
Jeffersonville, . Ind	15 00	10 50	5 75
Joliet, ... Ills	16 00	8 75	5 50
Kenosha, ... Wis	15 50	8 00	5 00
Kansas City, . . K. T	33 00	20 00	17 00
Keokuk, ... Mo	23 00	12 00	9 00
Kilbourne, ... Wis	18 90	11 90	8 90
La Porte, ... Ind	13 75		5 00
La Crosse, ... Wis	21 25	12 15	9 15
La Fayette, ... Ind	14 00		6 00
La Salle, ... Ills	17 75	9 75	6 25
Lansing, ... Io	21 25	12 15	9 15
Lawrence, ... K. T	35 50	22 50	19 50
Lecompton, . . K. T	36 00	23 00	20 00
Lexington, ... Ky	15 35		
Logansport, . . . Ind	13 00		5 50
Louisville, ... Ky	15 25	10 75	6 00
Madison, ... Ind	15 50		5 50
Madison, ... Wis	18 00	11 00	8 00
Manitowoc, . . . Wis	16 25	8 65	6 62
Marengo, ... Io	23 45	13 50	9 00
Marion, ... Ohio	8 00		2 85
Mattoon, ... Ills	17 75	11 00	5 95
Memphis, . . . T nn	28 50	16 37	9 87
Mendota, ... Ills	17 70	9 50	6 35
Mt. Pleasant, . Wis	22 15	11 65	8 65
Mt. Vernon, . . Ohio	8 90		3 00
Milwaukee, . . . Wis	16 00	8 00	5 00
Mineral Point, Wis	20 60	11 90	8 90
MinnJuncti'n, Wis	16 75	9 75	6 75
Muscatine, ... Io	21 55	12 25	7 80
Naples, ... Ills	20 50	13 25	7 75
Nashville, . . . Tenn	22 25		
Natches, ... Miss	36 00		12 00
Newark, ... Ohio	9 15		3 25
New London, . Wis	19 75		9 75
New Orleans, . . La (Rail Road from Cairo.)	42 00		
New Orleans, . . La (River from Cairo.)	36 00	18 50	12 00
Oshkosh, ... Wis	18 50	11 50	8 50
Oskaloosa, . . . Iowa	25 75	15 25	12 25
Ottumwa, ... "	24 00	13 00	10 00
Painesville, . Ohio	4 35		
Pittsburgh, ... Pa	9 25		3 85
Pana, ... Ills	19 20		6 35
Paducah, ... Ky	20 00		16 40
Peoria, ... "	18 00	11 50	7 00
Peru, ... Ind	12 50		5 25
Piqua, ... Ohio	10 70		3 80
Pra. du Chien, Wis	21 00	12 00	9 00
P't Washington "	15 75	8 50	5 50
Prescott, or Hastings, . . Min	24 20	12 90	9 90
Princeton, ... Ills	18 30	9 65	6 65
Quincy, ... Ills	22 50	12 00	9 00
Racine, ... Wis	15 50	8 00	5 00
Read's Land'g Min	23 45	12 40	9 40
Red Wing, . . . "	23 95	12 90	9 90
Ripon, ... Wis	17 85	10 85	7 85
Rockford, ... Ills	17 75	9 85	6 85
Rock Island, . . "	20 00	11 50	7 00
Sandusky, . . Ohio	6 85	4 25	2 45
Sidney, ... Ohio	9 80		3 50
Springfield, . . Ohio	9 90		3 50
St. Joseph, ... Mo	31 25	18 70	15 70
St. Louis, ... "	22 00	14 50	7 25
St. Paul, ... Min	24 45	13 15	10 15
Sheboygan, . . . Wis	13 00	8 50	5 50
Sparta, ... Wis	20 65	13 65	10 65
Springfield, . . . Ills	18 85	12 25	7 00
Stevens Point, Wis	21 00	14 00	11 00
Terre Haute, Ind	15 75		5 40
Tiffin, ... Ohio	7 75		3 00
Tipton, ... Mo	28 50	17 00	9 75
Toledo, ... Ohio	8 00	5 00	3 00
Tolono, ... Ills	16 55		6 75
Topeka, ... K. T	38 00	25 00	22 00
Urbana, ... Ohio	9 70		3 85
Vicksburgh, Miss	35 00	17 50	11 00
Vincennes, ... Ind	17 75		6 00
Wabash, ... Ind	12 00	8 75	5 00
Washington, . Iowa	23 00	13 00	8 50
Watertown, . . Wis	16 50	9 50	6 50
Waukegan, ... "	15 00	3 85	5 00
Waupun, ... "	17 30	8 00	7 30
Weston, ... Mo	33 00	10 30	8 50
Wheeling, ... Va	9 25		3 85
White Water, Wis	16 50		6 50
Winona, ... "	22 75	12 15	9 15
Woodstock, . . "	16 50	9 50	6 50
Xenia, ... Ohio	10 65		3 85
Zanesville, . . Ohio	9 95		3 50

FIGURE 4.13 Detail of map in *Condensed Time Table of the Lake Shore and Mich'n Southern R.R. Line* (1861). Courtesy of the Indiana Historical Society.

Passenger maps printed in public timetables became quite stylized. They emphasized routes of passenger service, usually with very thick lines, and often presented the lines as essentially straight, regardless of their actual path. They also omitted or deemphasized extraneous detail, such as branch lines that did not carry passenger trains. The 1889 Louisville, New Albany & Chicago Railway or Monon Route timetable map exemplifies thousands of railroad maps from the 1870s to the 1970s (figure 4.14).[72] The Monon's two main routes are drawn in exaggeratedly heavy lines, other railroads are deemphasized, little topographical information is provided, and stations along the two routes are shown as being roughly equidistant from each other in order to accommodate the map's typography.

3. "CLOUD-CAPPED GRANITE HILLS": THE ERA OF TRANSCONTINENTAL RAILROADS & SIGHTSEEING

Until the end of the Civil War America's focus had been on building railroads in the eastern half of the nation, including the tier of states just west of the Mississippi. This involved building short line railroads, connecting them, and then consolidating the shorter lines to create a web of connecting branch lines.

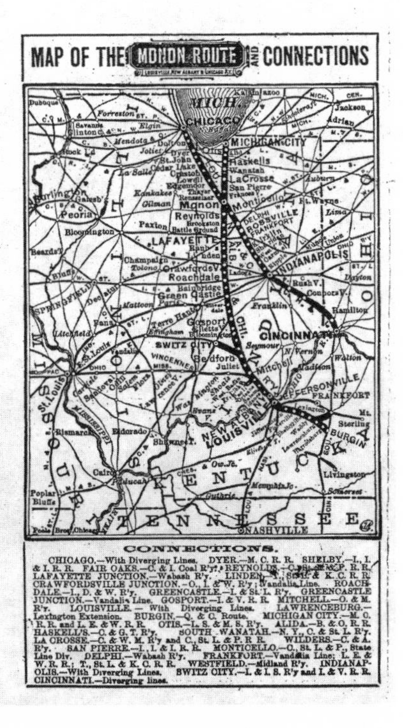

FIGURE 4.14 "Map of the Monon Route and Connections" (Chicago: Poole Bros., 1889), in *September, 1889 Local Time Tables, Monon Route, Louisville, New Albany & Chicago Ry.* (Chicago: Empire Snow Printing Co. for the Louisville, New Albany & Chicago Railway, 1889). Personal collection of the author.

Spanning the West As early as 1845, however, serious interest arose in building a transcontinental railroad to the Pacific. One impetus was the California gold rush of the 1850s. So many people wanted to go west, and thus needed cheap transportation and a means by which goods could be shipped to them, that newspapers, Congress, and average citizens were caught up in a decades-long interest in a transcontinental line. Various routes to the Pacific were considered—each supported by a very interested midwestern terminus, such as St. Louis or Chicago, and each backed by its own set of investors.[73] A number of speculative and survey maps were produced, each showing how a transcontinental line might connect with Chicago, St. Louis, Milwaukee, or Minneapolis and Duluth. As early as 1853 a bill was introduced in Congress to support Asa Whitney's plan to build a transcontinental line, followed by several other bills in the next three years.[74]

The War Department conducted the most important surveys to determine the feasibility of such an enterprise—the well-known U.S. Pacific Railroad Surveys of the mid-1850s is a twelve-volume assessment of the terrain, scenery, flora and fauna, and Native Americans of the West.[75] Gouveneur K. Warren's important *Map of Routes for a Pacific Railroad*, dated 1855, identifies the several routes under study.[76] Such maps helped whet the nation's appetite for building a transcontinental line, an accomplishment that contributed to an eventual interest in tourism and the production of tourist maps for rail passengers. Nonetheless, relatively little actual building took place until the close of the Civil War. Nebraska and Kansas, for example, each reported only a little more than 100 miles of railroads in 1865.

Rail Travel of the Late Decades of the Nineteenth Century By the last quarter of the century, railroads were perfecting their equipment to make rail travel rapid and comfortable. Cars were lengthened. As the English observer Edwin A. Pratt noted, "Americans were traveling greater distances and thus demanded cars that permitted them greater freedom to move about."[77] Pratt went on to note that the inside length of the typical passenger car body measured 45 feet 10 inches during the period 1862 through 1885, lengthened to 52 feet 10 inches in the 1890s and reached 69 feet by 1902.[78] Sleepers, including Pullman's fleet, became a success in the 1870s, while dining cars were introduced in substantial numbers by the 1880s. By the end of the century, open-platform observation cars were a feature of name trains, offering passengers an unobstructed view of the scenery. By the late 1880s, the vestibuled train became popular: enclosed vestibules connected cars so that passengers could safely walk from one car to another, allowing them to visit the diner, observation car, or lounge car.

Locomotives were becoming speedier and more powerful. The New York Central's Locomotive Number 999, a 4-4-0 fitted with eighty-six-inch drivers, achieved 112 miles an hour on a speed trial in 1893. And still larger engines were coming on line. The original wooden passenger cars were short and lightweight enough to be hauled by 4-4-0 Americans. But the new longer cars, and especially the newer steel cars, were heavier.[79] To achieve the necessary power to haul heavier trains, locomotives had to generate more steam, which meant a larger firebox, a larger boiler, and additional driving wheels. The result was either greater speed or greater power. All of this significantly enhanced the railroads' ability to provide the kind of long-distance service that attracted tourists.

Maps for Tourists Through the 1870s railroad maps designed for travelers chiefly focused on routes between various points, either to enlighten travelers about connections or to persuade them that a given railroad offered shorter routes to various destinations than did its competitors.

There had been earlier efforts by railroads to promote travel to tourist sights such as Niagara Falls. However, it was in the 1870s and 1880s that railroads began aggressively to promote travel for sightseeing purposes. The greater speed and comfort of trains helped to open the trans-Mississippi West and its magnificent scenery to railroad tourists.

Alfred Runte speaks of the trans-Mississippi west as the "romantic terminus." "Well into the twentieth century" he writes "the railroads of the West enjoyed a marketing advantage that was second to none. Theirs was the 'romantic terminus.' Americans have been drawn to the West not only by its history but also by its topography."[80] The Northern Pacific lobbied for the establishment of Yellowstone National Park. Other railroads followed suit by promoting the establishment of resorts and other national parks. Most importantly for our purposes, this interest in resorts and parks, according to Modelski, "created additional demand for maps to illustrate reports, promotional literature, displays, and timetables from the thousands of railroad and promotional firms which sprang up in the nineteenth century."[81]

The resulting promotional material took the form of an expanded timetable, a brochure with folded panels, or a bound booklet or even book. The 1885 Kankakee and Seneca Route timetable addressed the idea of tourism directly (figure 4.15).[82] The Kankakee and Seneca route ran on two cooperating railroads, the Cincinnati, Indianapolis, St. Louis & Chicago to the east and the Chicago, Rock Island & Pacific to the west. The map, occupying the timetable's inner four-panels, includes promotional text at all four edges. The text to the right speaks of the comfort of the line, while the left-hand text notes that travelers "can visit the Pellucid

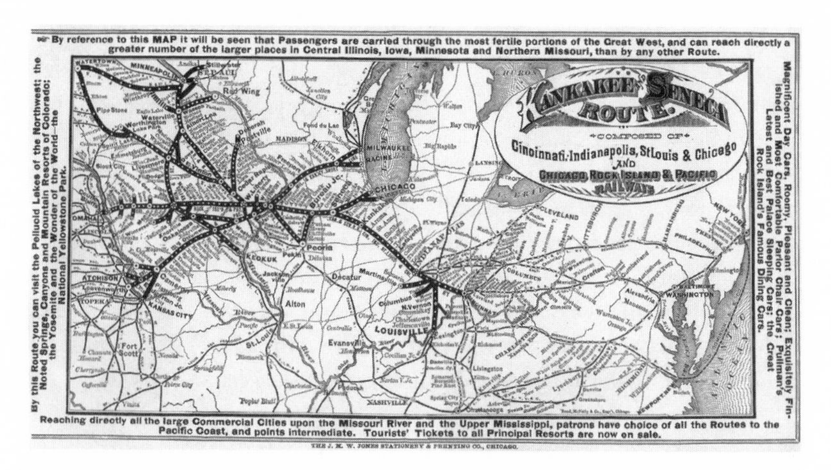

FIGURE 4.15 "Kankakee and Seneca Route" (Chicago: Rand McNally for the Cincinnati, Indianapolis, St. Louis & Chicago Railway and the Chicago, Rock Island & Pacific Railway, 1885), in *Kankakee and Seneca Route* (Chicago: J. M. W. Jones Stationery and Printing Co., 1885). Courtesy of the Indiana Historical Society.

Lakes of the Northwest; the Noted Springs, Canyons and Mountain Resorts of Colorado; the Yosemite and the Wonder of the World—the National Yellowstone Park." The timetable also invites readers to contact the railroads for separate maps, pamphlets, and tourists' guides describing cheap lands and health and pleasure resorts in the West and Northwest.

In addition to timetables incorporating more tourist enticements, various railroads introduced tourism brochures. The twenty-page, pocket-sized booklet, *Two Privileges of Summer Travel via The Lake Shore & Michigan Southern Railway: Niagara Falls and the Hudson River,* was printed in 1897 with numerous illustrations of scenery and travel accommodations. The "privileges" offered to travelers with tickets for east or westbound travel included a stopover at Niagara Falls without extra fare, or a trip between Albany and New York on the "palatial steamers" of the Hudson River Day Line. The booklet's final page is a small map. While the map image bears the title *Map of the Lake Shore & Michigan Southern Railway,* the page bears the title "Map of through car line between Chicago, Cleveland, Buffalo, New York and Boston, composed of the Lake Shore & Michigan Southern, New York Central and Boston & Albany Railways." A heavier line is used for the LS&MS road, but the other two railroads are included to suggest how travelers from many eastern and midwestern points can use the LS&MS Railway.[83]

Another example is from the Niagara Belt Line, a trolley line providing a two-hour, twenty-mile ride past several vistas from which to view the falls. This folding brochure has a color bird's-eye view of the falls and the Belt Line's route (plate 4). The Belt Line's double track route and its trolley cars are exaggerated for emphasis and clarity, with the steam railroads that brought tourists to Niagara Falls shown in the foreground. Besides describing the dramatic sights, the writer notes how important the railroad—in this instance, the trolley-system—is in realizing the grandeur of the Falls: "Before the great lines of electric traction were established at the Falls it was impossible to gain access to the many points of interest now reached with ease by the wonderful trolley system now encircling the Gorge which, for a distance of over twenty miles, continually presents to the tourist an ever-changing panorama of wonderful scenery."[84]

Just as the Niagara Belt Line highlighted its use of "large observation cars," so many railroads of the era acquired special excursion cars to further tourism. The Milwaukee, Lake Shore, & Western promoted travel by sportsmen to northern Wisconsin and the upper peninsula of Michigan. Their 1886 booklet *Gegobic,* featuring a map of the line's route and fishing and hunting sites along that route, included a diagram of the line's two special excursion cars. These cars, outfitted for fishing and hunting parties, staffed by a company cook, and designed to be attached to a regular passenger train, could be rented for $12 a day for a party of six.[85]

The first two categories of publications that railroads used to promote tourism—enhanced timetables and promotional brochures—gave equal or greater weight to the railroad itself. The third promotional medium, bound books, placed much more emphasis on the tourist destination. While produced by the interested railroad, such publications might devote only a few pages of advertisements to the sponsoring railroad and include a map showing how that railroad served the area. The Boston & Maine Railroad's 1887 *Down East Latch Strings; Or Sea Shore, Lakes and Mountains,* a 256-page softcover book, states its purpose as "presenting to the intending summer tourist a description of the scenery along the line of, and reached by the Boston & Maine Railroad and its immediate connections." The author goes on: "Here along [New England's] beaches or among its rocky promontories, with the broad Atlantic's breezes cooling the heated air, its piney woods and lakes furnishing health and sport to the Nimrods and Isaak Waltons of our land, or its pastoral valleys and 'cloud-capped granite hills,' may be found entertainment, health and pleasure for all tastes, however developed or inclined."[86] Twenty-three chapters of text provide detailed descriptions of such topics as "Portland and the Maine Beaches," "The Valley of the Kennebec," and the "Connecticut Valley." Along with the text are sixty-five black-and-white illustrations and seven maps, six of which are colored fold-out maps (figure 4.16). Two of the maps are bird's-eye views, while the other five show rail routes in relation to the area's attractions. The Boston

FIGURE 4.16 "Fishing & Hunting Resorts of Maine, Northern New Hampshire and Part of Canada and the Provinces . . . as Reached by the Boston & Maine R.R. and Connections" (Boston: Rand Avery for Passenger Department Boston & Maine Railroad, n.d.), in Ernest Ingersoll, *Down East Latch Strings* (Boston: Passenger Department, Boston & Maine Railroad, 1887). Courtesy of the Newberry Library.

& Maine passenger routes are depicted with a very heavy black line, while connecting railroads are represented by more modest lines. Various roads are shown, but in even fainter lines.

In the early part of the twentieth century a number of railroads moved away from the long promotional book of several hundred pages and instead offered the public a series of sixteen- to forty-eight-page booklets. Where the long books covered many destinations, the new, shorter works focused on one area served by a rail line. The forty-eight-page 1930 Union Pacific booklet *The Pacific Northwest and Alaska* includes photos and descriptive text of scenic areas of Alaska, Washington, Oregon, and Idaho served by the Union Pacific System—a collaborative of the Union Pacific, the Oregon Short Line, the Oregon-Washington Railroad & Navigation Co., and the

Los Angeles & Salt Lake Railroad. The final three pages are devoted to three maps. The first is of Washington, Oregon, and Idaho, with the Union Pacific System routes presented in heavy line. The second is of Alaska, again depicting the Union Pacific System in heavy line and showing the steamship routes with broken lines. The final map is of the United States west of Chicago and denotes the route of the Chicago & North Western, the line carrying travelers from Chicago to a connection with the Union Pacific in Council Bluffs, Iowa, and the Union Pacific System. The Union Pacific System is shown in heavy line for emphasis; the C&NW Railroad is shown in double line; and all other railroads earn a thin single line. As noted earlier, "the scale and directional relationships were carefully arranged to show the [railroad] lines and the region served by the company to the best advantage."[87] Since the Union Pacific did not directly serve Wisconsin, Illinois, and Mississippi, they are distorted to appear much narrower, and thus less important, than they were in fact.

The Chicago, Milwaukee & St. Paul Railway published *The Trail of the Olympian: Two Thousand Miles of Scenic Splendor, Chicago to Puget Sound* in 1924. This forty-page document includes thirty-three photographs of trackside scenes from Wisconsin to Washington state. One such photograph features the Olympian train and its open observation car. As the text notes, "In order that the scenery may be viewed to the best advantage and enjoyed without interruption, open observation cars are attached to the trains during the summer months while passing through the electrified portions of the route." This occurs in "the complete absence of smoke and cinders."[88] Also included is a map, highly stylized and lacking unessential detail (figure 4.17).

Promotional timetables and brochures, as well as tourist books of this type, were produced in sizable runs of tens of thousands of copies. The *Trail of the Olympian,* for example, enjoyed an initial press run of 25,000 copies. Printed promotional material that included maps was an important tool in railroads' effort to promote rail travel for the sake of travel. At the same time that maps proliferated as part of this promotional campaign, they continued to be simplified. In many instances, they became smaller, as the main purpose of the document in which they appeared was to highlight pictures of scenery or of comfortable railroad equipment. That railroads were beginning to relegate travel maps to a position of secondary importance is apparent from the fact the *Trail of the Olympian* booklet's map is quite small, measuring only 3 × 7 ¼ inches, and that the main burden of the enticement is borne by the thirty-three pages of photographs.

The point of the maps had become simply to show potential tourists that the railroad in question connected them to such locales as the New England

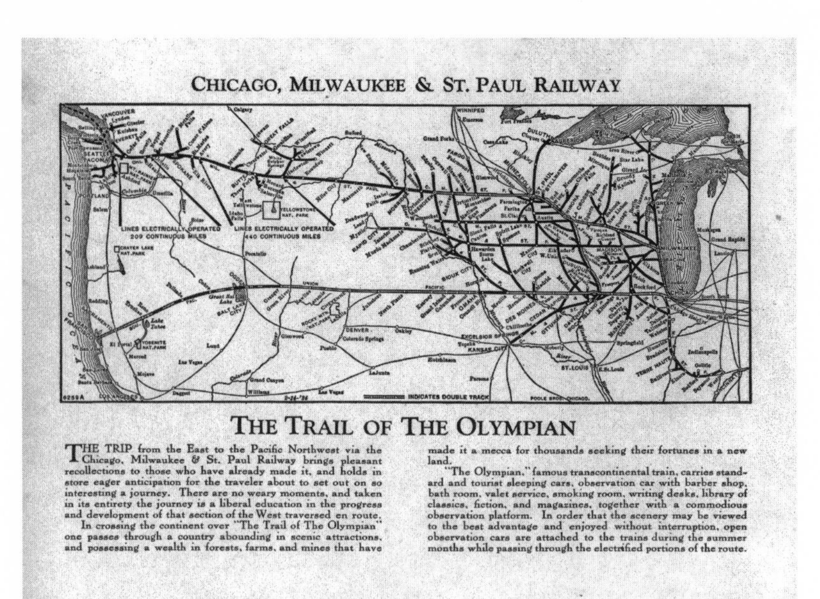

FIGURE 4.17 "Chicago, Milwaukee & St. Paul Railway, The Trail of the Olympian," in *The Trail of the Olympian: Two Thousand Miles of Scenic Splendor, Chicago to Puget Sound* (Chicago: Poole Bros. for Chicago, Milwaukee & St. Paul Railway, 1924). Courtesy of the Newberry Library.

mountains, Niagara Falls, the north woods and lakes of the upper Midwest, the mountains of the West, or various spas and resorts—not to meticulously delineate the course of travel. Indeed, including too much detail about the trip might make the journey seem a bit arduous, which would defeat the impression that one could quickly and comfortably be whisked away to some healthful resort or magnificent scenic area. Thus, the trend toward simplification and distortion in rail travel maps continued.

4. "COME WHERE THE BIRDIES SING": EXURBAN AMERICA

In a sense railroads had succeeded in creating a transcontinental nation by 1912 when Arizona, the last of the lower forty-eight states, joined the Union. Now all major regions of America were tied together by rail.

Interurban Railroads One last railroad development took place during this period: the expansion of suburban commuter service brought on by the new electric interurban railroads of the 1890s. Electric interurbans had a dramatic impact on short distance passenger traffic and in rural travel before automobile competition largely did them in.

While virtually every American community of any appreciable size sported an electric streetcar system, interurban lines quickly spread out over the national landscape as well. Nearly 1,000 route miles of track were operating in 1897; by 1905 there were 8,000. The country's interurban network peaked at slightly over 15,000 miles in 1915. Indeed, interurban companies built so much trackage and served so many communities that they saw themselves creating a network for long distance travel. A person could travel from Indianapolis to Chicago or to St. Louis by transferring from one interurban to another.

Interurbans arrived on the scene at a time perfectly suited to their development. As Roger Grant notes, smaller communities, especially those removed from large central cities, needed electricity. Quoting Grant: "By the turn of the century the electric railway had been largely perfected. Companies . . . built and operated networks of power stations and substations to supply electricity for themselves and frequently for commercial and residential users, as well."[89] Secondly, Americans had become more health conscious: diets tended toward fresh vegetables and city dwellers sought a way to move out of smog-filled inner cities. The scourge of tuberculosis led to suggestions that people protect themselves by breathing fresh air. Many homes at the turn-of-the-century were built with sleeping porches for this purpose, and some urbanites moved to the pastoral countryside for the outdoors and the opportunity to raise their own gardens. George Douglas writes of this general interest in suburbs: "An extremely common phenomenon in the last three decades of the nineteenth century was for a developer to identify some geographical area that was clearly suitable for suburban development and then buy up large parcels of land for purposes of subdivision into residential properties. Almost invariably, it was the presence of the railroad that gave impetus to these new developments, and more often than not the community was laid out with the railroad station and other commercial buildings being grouped together and constituting 'downtown.'"[90] The interurban especially fed the desire to move to healthful suburban and exurban communities. They projected the image of a cleaner way to travel in comparison to sooty steam trains. The Aurora, Elgin & Chicago and the developers of the community of High Lake, thirty miles west of Chicago, even commissioned a song called "Come Where the Birdies Sing" to call attention to the pastoral charms of this new subdivision.

Because they were lightweight and powered by electric traction motors, interurban cars could stop more rapidly and start with faster acceleration than could steam trains. Thus, they could much more effectively serve suburban and exurban communities, which demanded frequent stops. Turner,

Illinois, now West Chicago, serves as an illustration. It was a town of two or three thousand people at the turn of the century. It had a Chicago & North Western depot in the center of town for commuter traffic, but the new competing Aurora, Elgin & Chicago electric interurban had three stops in town—one in the center of town and one each on the eastern and western edges of the community. Indeed, the Aurora, Elgin & Chicago had five stops in the eight miles between the center of West Chicago and the center of Geneva, Illinois; the steam railroad C&NW had none. Developers of suburban and exurban communities recognized the advantages of the interurban, for commuters had no easy way to get from a steam railroad depot in the center of a town to their outlying community or subdivision. Such developers thus began to work closely with the new interurbans.

John Stilgoe notes that electric interurbans also offered advantages to rural areas:

> Farm regions that had never known railroad service suddenly discovered that they might have the advantages of a "car line." . . . Life on a rural car line struck many turn-of-the-century observers as remarkably pleasant. Cars ran regularly, often on twenty- or thirty-minute headways; away from villages, they proved quite speedy, humming along at thirty miles an hour on most lines. . . . Cheap fares—often only five cents, less than one-fifth the comparable railroad fare—encouraged frequent use. . . . Trolleys offered fast, regular, cheap, silent—and personalized—transportation to the nation's rural population, and to the rural people of the Northeast and upper Midwest in particular.[91]

In addition, the country had suffered through a depression starting in 1893, and companies operating one-third of the nation's steam railroad trackage entered bankruptcy. Thus, in the waning years of the 1890s, steam railroads were not prepared to meet the competition for commuter traffic that the new interurbans posed. They allowed their older commuter facilities to become seedy at the very time interurbans were building new, though not lavish, facilities.

Steam railroads eventually woke up to the threat. The Chicago & North Western went on a building spree in the middle of the first decade of the twentieth century. Many of their older, wooden suburban stations were replaced by new brick ones, and they built a new $20 million passenger station in Chicago. Similarly, Indianapolis steam railroads, confronted with the building of the nation's largest interurban station in their backyard, embarked on the first comprehensive renovation of their 1886 Union Station.

Interurban and Commuter Maps Both steam railroads and electric interurbans paid considerable attention to their commuter traffic and to the new commuter suburbs and exurban communities. This included the production and distribution of commuter maps, though this was delayed till the early twentieth century. Many interurbans had their start during the decade of the 1893 depression that had a restrictive effect on steam railroads' commuter operations. This same shortness of financing also meant that the new interurbans often did not have the funds to produce maps. As a result, many interurban traveler maps did not appear until the early years of the twentieth century.

Once they arrived, such maps shared many characteristics with conventional passenger maps. They were almost always contained within timetables or promotional brochures. They were stylized, with the main line drawn in a heavy, straight line. Some geographic distortion occurred, with stops shown as evenly spaced, in part for reasons of design and typography. And competing lines could be deemphasized or omitted altogether. In addition, pictures and text might extol the virtues of the communities along the route. Commuter railroads—both steam and electric—were, after all, trying to encourage people to locate along their routes, so promotion of the towns themselves became part of the purpose of the timetables and their maps.

The map contained within the twenty-four-page Electric Railway Guide Company's 1910 *Detroit United Railway* timetable encompasses the trains of seven connecting lines, including the Detroit United Railway; the Detroit, Jackson & Chicago; and the Detroit, Monroe & Toledo (figure 4.18). The guide serves two purposes of importance to interurbans. First, the various suburban and small outlying towns served are noted for commuters. And, as is suggested by the names of some of the interurbans represented in the guide, the map also highlights the fact that interurbans connect such cities as Detroit and Toledo or Detroit and Chicago. Because it was reproduced by a travel guide company rather than by any individual electric railroad, this map repeats the characteristics of early to mid-nineteenth century maps. In addition, the map's quality suggests the fiscal constraint of the time. The map is rather indifferently printed, with ink skips, and no map creator is identified.

Just as steam railroads used maps to assist in promoting tourism starting in the last third of the nineteenth century, so interurbans promoted tourism as well. *Trolley Trips in the Hudson-Mohawk Valleys,* a twenty-page illustrated brochure of the Schenectady Railway Company, features many photographs of natural scenes, including waterfalls, village scenes, area architecture, and cultural institutions such as Union College. While the

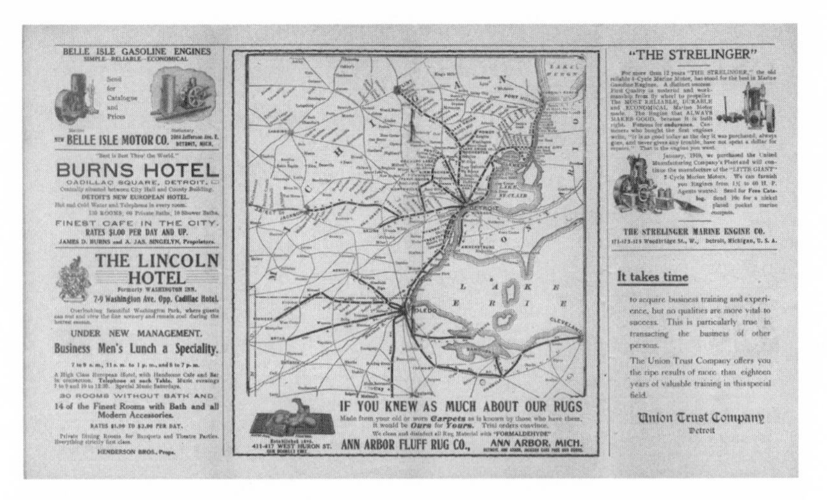

FIGURE 4.18 Untitled map of interurban rail lines in southeastern Michigan and northwestern Ohio, in *Electric Railway Service* (Electric Railway Guide Company for the Detroit United Railway, 1910). Personal collection of the author.

photographs are black and white for purposes of economy, the brochure's wraps and its five maps are in color. This use of color greatly enhances the brochure's appeal, and the color maps are particularly handsome, especially the center map, *Schenectady Railway and Connections,* which shows the area from Albany to Schenectady to Saratoga Springs.[92] The lines of the Schenectady Railway are shown in red, while the rivers and lakes are in blue. Topographical features, such as a range of hills, are shown in light olive, with city blocks in olive and major parks in dark olive. Other maps, also in color, show the Lake George area and the Hudson Valley.

A similar promotional brochure, dating from about 1907, is *Trolley Trips,* a twenty-four-page brochure issued by the Passenger Departments of the Boston & Northern and the Old Colony Street Railway Companies. Included is a colored folding map that shows the two interurban railroads' routes from Providence and Newport, Rhode Island, on the south through Boston, to the southernmost communities of New Hampshire (figure 4.19).[93] While the principal purpose of the brochure and its map was to encourage day trips, the brochure also noted that the two railroads offered special chartered trolley car service. "The special car service is especially adapted to picnics, outings of all kinds, lodge visitations, visits to social, religious gatherings, entertainments in neighboring towns, historical explorations."[94] Such service, with the ability to stop almost anywhere along the line, was a significant advantage over steam railroads enjoyed

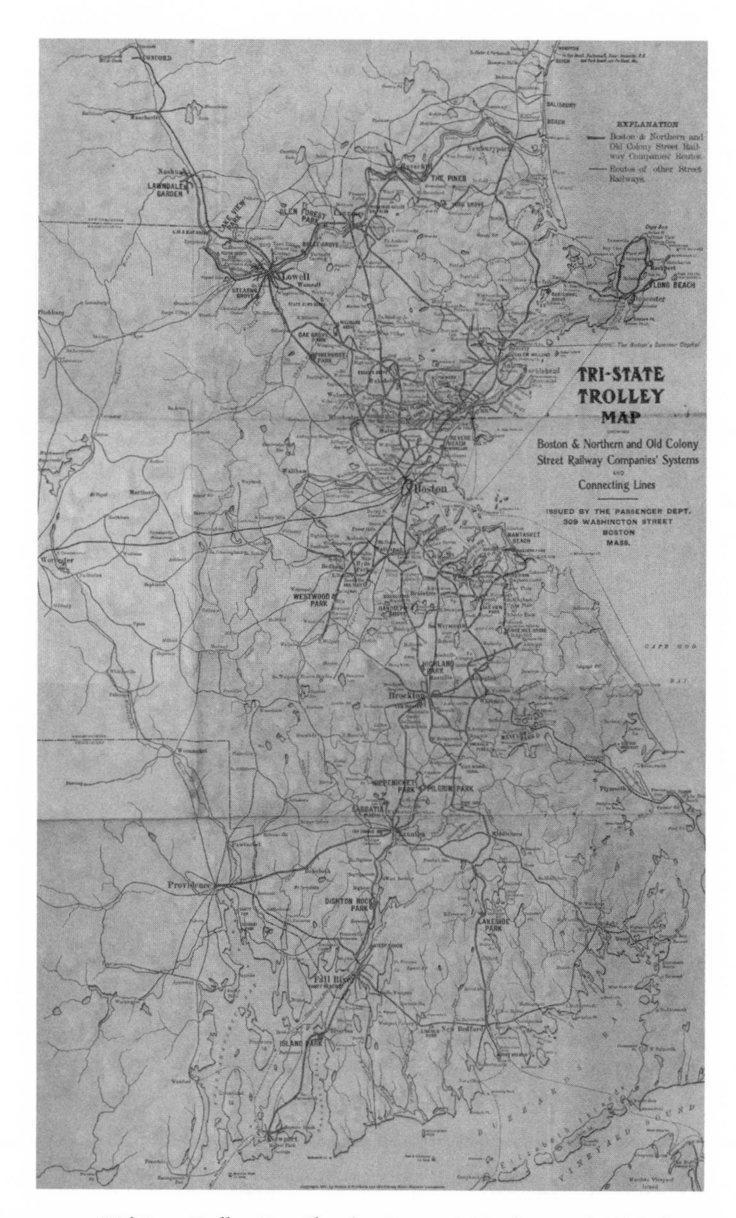

FIGURE 4.19 "Tri-State Trolley Map Showing Boston & Northern and Old Colony Street Railway Companies' Systems and Connecting Lines" in *Trolley Trips* (Boston: C. J. Peters & Son for Boston & Northern and Old Colony Street Railway Companies, 1907). Personal collection of the author.

by electric interurbans. And a handsome map assisted parties in planning their outings.

Curiously, while steam railroads responded to the new competition from interurbans by upgrading their stations and their service, they do not appear to have produced many commuter maps. This is certainly true of their commuter timetables, at least until after World War II. Perhaps this difference can be accounted for by recognizing the difference in the number of stops made by each of the two types of railroads. Steam railroads stopped only in the center of each town they served. Interurbans, by contrast, made frequent regular stops, including at the outskirts of many towns. Commuters, or potential commuters, might not need a map showing the towns along a commuter line, but they may very well need a map to help them understand all the stops between towns. Added to the fact that interurbans actively promoted settlement in outlying regions, the need to illustrate where trains stopped prompted interurbans to produce more commuter maps than did steam railroads.

In the late 1920s the Illinois Central (IC), a traditional steam railroad, electrified its suburban Chicago service to compete directly with the interurban lines. Employing a six-panel brochure to announce this new service, the company trumpeted the fact that the "Illinois Central Electrified serves added territory equal to 7040 City Blocks." The promotional brochure includes a map, produced in April 1927 by Rand McNally of Chicago (plate 5). Copying the promotional claims its interurban rivals directed at suburbanites, the IC includes a caption to its map that reads "A study of this map and time tables will show prospective homebuilders the many advantages of locating within reach of this service." Typically, the brochure does not promote or illustrate the steam railroad's traditional commuter service but rather features the IC's new electric or quasi-interurban service.

Conclusion

By the end of the 1920s—as automobiles, trucks, and airlines began to replace railroads as the primary transportation system—railroads had completed the work of creating a continental or transcontinental nation. As George Douglas notes, even as early as 1870, "it had been commonplace to say that the railroad was the agency that had made the United States a unified nation, that it had been the railroad which had drawn so many lonely and isolated communities into one national community."[95] Americans, a people seemingly in perpetual transit, could now move anywhere—reach any state and any city, travel to nearly any scenic area, leave cities to reach most outlying suburbs and rural communities—cheaply, safely, and conveniently by rail. Tens of thousands of railroad passenger maps first foreshadowed and then chronicled this movement. In the decades following 1830 they showed how railroads connected with rivers, lakes, and canals to offer a fairly primitive transportation network. Early maps showed how railroads tied together major cities, usually port cities, with their outlying farm areas. By the 1850s, railroad companies were creating a new transportation network, one that relied chiefly on rail and less on water transportation. This development dictated the production of a new set of maps, initially produced by independent publishers but soon by the railroads themselves. The establishment of new railroads and the extension of existing ones were so intense that maps had a relatively short useful life. With all the new rail lines and with the new towns such lines fostered, railroads had to continuously update their maps, resulting in the production of thousands of railroad maps.

In 1869, the first transcontinental railroad was completed. The nation now turned its attention to development west of the Mississippi, which in turn led to the distribution of more maps. At the same time, railroads both east and west of the Mississippi recognized the potential of tourist traffic and produced numerous new promotional brochures encouraging tourism. While these brochures needed to include new maps, the maps now played a secondary role. Photographs and enticing text were emphasized, with simplified maps relegated to the back page of the brochures.

As Frederick Jackson Turner famously noted, the American frontier closed by 1890, a development in which railroads played a principal role. But there was still one more development to occur in the settling of the American continent—the growth of suburbs and exurban areas around major cities—and once again railroads played a key role. In the 1890s the electric interurban took hold, especially east of the Mississippi. Just as steam railroads had worked to build the concept of travel, the interurbans worked to encourage suburban real estate developments with the knowledge that future residents would become riders of interurban cars. To promote the idea of such developments, interurban companies, like their predecessors, developed maps to educate the public.

The early twentieth century also saw the advent of the automobile. That railroads no longer ruled the arena of passenger travel is suggested by the publication of such map series as Clason's Green Guides, "with Road and Railway Maps." Clason's Green Guides included fold-out sheets featuring a map of a given state's highway system "Showing Paved Roads, All Weather Roads and Other Thoroughfares," while the verso comprised a map of the state's railroads.[96] Significantly, the index of towns and the publication and copyright information is printed on the page featuring the map of roads, not on the railroad map page.

For one hundred years, railroads were the dominant means of transportation in America. During that time they not only produced thousand of maps, they also strengthened the map publishing industry and helped educate the public about geography by providing free maps to whoever wanted them.

PLATE 1 Five hundred dollar bond, the New York Central Rail Road Company [1853]. Personal collection of the author.

PLATE 2 "Map of the Great Southwest" in 5,000,000 *Acres, Fine Farming Lands in Northern and Western Texas for Sale by Texas & Pacific Railway Co.* (St. Louis: Woodward & Tiernan for the Texas & Pacific Railway Co., 1882). Courtesy of the Newberry Library.

PLATE 3 J. Calvin Smith, "Guide Through Ohio, Michigan, Indiana, Illinois, Missouri, Wisconsin, and Iowa," in *The Western Tourist and Emigrant's Guide* (New York: J. H. Colton, 1840). Courtesy of the Newberry Library.

PLATE 4 Detail view of the Niagara Belt Line, in *The Niagara Belt Line—around and through Niagara Gorge* (Buffalo: Matthews-Northrup for the Niagara Belt Line, n.d.). Courtesy of Jeff Darbee.

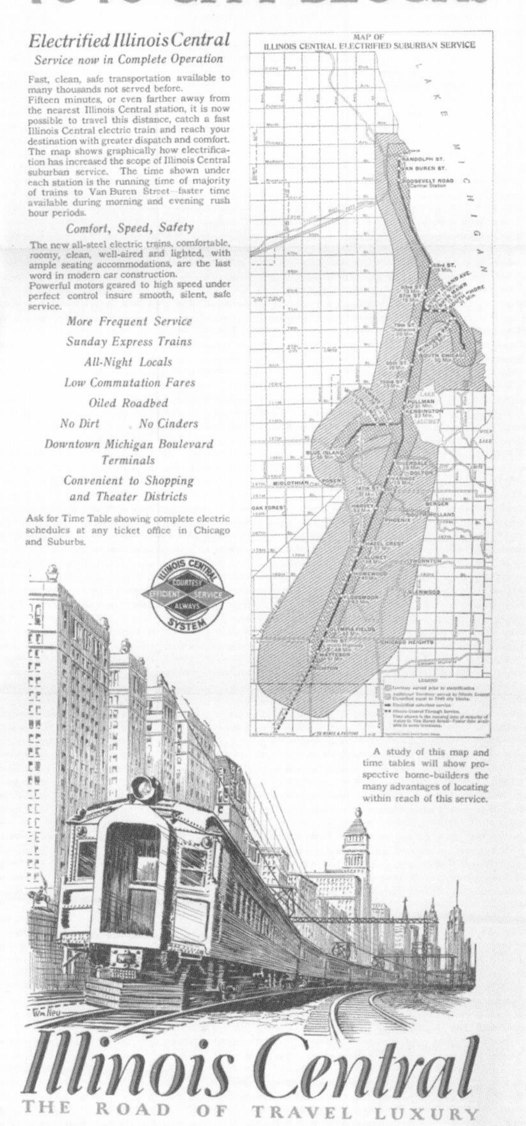

PLATE 5 Untitled map of Illinois Central electrified suburban service, in *Illinois Central Electrified Serves Added Territory Equal to 7040 City Blocks* (Chicago: Rand McNally for the Illinois Central Railroad, 1927). Courtesy of the Newberry Library.

PLATE 6 Detail of "Pictorial Map of the United States with Trip-Planning Guide" (Convent Station, NJ: General Drafting for American News Co., 1952). General Drafting Collection, the Newberry Library. American Map Corporation; reprinted with permission.

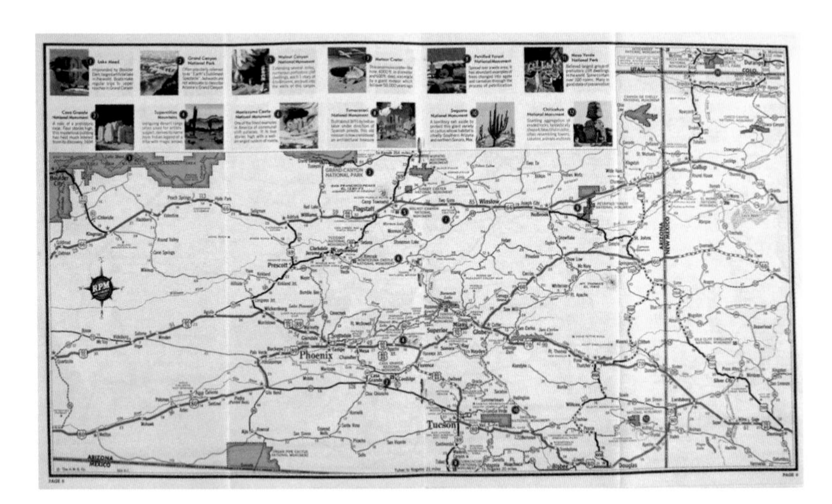

PLATE 7 Detail of Arizona from "U.S. 60-66-70-80 Southern Routes West-East Standard Oil Interstate Route Map" (Chicago and San Jose: H. M. Gousha for Standard Oil Company of California, 1941). Personal collection of James Akerman. Map by Rand McNally, R.L. 04-S-108.

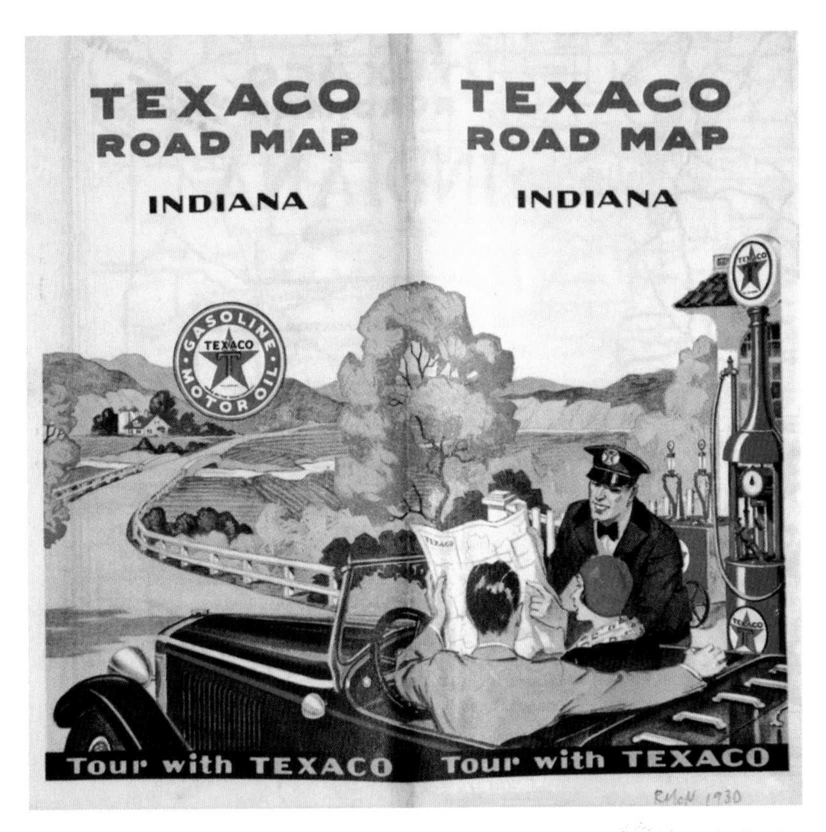

PLATE 8 Cover illustration, "Texaco Road Map, Indiana" (Chicago: Rand McNally for The Texas Company, 1930). Rand McNally Collection, courtesy of the Newberry Library.

PLATE 9 Cover illustration, "Florida, Sinclair Road Map" (Chicago: Rand McNally for Sinclair, 1936). Rand McNally Collection, courtesy of the Newberry Library. The prosperous suburban setting of this Depression-era cover illustration associated gasoline consumption with economic comfort.

PLATE 10 Cover panel, "New England Road Map, Sunoco Gas, Oils" (Chicago: Rand McNally for Sunoco, 1938). Courtesy of the Newberry Library.

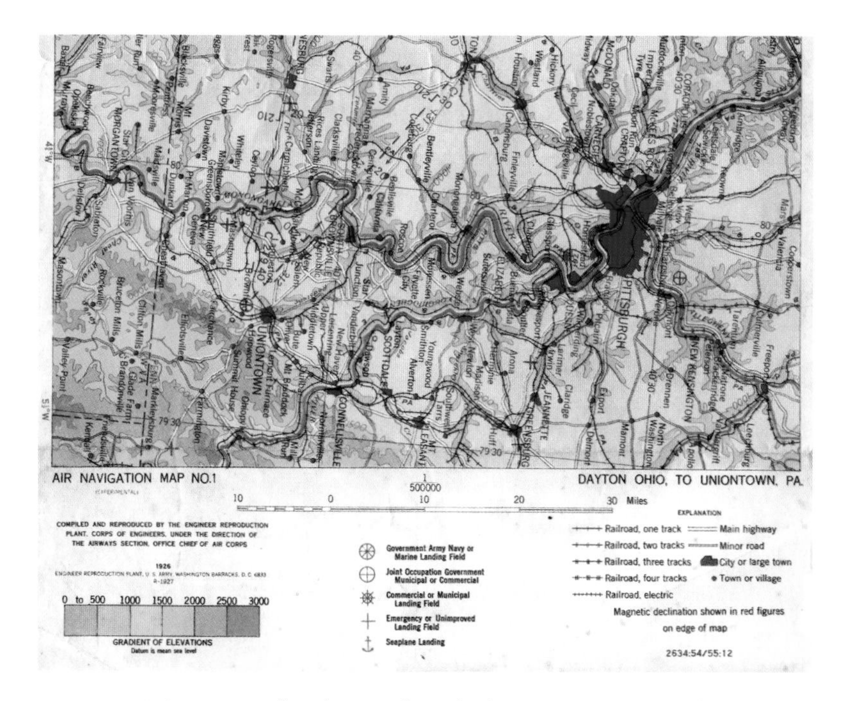

PLATE 11 U.S. Army Corps of Engineers, "Air Navigation Map, No. 1, Dayton Ohio, to Uniontown, Pa." 1926, revised 1927. One of fifty-two air navigation strip charts produced by the U.S. Geological Survey and the Corps of Engineers under the direction of the Army Air Service/ Air Corps from 1923 to 1933. Private collection of Ralph Ehrenberg.

5

Twentieth-Century American Road Maps and the Making of a National Motorized Space

JAMES R. AKERMAN

In his poetic study of the American railroad landscape of the late nineteenth and early twentieth centuries, *The Metropolitan Corridor*, John Stilgoe writes evocatively of the almost cinematic way in which the American landscape appeared to whisk by turn-of-the-century train passengers. The thrill of motion at speed was, like the experience of a movie, mitigated by train passengers' inability to control the movement of the vehicle and their isolation from the landscape enforced by the great "plate glass window of the Pullman car." "For its occupants [the express train] became the world, and the world beyond it became a series of snapshots flashed by a cinemagraph performance."[1] What riders on a train saw of the silent world outside was also effectively limited to the 180-degree field of vision on their side of the car and by the character of the corridor landscape itself, which faced them at stations, but turned its back on them elsewhere in what were still pedestrian-oriented urbanized areas.

Slow, weakly powered, and hampered by the poor quality of American highways, the first automobiles could not match the range and speed of express trains. Automobile travel nevertheless appealed to early twentieth-century Americans because it held out the possibility of escaping the tyranny of the rails and interacting with the landscape on their own terms, going wherever an accommodating road allowed, at one's own pace, with the option of disembarking to rest or to explore on foot wherever one pleased. The near total control of one's mobility the car offered was superbly suited to Americans' legendary restlessness and love of frequent, if not constant motion.[2] Douglas Brinkley observes:

> To find liberty in America—that notion enshrined in our nation's birth announcement, that sought-after sensation of freedom—hop in your car and drive in any direction for five hours, at any speed on any highway or road, with no objective, no one to meet, alone behind the wheel. Then, when the tank nears empty, fuel up at a gas station, look into the sky, feel the wind, sun, rain, or snow in your face, and freedom is yours.[3]

Over the course of the twentieth century, Americans came to see unfettered exploration of the countryside by automobile as the quintessential expression of American identity, appealing, in the words of Henry B. Joy, highway promoter and early president of Packard, to "the pioneer instinct . . . in all of us."[4] For David Halberstam's immigrant grandfather, the car "had not merely transported him to different places, it made him feel more *American*, more a citizen of this land than a refugee from the old one."[5]

The train endures in American culture and mythology as a symbol of continental conquest and modern humanity's mastery of nature, but the automobile's ability to infiltrate the landscape it occupies ultimately has made its geographical and its cultural impact more far-reaching. John Jakle observes that "no other technological innovation has so transformed the geography of the United States as the automobile."[6] Over the course of the twentieth century it has transformed—some would say destroyed—the structure and fabric of our urban areas, while making even the most remote rural areas accessible. Self-styled *Road Scholar* Andrei Codrescu understands, however, that automobility's reach extends into our very souls and identities:

> An American without a car is a sick creature, a snail that has lost its shell. Living without a car is the worst form of destitution, more shameful by far than not having a home. A carless person is a stationary object, a prisoner, not really a grownup. A homeless person, by contrast, may be an adventurer, a vagabond, a lover of the open sky. The only form of identification an American needs is a driver's license.[7]

I would add that the only map an American of the automobile age really needs is a road map. Americans' sense of their national geography over the last three centuries is epitomized by the maps made for their general use. Waterways, the principal paths by which Europeans came to know, penetrate, and conquer the eastern half of North America, visually dominated eighteenth-century maps and many early nineteenth-century maps of the continent. Maps published by American commercial cartographers during the first half of the nineteenth century stressed counties, territories, and states: the political organization of a nation constantly acquiring and restructuring western territories.[8] In late nineteenth-century maps, the saturated solid colors of the various states and counties receded into the background as the bold lines representing steel-railed sinews of the nation came to the fore, appealing to a nation concerned with the social and economic integration and consolidation of the previous century's conquests.[9]

If the United States was in the twentieth century indeed a nation defined by its obsession with and reliance on personal motorized transportation, then the common automobile road map was America's national blueprint. Cursed for its occasional inaccuracies, abused when in use, discarded when out of date, this humble paper artifact also became the essential guide to an American landscape imprinted almost everywhere by the needs and demands of the car. The web of highways on the road map became, alongside the boundaries of the fifty states, a critical frame of reference for all Americans, an image of their country they committed to memory as they did the national flag and the Statue of Liberty. Moreover, the road map and the highway space it represented gave driving Americans the tools to explore the national territory in numbers and over distances without parallel in any other nation's experience. It widened the horizons of average Americans by leading them to and through radically different physical and cultural landscapes from those they knew at home. At the same time, maps and highways enlarged Americans' concept of home. The automobile journey, guided by the road map, became for many Americans a means of discovering a common national geography, history, and culture.

Introducing their appreciative album of oil company road map cover art, Douglas Yorke, John Margolies, and Eric Baker observe that "Oil company road maps were the most universally loved and the most genuinely helpful by-products of America's abiding love affair with the automobile. For generations of motorists they were the charts of dreams and future adventures on the open road. And they were powerful symbols of mobility and prosperity in the United States."[10] Maps were incorporated into travel scrapbooks to help tourists recall the places and spaces they had experienced; in some instances, the maps themselves were annotated and retained as narratives of past vacations. Chicago's Newberry Library possesses several such maps, including an early 1920s map of Minnesota annotated with the route and sites of a hunting and fishing trip.[11] I possess several copies of AAA Triptiks prepared by the Chicago Motor Club for a fastidious traveler who annotated each with travel times, trip mileage, gasoline expended, and the location and character of stops for meals.[12]

The promotional road maps issued in profusion by the purveyors of roadside goods and services more publicly exploited the sentimental and cultural content consumers packed into road maps. These maps targeted leisure travelers who needed guidance on roads that extended beyond the bounds of their usual geographical experience and competence. Free road maps fulfilled this need, and in so doing encouraged the growth and health of automobile-related consumption. "Free" maps also served as simple advertisements for whatever their sponsors had to offer. Oil company maps included advertising copy and artwork asserting the superiority of the branded gasolines and services. Hotel and motel maps likewise touted the value and convenience of their sponsor's accommodations. State governments used road maps to assert the modernity and scenic quality of their highways. AAA maps, in tandem with their regional guidebooks, helped motorists locate subscribing motels and restaurants.[13] Implicitly, and often explicitly, underlying these various commercial and regional agendas was an industry-wide promotional argument that discretionary automobile travel was not merely a pleasant diversion, but was in fact an essential act of American citizenship in the twentieth century—citizenship, that is, in a national motorized space.

One of the most straightforward statements of this rhetorical tack may be found the *Pictorial Map of the United States with Trip-Planning Guide* issued in 1952 by the General Drafting Company, one of the so-called Big Three of American automobile road map publishing (along with Rand McNally and H. M. Gousha).[14] On the back panel of this map, a tableau of diverse but quintessentially American scenes and landmarks (including the Capitol, Independence Hall, Niagara Falls, a New England church, pink flamingoes, a lighthouse, and a broad plain bisected by a highway populated by oversize cars and receding into purple mountains) frames what might be interpreted as a mission statement for automobile tourism:

> Know your country! Wider horizons make better citizens. Your community is part of America—the most important part for you. But America is also the campus of some distant university, a wheat field in Kansas, lofty Mt. Whitney in California, the Florida Everglades and the Detroit automotive plants.

Inside, a large map of the United States filled coast-to-coast with more than 700 drawings showing "a sampling of America's natural beauty and culture, its history, industry and products," was intended to help travelers "learn more about our country's beauties and its wealth of resources" (plate 6). The illustrations in South Dakota, for example, include images of the state's principal agricultural products, its gold and silver mines, Mt. Rushmore, and the Badlands. Alabama illustrations include the Wilson Dam, the iron and steel mills of Birmingham, "old French settlements" near Mobile, the Carver Museum, the "capital of the Confederacy," and a black cotton worker. On the other side of the unfolded sheet is a road map of the United States framed with photographs of American scenes topped with a banner proclaiming that "the heritage of every American reaches across a continent."[15]

This promotional tack had its origin in the See America First movement that flavored much late-nineteenth and early twentieth-century railroad tourist promotions.[16] But automobile tourism added something with which railroad tourism could not effectively compete, namely the promise of a more intimate and democratic means of knowing the American landscape. Railroad-dominated tourism promotion in the nineteenth and early twentieth centuries elevated the stature of particular, geographically restricted landscapes as quintessential expressions of an American identity, rivaling and surpassing comparable European landscapes in their beauty and savage grandeur.[17] This approach suited well the dependence of railroad tourism on the promotion of specific places, particularly the emerging system of national parks, to which individual railroads enjoyed near exclusive access. As we shall see, the automobile tourist industry—and the maps it produced—also promoted these "sacred places"[18] and landscapes as touchstones of American identity. But auto interests also stressed the more democratic aspects of the highway, the ability of the car to take its operators wherever they wished to go at speeds that nearly matched the train's, without the routing limitations railroads imposed on their passengers. Automobile promotions urged tourists to develop their citizenship by discovering the country in its entirety, not merely by traveling from one coast to the other, but by experiencing regional landscapes in intimate and exquisite detail.

Mapping was the ideal medium for making this point. A small road atlas published by the H. M. Gousha Company for Standard Oil of California consists of four sectional maps of the corridor stretching from southern California to Oklahoma and eastern Texas, served by four major east-west U.S. highways (U.S. Routes 60, 66, 70, and 80). A panel running across the top of each section pictures and briefly describes twelve or sixteen places of

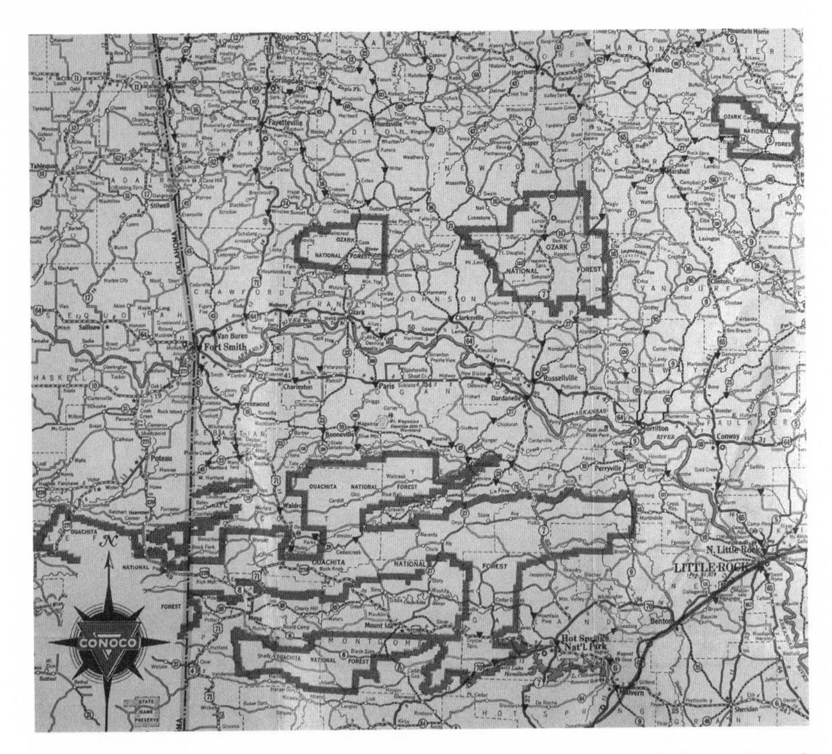

FIGURE 5.1 Detail of "Travel Arkansas with Conoco" (Chicago: H. M. Gousha for Continental Oil, 1933). H. M. Gousha Collection, the Newberry Library. Map © by Rand McNally, R.L. 04-S-108. While guiding tourists to sites in the Ozark Mountains, the map also pinpoints, by means of small green triangles, the locations of Conoco service stations.

interest, identifying these by numbers that may be found on the map below (plate 7).[19] These sites, along with the dozens of "points of interest," parks, and cities, marked on the map provide the motorist with an apparently unlimited choice of itineraries. The apparent freedom to appreciate the southwestern landscape on the motorist's own terms and schedule stands in stark contrast to what the comparable traveler on the Southern Pacific Railroad or the Santa Fe might experience from the confines of the Pullman car.

And yet, by posing itself as a portable tour guide, the atlas is certain to shade and possibly limit the motor tourist's reading of the landscape. The great irony of free automobile road map distribution was that while it worked in many practical ways to break down the barriers to personal geographical mobility, it also sought to regulate the movement of drivers, to bring their itineraries into conformity. At the simplest level, this meant directing motoring tourists to specific commercial establishments. Many oil companies, for example, saw to it that their maps clearly indicated the location of each service station in the mapped area (figure 5.1). The same technique was by applied on maps issued by motel, hotel, and restaurant chains and even clothing stores.[20] Seeking to draw tourists their way, associations of communities and roadside businesses strung along specific federal or state routes frequently banded together to publish glossy maps or map-saturated booklets promoting their route as the most scenic and/or direct. Among the most widely distributed of these was a simple three-color strip map (measuring 69 × 11 cm) published by the U.S. 441 Highway Association, one of several U.S. highways that funneled traffic from the Midwest to Florida (figure 5.2).[21] Typical of this genre, the map emphasizes several advantages of the route over its competitors: Route 441 is the only U.S. highway that passes through Great Smoky Mountain National Park. The road is the "fastest through Georgia," and it enjoys convenient highway connections from Chicago, Detroit, and Cleveland to its northern terminus in Lake City, Tennessee (most alternative north-south routes through Tennessee are carefully omitted). The points of interest catalogued on the map's reverse side include a selective list of the "best hotels and motor courts . . . along U.S. 441," which we may assume have paid fees to the association or in some other way have sponsored the map's publication.[22]

Even ubiquitous brochures promoting the myriad caverns, Mystery Spots, and Rock Cities that dotted roadsides incorporated maps to draw and direct motorists off the highway to their hidden treasures. Many of these, such as the $2\frac{1}{2} \times 3\frac{1}{2}$—inch map included in a circa 1970 brochure advertising *The Amish Farm and House* of Lancaster County, Pennsylvania,[23] are highly schematic, interested only providing simple guidance to the roadside attraction from major highways (in this instance, the Pennsylvania Turnpike). Others, such as the circa

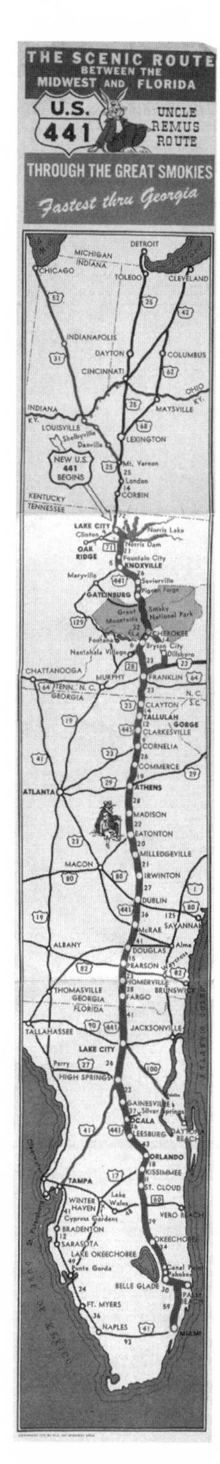

FIGURE 5.2 "The Scenic Route between the Midwest and Florida, U.S. 441" (Commerce, GA and Gatlinburg, TN: U.S. 441 Highway Association, 1951). Personal collection of the author.

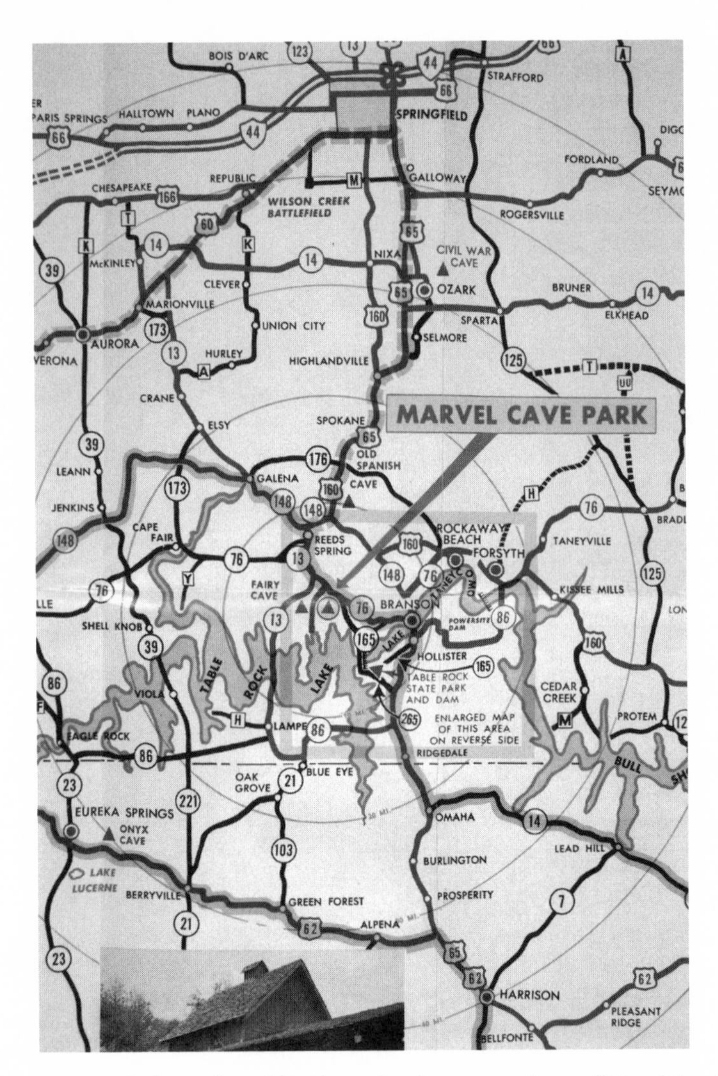

FIGURE 5.3 Detail of "Ozarks—White River Lakes Country . . . Home of Marvel Cave," in *White Lakes Area Map Compliment of Marvel Cave Park* (Aurora: MWM Color Press for Marvel Cave Park, c. 1965). Personal collection of the author.

1965 map "Ozarks—White River Lakes Country . . . Home of Marvel Cave," are full-fledged (if undistinguished) highway maps designed to enhance the allure of a particular attraction by putting it in the context of a well-known resort district (figure 5.3).[24]

In these instances, the cultural and commercial destinations confined to marginal vignettes, back panels, and cover art in more conventional road maps were allowed to mingle seamlessly with the maps' principal navigational content—the directionality and connections between roads, travel distances, and the location of salient features such as towns and rivers. But in truth, the separation between the interpretive and navigational elements of any road map is largely an artificial one. Whether or not they made this explicit, promotional road maps were intended to bring consumers to specific businesses and destinations.

Through their representations of directions and destinations road maps mediated the emergence of a national motorized space created by and for the American motoring tourist. This space—part of the larger geography of the car culture that also embraced the shopping mall, the bedroom suburb, and the urban expressway—was created for the pursuit of pleasure, but it also fostered citizenship and belonging. Physically, the space consisted of motels, roadside attractions, national parks, rest areas, historic markers, and the highways linking them. Conceptually, it was a geographical expression of national citizenship, developed and reinforced by what became, by the middle of the twentieth century, a common and almost mythological ritual of automobile travel. This space was national not merely because tourists crossed state lines when operating in it, but also because the passages of automobile tourists through this space defined places and experiences that—at least in theory—all citizens could know and should know, and ideally experience first-hand, as significant to their identity as Americans.

Origins

American highway mapping, like automobile tourism, drew its inspiration from two late nineteenth-century sources: railroad-oriented tourism and the bicycle craze that swept the country in the 1880s and 1890s. The starkly different cartography produced for these earlier modes of transportation reflected the character of each. As Gerald Musich has shown in the previous chapter, railroads produced maps to promote the interests of large corporations, whose operations might span the continent. They tended to gloss over geographical details, relegating to accompanying text the description of landscapes through which railroads passed. Late nineteenth-century railroad mapping offered a sweeping vision of a transcontinental nation state, bound together by a dense and rapidly growing network of steel roads capable of transporting passengers from coast to coast in a few days. Bicycle mapping, in contrast, had a far narrower geographical vision befitting the limited range of the machines; cycling maps depicted districts rather than entire regions or nations. They were large in scale and modest in geographical scope, concerned with the detailed description of localities, often in exquisite geographical detail. Their authors and publishers usually were local mapping companies or local bicycle clubs.

Each of these traditions defined traits that we recognize in the mapping produced for twentieth-century automobile tourists. Intensely promotional and commercialized railroad cartography bequeathed to automobile mapping its transcontinental vision, its emphasis of speed, and its fetishization of particular destinations and experiences. Bicycle mapping bequeathed—particularly to the early motorist—a mapping genre that emphasized self-reliance and unlimited possibilities.

Scenic tourism was already well established before the Civil War at eastern destinations such as Niagara Falls, the White Mountains, and Mammoth Cave.[25] The extension of the nation's railroad network to the Rocky Mountains and beyond, coupled with the steady growth of an American middle class, promised and delivered further spectacular growth. A vigorous publicity industry emerged by the end of the century devoted to the production of guidebooks, maps, and publicity for emerging resorts across the country, now accessible via a few days journey in the relative comfort of the new Pullman rail cars. Printing companies such as Rand McNally and Poole Brothers in Chicago and Rand Avery in Boston became specialists in cheaply produced, cartographically enriched brochures issued by railroads to publicize routes to emerging resorts or new ways to reach old ones.

The publication of maps and related materials for tourists intensified after the 1880s as various western railroads began to carve out market territories, forging alliances with newly created national parks and monuments, and investing in park infrastructure and services in return for gaining almost total control of the stream of park visitors. The brochures published by railroads emphasized both the exotic nature of the destinations as well as the comfort and simplicity of railroad visits to them. Small-scale maps attached to these brochures did their best to emphasize the directness of rail connections to the parks from major eastern and midwestern population centers, while large-scale maps stressed the accessibility provided by railroad gateways to the leading parks' sites and amenities.[26] Yet most of these maps stressed transportation and tours operated for rail passengers by local stage lines (and later, motor coach operators). With the benefit of hindsight, the look of park guide maps published during the period of railroad dominance suggests the inevitability of the transition from train to car as the ideal conveyance into the park landscape. A map published in 1902 by the Santa Fe (figure 5.4) shows that while a railroad spur could bring travelers to the brink of the Grand Canyon at one spot, full appreciation of the landscape could only be achieved on the system of trails that traversed the rim or led into the canyon.[27] Motorcars could not, of course, descend

FIGURE 5.4 "Map of Grand Canyon of Arizona" (Chicago: Poole Bros., n.d.), in *The Grand Canyon of Arizona* ([Chicago]: Passenger Department, Atchison, Topeka & Santa Fe Railroad, 1902). Courtesy of the Newberry Library.

into the chasm, but, like the stagecoaches that preceded them, they could traverse the rim more flexibly than trains. As soon as improvements in the road infrastructure both leading to and within the parks made this possible, the railroads' dominance of national park tourism—not to mention domestic tourism at large—was doomed.[28]

The motorization of national parks reflected the ideology of the National Park Service's first director, Stephen Mather, who believed that increased public access to the publicly funded parks was not only an ethical imperative, but was also essential to the development and survival of the park system and its conservational mission. Mather's vision was already on its way to being realized when he took his post in 1916. Yosemite National Park was opened to automobiles in 1913, and in 1916 it welcomed more visitors by automobile than any other form of transportation. Sixty percent of Yellowstone's 138,000 visitors in 1923 arrived by car. With Congress's endorsement, road mileage in the parks rose from a little over 1000 miles in 1924 to nearly 5400 miles in 1947. Over the same period, the number of cars entering national parks increased from half million to about twenty million.[29] For a time, during the 1920s and 1930s, railroad promotions accepted and even embraced the presence of automobiles in their former domain. The concessionaire subsidiary of the Union Pacific Railroad, the Utah Parks Company, even published a road map of southern Utah and northern Arizona "to assist those who are driving to [the] scenic wonderlands" of the region embracing the Zion, Bryce, and Grand Canyons.[30] Railroads operated motor coach tours of the most popular park areas, or encouraged travelers to rent or ship cars for use upon arrival to avoid the dangers and drudgery of driving long distances. Despite these efforts, by the mid-1950s 98 percent of all visitors to the national parks were private motorists.[31]

Railroads and their promotional cartography passed on to the automobile era a vision of transcontinental consumption. The first transcontinental rail trips were expensive: in 1873 John Erastus Lester noted that "it cost a New Yorker as much to see the Big Trees in California as it did to see Mount Blanc in Switzerland or St. Peter's in Rome."[32] Nevertheless, by the end of the 1870s, one hundred thousand took the trip on the Union and Central Pacific line from Omaha to San Francisco each year.[33] The later nineteenth and early twentieth centuries saw rapid growth both in the leisure time of the expanding middle class and in their financial opportunity to travel.[34] The attractive brochures, maps, and guidebooks published on behalf of the railroads hoped to convince this new legion of tourists that their geographic opportunities had expanded as well. Images and text fostered the notion that intimate contact with unspoiled wilderness would bring personal fulfillment—without sacrificing comfort, safety, or time. *The Pacific Tourist,* an 1876 guidebook to the Union Pacific and Central Pacific transcontinental route, boasted that:

> In no part of the world is travel made so easy and comfortable as on the Pacific Railroad. To travelers from the East it is a constant delight, and to ladies and families it is accompanied with absolutely no fatigue or discomfort. One lives at home in the Palace Car with as much true

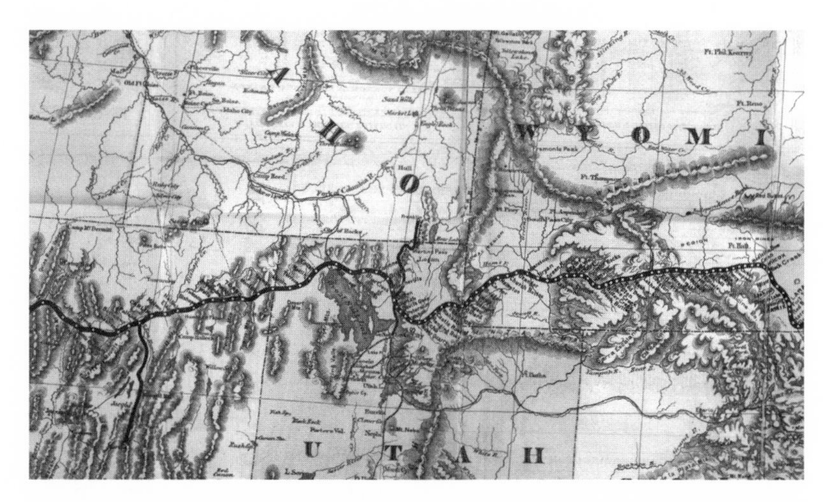

FIGURE 5.5 "Williams' New Trans-Continental Map of The Pacific R.R." (New York: Henry T. Williams, 1876), in *The Pacific Tourist* (New York: Henry T. Williams, 1876). Courtesy of the Newberry Library.

> enjoyment as in the home drawing room. . . . On the second day out [of Omaha], the traveler is fast ascending the high plains and summits of the Rocky Mountains. . . . He is alive with enjoyment, and yet can scarcely tell why. . . . Ah! It is this keen, beautiful, refreshing, oxygenated, invigorating, toning, beautiful, enlivening mountain air which is giving him the glow of nature, and quickening him into greater appreciation of this grand impressive country.[35]

The smooth and bold railroad lines inscribed on the maps attached to such texts reinforced the image of safety and comfort, and speed. A large fold-out map accompanying our example emphasizes the severity and complexity of the topography traversed by the Union Pacific route, but the railroad seems unyielding to the challenges posed by topography, even oblivious to it (figure 5.5).[36]

Topography was a mere source of amusement for the railroad traveler, something that could be casually and passively observed in transit, and packaged for restful contemplation en route or upon reaching one's destination. For the bicyclist, however, knowledge of topography was a practical necessity. Cyclists were exposed to the elements at every stage of the journey; they were their own drivers as well as their own mules. Like the motorists that followed, bicyclists controlled both their routes and their timetable, but their strength and endurance limited their range of travel. Bicycle mapping had therefore to be keenly aware of topographical features that would frustrate cyclists or threaten their personal safety.

And, since cycling involved personal navigation without the guidance of rails, it also required advance knowledge both of the road network's general layout and the suitability of individual roads to bicycle travel.

Not surprisingly, then, some of the earliest bicycle maps were modified topographical maps. In Britain, large-scale topographic base maps produced by the Ordnance Survey were already widely used by travelers in the 1870s; in the 1880s and 1890s, commercial publishers such as Bartholomew, W. & A. K. Johnston, G. W. Bacon, and Gall & Inglis adapted these maps to show roads suitable for cyclists, often by adding notations on the character and quality of road surfaces.[37] The bicycle was introduced to the United States in 1877, and in the 1880s a small number of commercial publishers produced larger scale maps for cyclists, notably the Boston-based publishers George H. Walker and The Scarborough Map Company. The main impetus for the production of bicycle maps in the United States came from local and state cycling clubs and especially their national organization, the League of American Wheelmen (LAW).

Founded in Boston in 1880 to coordinate the promotion of cycling nationwide, the LAW is best remembered for its role in the early good roads movement. The LAW and it member clubs cut their political teeth combating a rising tide of local and state ordinances that restricted or taxed bicycle use on public rights of way. By the late 1880s its leadership reacted to members' complaints about local road conditions, perceiving that technical impediments to free bicycle use were as worthy of its attention as legal ones. By 1892 the league had begun its own *Good Roads* magazine. The articles published there, and in the LAW's pamphlets, played a significant role in the 1893 creation of the federal Office of Public Road Inquiry, the forerunner of the Federal Highway Administration.[38]

The LAW's mapping activities consisted primarily of the compilation of road books, pocket-sized guides to the best roads. The earliest of these may be the *Road-Book of Long Island,* compiled by Albert B. Barkman and published for the Brooklyn Bicycle Club in 1885. Barkman's book for New York state, as well as a LAW Michigan road book, appeared in 1887, followed by Massachusetts and Connecticut books in 1888. Within a decade road books or road maps had appeared for all of the states east of the Mississippi River and north of the Ohio River, the New England and Mid-Atlantic states, as well as Northern and Southern California, Oregon, Missouri, the District of Columbia, and Metropolitan New York City. Southern and most other western states were conspicuous by their absence.[39]

The LAW books were compiled by individuals or by the collective membership of state and city clubs who contributed information on local road conditions to coordinating road book committees. That road information was often derived from base maps supplied by commercial publishers. J. B. Beers supplied the base maps for the Illinois and New Jersey books; Rand McNally supplied maps for Oregon; and George H. Walker did the cartography for Massachusetts. The editors of the 1898 Pennsylvania book, however, noted with pride that "there was not in existence a good, up-to-date map of the *whole* state, showing all the roads, which could be used as a basis. Your Committee therefore not only had the labor of compiling the nine hundred miles of reported roads on these maps, but they had also to actually created the maps on which those reports are given by comparison with the best existing sources."[40] Particular attention was paid to the quality of road surfaces and gradients, matters of vital importance to the riders of these fragile machines. The 1898 Pennsylvania edition recorded nine different kinds of road surfaces (clay, gravel, sand, cinders, shale, loam, slate, plank, and dirt) and distinguished between steep and gradual gradients. The Illinois book recorded four surface qualities, from poor to fine, and four classes of gradients, from level to mountainous.[41]

The LAW abandoned its brief role as road cartographer after 1901, reflecting both the increasing success of commercially produced road maps and the waning national interest in cycling as a form of travel and recreation. In 1896, participants in a transcontinental bicycle relay to promote postal and military uses of bicycles succeeded in covering the distance from New York to San Francisco in thirteen days. Bicycle travelogues such as Thomas Stevens's *Around the World in a Bicycle*[42] claimed a global reach for the two-wheeler. But however much roads might improve, the geographical range of the bicycle and its ability to carry freight or passengers was limited by the strength and endurance of its human power source. Almost everywhere in the world except the United States bicycles endured alongside the automobile as working vehicles for adults. In the United States, for most of the twentieth century, bicycles were redefined essentially as children's toys or as vehicles of last resort for those unable to obtain or drive a car. The sport and leisure magazine *Outing,* which had absorbed LAW's publication *The Wheelman,* regularly ran feature articles on cycling in the 1880s and 1890s. In 1900 it began publishing articles on the new sport of automobiling; it ceased publishing articles on cycling by 1902, about the time that the LAW stopped publishing cycling maps and commercial publishers rechristened their cycling maps as automobile road maps.[43]

Automobile Mapping in the Good Roads Era

Early automobiles relied on many of the technical innovations pioneered by bicycle manufacturing, and many of the earliest American (as well as

European) automobile manufacturers started as producers of bicycles. But automobile historian James Flink argues that the "greatest contribution of the bicycle . . . was that it created an enormous demand for individualized long-distance transportation that could only be satisfied by the mass adoption of motor vehicles."[44]

Nevertheless, the automobile's potential for long-distance travel was not instantly realized. The "do-it-yourself" aspect of LAW mapping thus had intriguing echoes in the method of early automobile mapping, partly because state and federal efforts to support road construction, or even to identify the most suitable roads for automobile use, remained negligible until the 1910s. Early motorists, and the motor clubs that supported their interests, were consequently required to survey the best roads themselves, just as bicyclists had. The earliest automobiles also had limited horsepower, climbed hills with difficulty, and struggled like bicycles to negotiate poor road surfaces. Most of the early maps produced for motorists consequently showed somewhat small districts where auto use was relatively high, as in the Northeast. Often they were based directly on earlier cycling maps and retained their emphasis of surface qualities. The first Rand McNally map targeted at motorists was simply an edition of a regional New York City map first published for cyclists nine years earlier.[45] Boston map publisher George H. Walker, whose maps had been favored by northeastern bicycle clubs, was marketing the same base information on the same scale (two miles to the inch) as automobile maps by 1910 (figure 5.6).[46] Other New England publishers, such as The Scarborough Company and F. S. Blanchard, sought to appeal simultaneously to users of several transportation modes at once. In 1905, Blanchard published a map titled "The New England Commercial and Route Survey" that depicted steam and electric railways as well as roads, classified as "good roads" and "good roads following electric railways."[47] Likewise, The Scarborough Company published a road atlas of Massachusetts and Rhode Island in 1905 (scale two miles to the inch), on which roads and railroads vie for prominence with hachured relief.[48]

The publication record of the first decade of the twentieth century suggests that established map publishers were not motivated to develop truly innovative product lines for automobilists so long as the market remained small and geographically confined. There was only one motor vehicle registered for every 196 Americans in 1910, and ownership rates were consistently highest in the Northeast,[49] where most early road map publishing was done. The relatively large-scale maps published for the early northeastern market were adequate to the task so long as the typical automobile

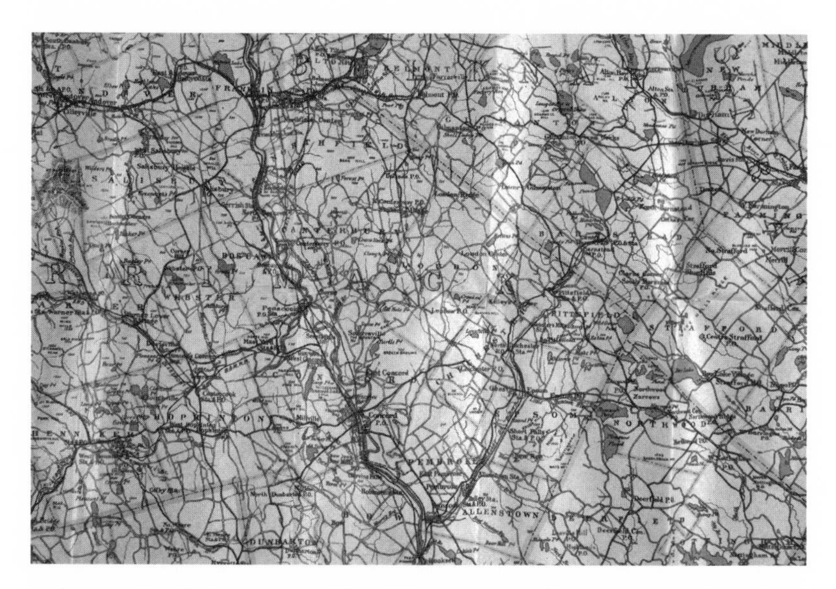

FIGURE 5.6 Detail of "Latest Map of Southern New Hampshire" (Boston: Walker Lith. & Pub. Co., 1917). Courtesy of the Newberry Library. Though identified as an automobile map, the design and content of this map has only been slightly modified from similar maps issued for bicyclists by George Walker in the 1890s.

jaunt stayed within the relatively familiar territory a hundred miles or so from one's home.

Travel over longer distances was another matter. Almost as soon as it hit American streets in 1896, automobile advocates began testing the geographical limits of the car. The first transcontinental journey by automobile—one of several well-publicized endurance tests staged by manufacturers and automobile clubs in the first decade of the twentieth century—was completed in 1903.[50] The tests of greatest renown were the annual reliability tours staged from 1904 to 1913 by the American Automobile Association, better known as the Glidden Tours, after the celebrated driver and automobile publicist Charles J. Glidden, who offered a trophy to the car that completed the tour in the fastest time. These covered as much as a thousand miles or more, bringing their contestants from eastern and midwestern urban centers to favored tourist spots. Despite their competitive aspect, the Glidden Tours were essentially publicity stunts and social events. A brochure promoting the ninth and last reliability tour in 1913, which ran from St. Paul to Glacier National Park, devotes as much space to describing sites to be seen along the way as it does to contest rules.[51] Even so, when newspaper columnist Emily Post traveled by car from New York to San Francisco in 1915 the transcontinental trip was still a novelty. New

Yorkers read her accounts of each stage of the journey with such interest that Post felt compelled to publish a book on her adventures the following year.[52] Carey Bliss has recorded the publication of no less than thirty-one accounts of transcontinental automobile journeys between 1910 and 1919, and accounts of lesser journeys were common fodder for general and sporting magazines as well as local newspapers well into the 1920s.[53]

An abundance of how-to articles advised the most adventurous automobile travelers on practical matters, usually including an extensive list of supplies and gear necessary for the sustenance both of man and machine. These lists frequently include some advice on obtaining good road guides or maps, but with the acknowledgement that they were not always suitable to the particular needs of long-distance motorists. In a 1905 contribution to *Outing* Hrolf Wisby observed that "most of the big automobile clubs the world over publish good maps, and the topographic charts of the various States are of great aid judging the lay of the country." But feeling these to be cumbersome in transit, he advised his readers to make their own "set of handy maps" by taking

> a good, plain geographic map. Place over it a transparent tracing-paper. . . . Trace with black ink the rivers, railroads, towns, etc., of the country you wish to pass through. Trace with red ink the route you decide to take in passing through. Then take a topographic map and trace, in the same manner, mountains, hills, elevations and valleys. . . . If your tour includes travel in but little known regions of this country, provide yourself with a tracing of a survey map such as the U.S. Geodetic Survey issues, which will give you what accurate information is available in advance. The data gleaned from these maps should enable anyone to make up a driving chart, its object being to show at a glance the combined information contained in all the other maps. To be practical, such a chart must be simple—a sort of ready-reference resumé of all needed road and camping information. Stake out your route in a heavy, red line . . . and after pasting it the map on chart canvas it should be rolled around a paper or wooden cylinder, placed in waterproof box with a glass pane, so that the chauffeur may have it continually before him and turn the cylinder as he progresses over the route shown on it. This arrangement will obviate that most wearisome operation of a long-distance trip, of having to stop to consult maps—a very exciting performance when it rains or blows—or to consult "people you meet," who, as a rule, don't know more about the country they live in than the average New Yorker knows about New York.[54]

H. B. Haines outlined a similar though perhaps less labor-intensive method for constructing a road guide that was also more trusting of local informants:

> The writer has always found it preferable to write out the route intended to be followed. He has also found it a convenient practice to have a leather case made with a celluloid front in which the road directions can be placed. The map or directions can be read through the celluloid, which, however, protects them in case of rain. The easiest way to get along when asking road directions of natives in various towns is to know the name of the next following town and then ask the best road to it, if the road book directions are not explicit. For instance, before starting on a trip I generally take a map of the country through which I am going to travel and ascertain the various small towns through which I have to pass. These are listed, and as I proceed I inquire the way from one to another. This method has been found satisfactory, for when one asks the road directions to the larger cities, which may be ten, twenty or more miles apart, it is difficult to find anyone who can direct you properly, but any boy or girl can tell you the best road to the next town, which will probably be found three to five miles away.[55]

Less ambitious and self-reliant travelers relied not on maps, but on published itineraries, or "logs," of intercity routes. Following the pattern established by bicycle enthusiasts, energetic private individuals, often in coordination with automobile clubs or commercial enterprises, determined the best automobile routes, logged and mapped them. A handful of these automobilists, such as William Rishel and A. L. Westgard, acquired local or national reputations for their work, becoming professional "pathfinders." Rishel, a Utah cyclist turned automobile enthusiast, blazed major touring routes leading from Salt Lake City, and published his logs and maps in an annual guidebook that became known as *Rishel's Routes*.[56]

During the 1900s and 1910s route logs were published by automobile companies, newspapers, motor clubs, and a host of semipublic and private organizations that stood to benefit from the increased traffic a log might bring. These varied in size, detail, and quality, from cheaply reproduced sheets[57] to expensive collections of logs that might run several hundred pages. Some publications were accompanied by simple diagrammatic maps of the route described, but this was hardly a requirement.

By the second decade of the century, the standard publication of this sort had become the multiple regional volumes published by the Automobile Blue Book Company. Its first volume, covering only the eastern United

States, was published in 1901. The *Automobile Blue Book* gained the endorsement of the American Automobile Association (AAA) in 1906, but the covers of later editions also prominently advertised commercial sponsors, such as Kelly Tires and Standard Oil of New York. By 1920, the *Blue Book* series had grown to eleven volumes, covering the entire United States. *Blue Book* logs typically began in the city center or another common destination identified by a landmark, such as a courthouse or post office. One column provided navigational instructions, guiding the motorist through each essential turn, crossroads, or fork in the road or identifying landmarks, such as bridges, notable buildings, side roads, scenic vistas, and entrances to parks and other attractions. The other column kept a precise running tally of the miles traveled from the starting point of the log. The (paraphrased) log of the route from Pittsfield, Massachusetts, to Waterbury, Connecticut, published in the *Automobile Official AAA 1910 Blue Book,* New England volume, instructs travelers to commence in Pittsfield at the intersection of North, South, East and West Streets. The traveler is then to bear south with a trolley on South Street, crossing a railroad bridge at 1.1 miles, and to continue past the end of the trolley line at the country club at 1.7 miles. The road passes the Aspinwall Hotel on the right at 6.0 miles, arriving at the monument in the "four corners" at the center of the town of Lenox at 6.6 miles. Turning left, the traveler will encounter an Episcopal Church at a fork in the road at 6.8 miles. Bearing right on a "winding macadam" road, the traveler will pass a number of "fine estates." And so it goes for 78.7 miles, before arriving at the central square of Waterbury.[58]

There were a number of ingenious but less successful variants on the Blue Book model. Prefiguring today's in-car navigation systems, Live-Maps employed disks representing 100-mile lengths of selected routes that were mounted on a rotating mechanism geared to a front wheel of the car. As the wheel turned, so did the disk, past a pointer indicating automatically the car's location at any moment and "what to do then and there." This passive form of navigation would eliminate "endless trouble and delay poring over maps and guide books."[59] Another variant supplemented the standard route log information with a series of photographs of forks and crossroads marked with arrows to indicate the direction a vehicle had to turn to stay on course (figure 5.7a, b). This attractive idea proved short-lived, perhaps because the photographs were quickly outdated and added bulk to the publication. The 1909 edition of *Photo-Auto Maps . . . New York to Chicago, Chicago to New York* required 796 photographs and comprised several hundred pages.[60]

Route logs and their variants were essentially itineraries (like those described in the previous chapter) adapted to new circumstances—though

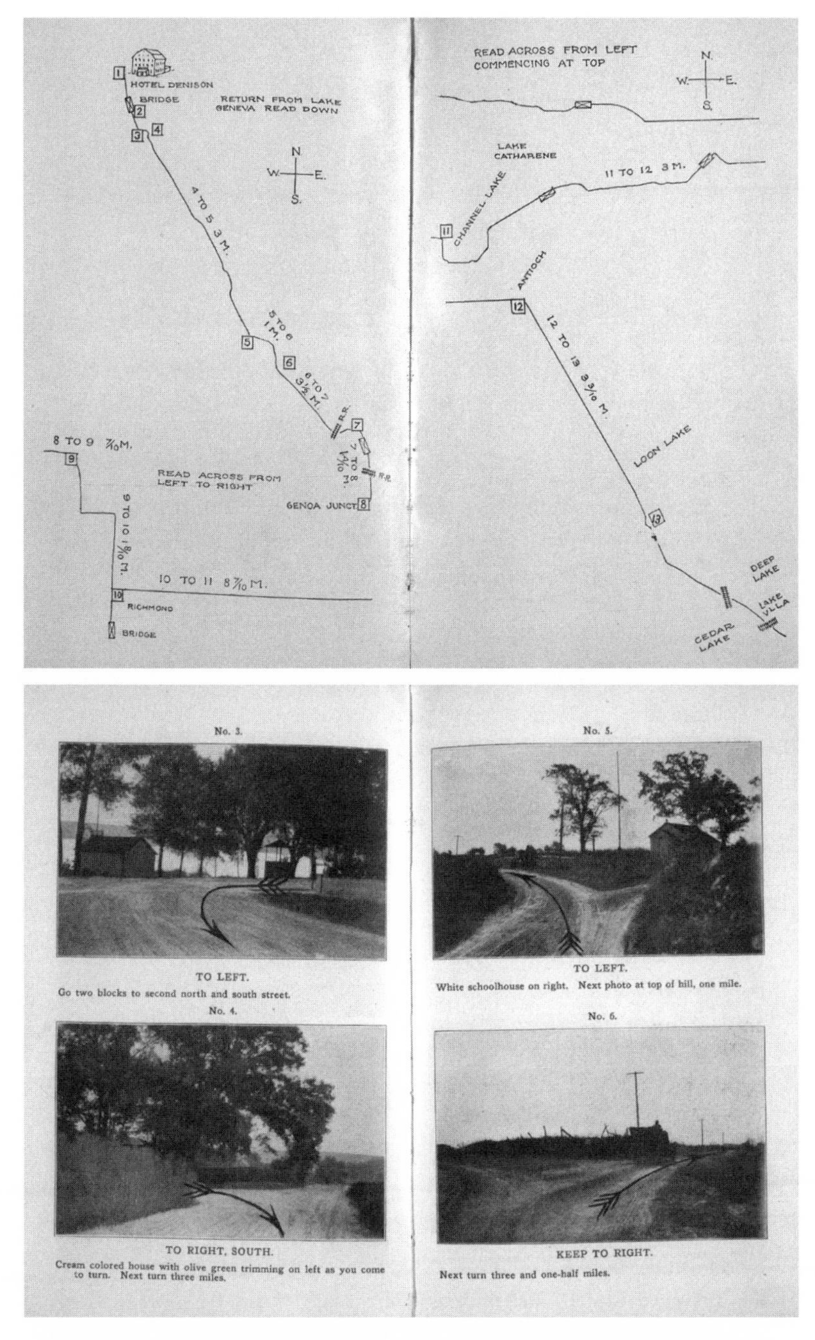

FIGURE 5.7a,b Map and accompanying photographs showing a portion of the route from Lake Geneva, Wisconsin to Chicago, Illinois, from *Photographic Automobile Map, Chicago to Lake Geneva, Delevan Lake, Delevan, Beloit, Returning via Antioch and Fox Lake Country* (Chicago: H. Sargent Michaels, 1905). Andrew McNally III Collection, courtesy of the Newberry Library.

ultimately, perhaps, poorly adapted. Route books were bulky; the 1910 New England volume of the *Automobile Blue Book* required more than 900 pages to describe the most common automobile routes in the six New England states. Motorists were expected to follow these instructions while keeping an eye on their odometers and the road ahead. One motorist observed that "Whoever had the seat of honor beside the driver got the Blue Book job and spent the day with his nose glued to the fine-typed pages and read aloud each direction, but never quite in time to prevent the wrong turn."[61] If any one of the instructions describing a critical turn was missed or misunderstood motorists might easily travel some distance before the mistake was realized, and have difficulty finding the route again without the benefit of written guidance.

In many instances reliable route information, like that provided by automobile blue books, was either hard to come by or simply unknown to travelers. Returning from California to their Indiana home after a failed attempt to resettle in the West, Guy and Estella Copeland were required to engage in extensive research about the best possible routes. The 1912 edition of the "General Map of Transcontinental Routes" published by the AAA offered some guidance on the possible routes across the southwestern deserts and mountains, but detailed navigational information about their chosen southern route through San Diego, Yuma, Phoenix, Albuquerque, Santa Fe, and Trinidad, Colorado had to be cobbled together from various sources:

> The Copelands had assembled a fairly complete collection of local road information for the southern route. These materials were road maps in the modern sense but rough sketches showing distances between towns and landmarks to confirm that the tracks an automobile was following were the correct ones. For the first leg of the trip the Copelands acquired the *Pacific Road Guide* published in 1912 by a San Francisco insurance agent with maps drawn by E. Rawlins. It covered roads from Seattle to Yuma. A hard-covered volume published by the Arizona Good Roads Association led them across Arizona. From eastern Arizona to Albuquerque and the Santa Fe Trail the Copelands had only a simple "log" published by the Butler Auto Company of Albuquerque, giving mileages between landmarks and terse instructions to bear left or right. . . . From Albuquerque to Newton, Kansas, approximately 780 miles, the Copelands apparently had a guidebook for the Santa Fe Trail, which was also marked in places by colored rings on telephone poles. . . . From Newton, Kansas, they turned south on the Chisholm Trail for their journey to Olustee, Oklahoma. They had no printed guides or logs for most of this section of the

> trip, and they simply asked directions from Olustee to Wichita. They had acquired maps from Indiahoma, Oklahoma to El Reno, Oklahoma, and from El Reno, to Wichita. They found good roads and apparently adequate markers from Newton, Kansas, to Kansas City, and had a *Goodrich Route Book* from Kansas City to St. Louis. At St. Louis they acquired *A Guide and Directory for Automobile Tourists for 1913,* which consisted mostly of advertisements and included only three very generalized maps and a log of distances between towns. For Indiana they had excellent maps and route guides in the *Goodrich Route Book* for Ohio and Indiana.[62]

The well-heeled New Yorker, Emily Post, was astonished to discover that her precipitous decision to begin a transcontinental trip from New York to San Francisco in the spring of 1915 meant that the most reliable guide of the day, the AAA sponsored *Automobile Blue Book,* was not yet available for the year, and the 1914 edition was out of print. The information supplied her by the Lincoln Highway Association was optimistic but not helpful practically. The man at the automobile club confessed that "we seem to be out of our Western maps," and attempted to persuade Post to take a tour of New England, where the roads were "very much better." Doubting her sanity, the agent suggested that if she must make a motor trip across the country, the best course might simply be to rely on local advice she got along the way. Finally, at Brentano's bookstore, she found a map of the United States, which showed "four routes crossing it, equally black and straight and inviting."

> I promptly decided upon the one through the Allegheny Mountains to Pittsburgh and St. Louis when two women I knew came in, one of them Mrs. O., a conspicuous hostess in the New York social world, and a Californian by birth. . . .
>
> "Can you tell me," I asked her, "which is the best road to California?"
>
> Without hesitating she answered: "The Union Pacific."
>
> "No, I mean motor road."
>
> Compared with her expression the worst skeptics I had encountered were enthusiasts. "Motor road to California!" She looked at me pityingly. "There isn't any."[63]

The Copelands' and Post's struggles to find navigational information for separate transcontinental trips show that the problems of road quality and road navigation were closely related. Travel agents and cartographers could not agree on the best routes across the country because all roads were poor, and such route maps that existed were at best advisory.

In the face of state and federal recalcitrance, the various automotive and highway interests accelerated their demands for an integrated national highway system in the early 1910s. In its infancy in 1880s and 1890s, the good roads movement had perceived road improvements largely as a local matter. Farmers agitated for upgrading the roads they used to bring produce to market and by which they received mail. They formed an uneasy league with urban and suburban bicyclists, represented by the LAW. The latter shared the same general goals but were occasionally at odds with farmers, who regarded bicycles as nuisances and thought hard surfaces unhealthy to their horses and draft animals. Both groups were hampered by the existing system of road finance, which placed the burden of road construction and maintenance almost entirely on localities and counties. Now, the rise of an automotive industry capable of building cars affordable to the masses both magnified public interest in the good roads debate and transformed it into a contest between camps favoring local road improvements, on the one hand, and the construction of an integrated system of state and national highways on the other. The main automotive interest groups disregarded the need for local roads in favor of the development of a national highway system. To dramatize the potential benefits of improved cross-country highways and the increased automobile travel the highways would support, the automobile and highway industries allied with state highway officials, automobile clubs, civic groups, and like-minded local officials and entrepreneurs to form highway associations committed to the creation and improvement of specific highway routes.[64]

The best publicized of these association highways was the Lincoln Highway, first proposed to a consortium of midwestern automobile and tire manufacturers in 1912 by Carl G. Fisher, president of the Prest-O-Lite Company, a manufacturer of automobile headlamps. The group formed the Lincoln Highway Association (LHA) in 1913. Under the leadership of Henry Joy of Packard, a plan was forged to complete an improved and partially paved highway from New York to San Francisco by May 1915, in time to accommodate summer traffic to the Panama–Pacific International Exposition in San Francisco. But chronic financial problems reduced the association's ambitions to the marking of the route and paving of scattered "demonstration miles." Travelers using the Lincoln Highway in 1915 reported it to be excessively muddy, or simply nonexistent in places west of the Mississippi River. Jessie Wiant recalled that "one time they came to a place where the road faded out into a sandy field full of clumps of grass. There seemed to be two faint tracks like a branch in the road, and they didn't know which way to go. They stopped and studied the situation, then having to choose, they fortunately chose the right land and eventually came to a Lincoln Highway sign."[65]

There were many other association highways, including hundreds of regional routes, and approximately forty important transnational routes (both east-west and north-south).[66] Some of these highways followed established and well-known routes. The Atlantic Highway (now U.S. Route 1) followed the Boston Post Road from New York to Boston. The National Old Trails Road followed the old National Road from Baltimore to Vandalia, Illinois, and portions of the Santa Fe Trail. Most of the highway associations were obliged, however, to cobble together their routes from an assortment of often poor local roads.

Despite this, or perhaps because of it, the highway associations were prolific producers of maps. Typically, these were promotional and not useful for navigation. A small map issued with a prospectus of the Colorado to Gulf Highway merely suggests the general line of the route and the major cities it serves between Denver and Galveston. The largest portion of the highway's sixty-eight-page booklet is given over to an extensive route log of the highway excerpted from the 1914 edition of the *Automobile Blue Book*.[67] Other association maps provided only minimal guidance. A map published in 1914 by the Yellowstone Trail Association confidently shows the route of the Yellowstone Trail linking the Northeast to Yellowstone Park and the Pacific Northwest via South Dakota, but accompanying text reveals that the sponsoring association is "lacking in accurate information and data.... We would appreciate several logs, and ask those making the run to mark on a map the exact distances between towns as shown by the speedometer, and to forward such maps to us."[68] The continuous black line representing the Appalachian Scenic Highway on a crude 1925 map published by the president of the Appalachian Scenic Highway Association might have suggested to the unwary motorist the existence of a modern highway connecting New Orleans and Quebec.[69] The authoritative *Auto Trails* maps issued by Rand McNally in that year show that many of the roads this route supposedly followed were still unimproved or unpaved. Worse, the maps do not even recognize the existence of the Appalachian Scenic Highway, suggesting that it was poorly marked, if at all.[70]

Highways such as the Colorado to Gulf, the Appalachian Scenic, and the Yellowstone Trail aspired to national significance, but they, like most association highways, were motivated by regional, even local concerns. As such, they were poorly capitalized and—as many of their brochures and maps emphasized—they were *not* in the business of highway construction, only highway marking and promotion. The circa 1914 Colorado to Gulf Highway Association booklet noted that

> Some have gotten the erroneous idea that this Association expects to at once construct a macadamized highway from Galveston to Denver. But as much as we desire this, we are not foolish enough to expect such happy results in the immediate future; nor was such the primary object of the organization; however, we did intend to put into motion the forces that would connect up a continuous roadway, that would be passable for tourists between these points. . . . So we found that the first work of this Association was to map out a line and to secure the co-operation of counties through which it passed.[71]

Association highway promotional maps thus served the dual purpose of attracting or routing early motorists through specific corridors, thereby creating both short-term income for local interests and long-term justification for the construction of regional and interstate highways, specifically along routes designated by the associations. To build and sustain momentum for these efforts, the maps had subtly to persuade potential tourists that the primary goal—a continuous improved motorway—had already been achieved. Many did this by adopting designs reminiscent of railroad maps (see figure 5.8). The featured association highway appeared as a bold and straight or gently curving line that seemed to cut confidently through the regions they served, like railroads apparently heedless of physical barriers.

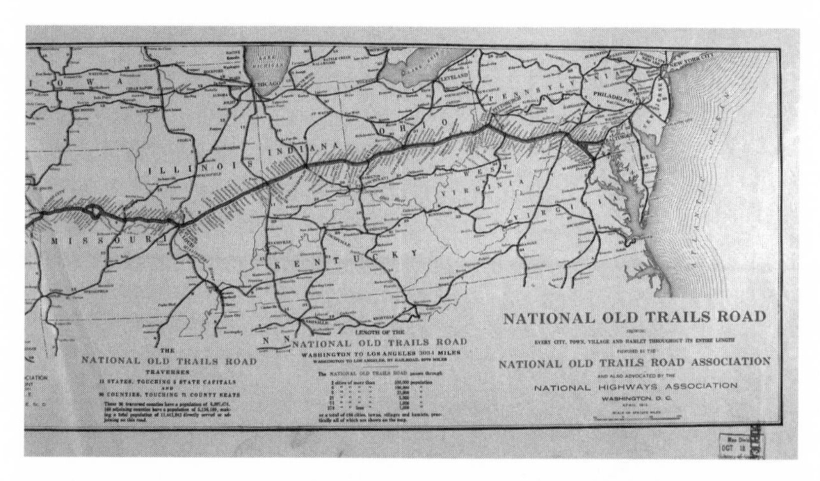

FIGURE 5.8 Detail of "National Old Trails Road" (Washington: National Highways Association, 1915). Courtesy of the Newberry Library. Highway association maps of this type frequently imitated the style of contemporary railroad maps.

Towns along the route are named in profusion at right angles to the line of the highway, like station stops on an express passenger railroad route.

The most prolific publishers of highway association maps, the National Highways Association (NHA), made particularly effective use of persuasive cartographic techniques in the various series of maps it published during the 1910s and 1920s. T. Coleman Du Pont (industrialist, financier, and senator) and Charles H. Davis (inventor, civil engineer, and president of the American Road Machine Company, a manufacturer of road construction equipment) founded the NHA in 1912. Through the zeal and organizational ability, Davis sustained a prolonged good roads campaign that included the publication of pamphlets and maps from 1912 to the late 1920s. Throughout its history the NHA favored the construction of a fully integrated national system of highways, adopting the extreme position that the system should be entirely funded by the federal government. To that end, it embraced the plans of the individual highway associations, soliciting their support as member organizations, and publishing maps showing how individual highways could serve as the basis for a coherent national system. Davis and the NHA's maps implicitly opposed the federal-aid camp of the good roads movement, which sought to create a state and farm road system built with a combination of state and federal funds, but largely supervised by states. Mindful of this, a caption across the bottom of the NHA's 1914 map of Maryland noted that 700 road miles designated as National Highways represented only 4.17 percent of the state's 16,773 miles; but that the counties through which the National Highways passed represented 97 percent of the state's population. The caption thus cleverly suggested that an overwhelming majority of the state would be served by the NHA's routes.[72] Such arguments bordered on the absurd in states further west, where farmers would supposedly be served by a proposed NHA highway as much as fifty miles away. The NHA also issued attractive maps of the individual association highways it supported, adopting the classic railroad design favored by the highway associations (figure 5.8).[73]

The local road camp won the first round of this contest when the 1916 Good Roads Act, the first to authorize significant federal funds for road construction, allocated $75 million over five years primarily for rural post roads selected cooperatively by the state and federal governments. But during the First World War, American railroads failed to meet the extraordinary demands placed on them to transport war materiel and foodstuffs from the American interior to the East Coast for shipment to Europe, swaying public opinion decisively toward the creation of a national highway system. Experiments with military truck convoys both during and after the war dramatized the need for interstate highway improvements, vindicating the arguments of the NHA and its allies that locally coordinated federal aid could meet the needs of long distance auto and truck travel. Accordingly, the Federal Highway Act of 1921, while still rejecting the idea of a federally owned national highway system, nevertheless required states to designate up to 7 percent of their federally aided roads as a state trunk highway system. Three-sevenths of that system would be "interstate in character"—that is, usefully connected with federal aid highways in adjacent states—and would corral sixty percent of all federal road aid funds.[74] Planning for the state and federal highway systems continued through the mid-1920s, and in 1926 approximately one hundred thousand miles of these federal aid highways were formally incorporated in a system of numbered federal highways—a system similar in scale (and in many details) to that originally envisioned by the NHA.

Commercialization and Consolidation of the Road Map Industry

In 1917 Henry Joy estimated that between five and ten thousand motorists had used the Lincoln Highway in 1916 to travel between points east of the Mississippi and the Pacific Coast. He called these "the vanguard of a tremendous army of Easterners who will seek the beauty of the West when out main transcontinental route is improved in its entirety to the standard which now prevails east of Chicago."[75] Joy's figures may have been inflated, but there is no question that the maps and articles published by good roads advocates in the years immediately preceding the First World War had succeeded in fostering interest in automobile touring. Rapid improvements in interstate highways during the 1920s and postwar prosperity boosted the new pastime still further. The motor camping enthusiast Elon Jessup estimated that twenty-thousand Americans completed a transcontinental journey in 1921, while only twelve had completed such a journey nine years earlier. Warren Belasco cites varying estimates of the annual number of "motor campers" in the 1920s ranging from ten to twenty million.[76] At first, motor tourists relied on downtown railroad hotels for accommodation, but as automobile travel spread to the middle and lower classes, such hotels proved to be both too expensive and inconvenient. Motorists came to rely on special camps set aside for their use on the outskirts of towns. In time, these became commercial operations that offered amenities (water, bath facilities, food, and supplies). Some provided modest cabins, which evolved into "cabin courts," and then motels.[77]

Automobile ownership rose precipitously in the 1920s. One in every 196 Americans owned a registered motor vehicle in 1910; by 1920 there was a car for every 11 Americans; by 1930, one car for every 4 Americans. Private

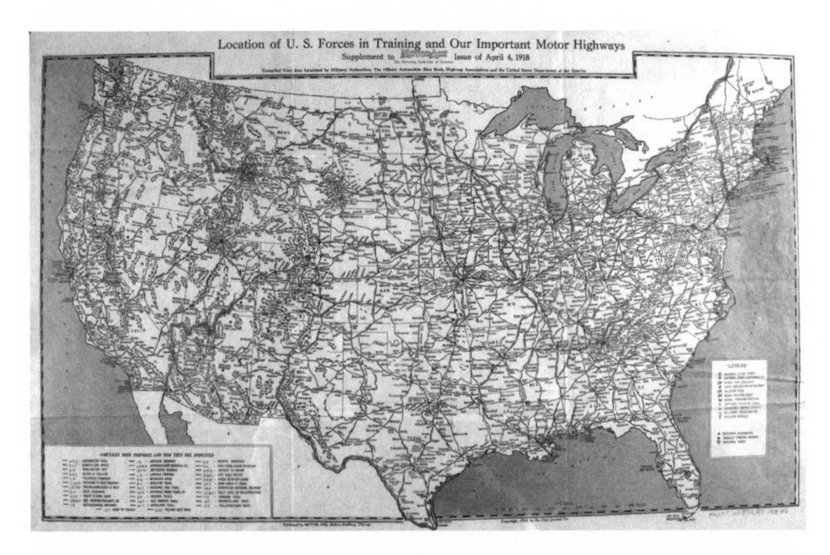

FIGURE 5.9 "Location of U.S. Forces in Training and Our Important Motor Highways" (Chicago: Motor Age, 1918). Courtesy of the Newberry Library.

automobile passenger miles totaled only one quarter of the miles traveled by train in 1922, but were four times greater than railroad passenger miles in 1929.[78]

During this period of expansion the idea of publishing a national automobile highway map began to make sense in practical terms. The highway associations, and particularly the National Highways Association, published a number of these before the First World War in order to promote the idea of a national highway system. Another national map, published by *Motor Age* magazine in 1918, invokes the military case for a national highway system by showing how such a system would effectively link the nation's military installations (figure 5.9).[79] The 1912 AAA *General Map of Transcontinental Routes,* consulted by the Copelands while planning their 1913 transcontinental journey, was probably among the earliest national maps published for the use of motorists. During this time a number of NHA maps promoted the idea of a national highway system, but the country's largest map publisher, Rand McNally, which had been publishing automobile road maps since 1904, did not see fit to publish its first national road map until 1921.[80]

It was also during this period of phenomenal growth in long-distance motor vehicle use that the idiomatic design of American highway maps—we'll call it the "network" road map—came into common use. Whether depicting states, regions, or the entire nation, the features of the network road map have remained largely unchanged from the late 1920s to the

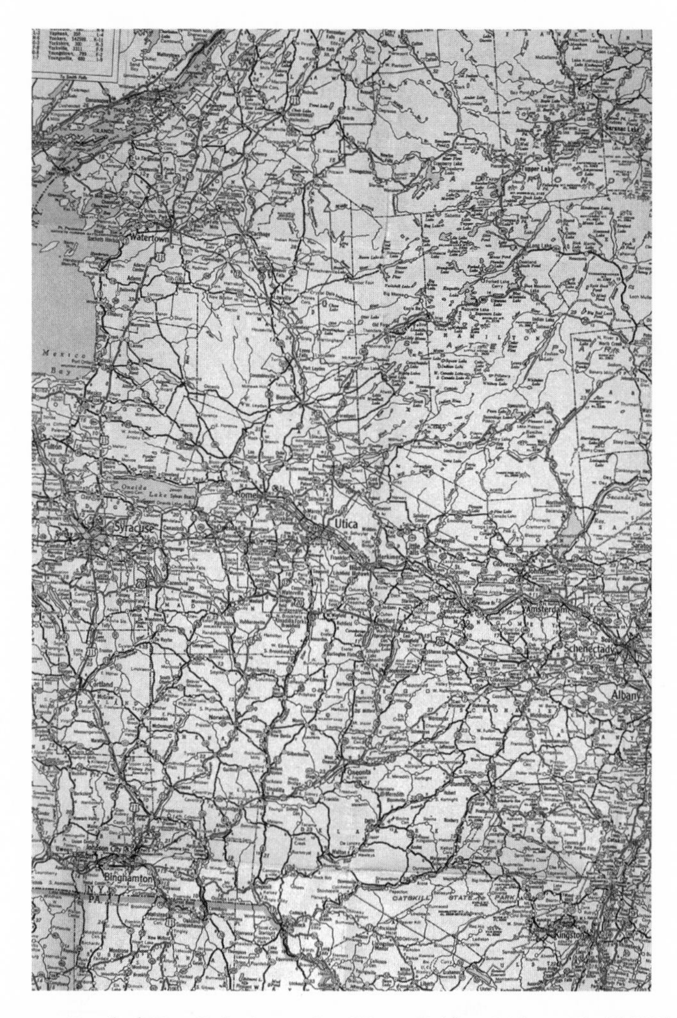

FIGURE 5.10 Detail of "New York, Amoco, Best Way to Go! from Maine to Florida" (Chicago: Rand McNally for American Oil Co., 1947). Map © by Rand McNally, R.L. 04-S-108. Rand McNally Collection, courtesy of the Newberry Library. Mountain ranges, such as the Adirondacks of upstate New York are not indicated on this typical oil company road map from the mid-twentieth century.

present day (figure 5.10). Paved highways and side roads appear as solid lines, graded by thickness or color according to their relative importance as bearers of traffic; their status as federal, state, or county highways; the number of traffic lanes; or (in later years) whether they are toll or free superhighways. Unpaved or poor quality roads appear as hollow or dashed lines. These highways traverse a largely featureless landscape populated chiefly by towns, crossroads, parks, and designated places of interest.

County boundaries are usually provided as a frame of reference, as are major rivers and streams, but the maps supply relatively little topographic information. Mountains, shown as in hachure or shaded relief, appear in some issues by Rand McNally and H. M. Gousha and on official state maps, particular of western states, where the mountainous character of the terrain itself is a tourist attraction. On the whole, the "flat" character of network maps suggested that rapid travel on paved highways was largely unaffected by terrain, as, in truth, it was by the mid-1930s.

Cheap and easy to reproduce in thousands of copies, easily updated and customized to the needs of advertisers, the network map quickly became the industry standard, in large measure because of its value as a promotional tool. Not surprisingly, tire and automobile manufacturers were among the first industrial concerns to publish road guides and maps under their own name as a means to encourage automobile travel and cultivate consumer loyalty to their products. Advertisements for automobiles, auto parts, hotels, and roadside businesses provided a major source of income for the publisher of the popular *Automobile Blue Books,* and from 1905 to 1913, a small number of tire companies and automobile companies—including the Hartford Rubber Company, the White Motor Car Company, Pierce-Arrow, and the United States Tire Company—dabbled in the production of their own route guides and road maps.[81] The most successful early industrial sponsor of route guide and road map production was probably tire manufacturer B. F. Goodrich, which from 1912 to 1917 issued an ambitious series of road guides describing routes marked by "Goodrich Guide Posts" placed by the company itself. Goodrich customers also had access to free maps from the mid-1910s until about 1930.[82] From the mid-1920s into the 1930s, the Mohawk Rubber Co. published an innovative and apparently popular series of *Grade and Surface Guides* profiling the major auto trails and numbered federal highways. Researched and prepared by Howard F. Hobbs, the guides sold for a very modest price in 1926 of seventeen cents and folded to a size comparable to a typical network road map. Most of the early verbal route guides included maps showing where the routes went, but as the entire road network improved, reliance on the detailed navigation instructions for individual routes limited the utility of these guides, diminishing their value as promotional premiums. After a few early experiments, comparable American firms quickly abandoned the notion of compiling and publishing their own maps for sale, in favor of the free distribution of network-type maps made for them by firms specializing in cartographic production.[83]

Gulf Oil, pioneer of the modern gas station, also claims to have issued the first free "oil company" road map, showing Allegheny County,

Pennsylvania, in 1914. But Monarch Oil, having sold a diminutive road atlas of California and Nevada, also issued a free map of the Pacific Coast states and Nevada in 1914. Rand McNally published a network type map in 1915 for free distribution with the Atlantic Refining Company, but Walter Ristow observes that "[n]either Rand McNally nor Atlantic seem to have followed up immediately on this early experiment with complimentary maps."[84] Instead, Rand McNally put its energy into the development of a network-type road map issued under its own name. By the end of the 1910s, George F. Cram, C. F. Hammond, and the Clason Map Company were among several established commercial publishers to have experimented with simple network designs. Several new cartographic publishers concerned primarily with producing inexpensive road and tourist maps emerged during this period, notably the General Drafting Company and The National Survey.[85] The AAA, which had dabbled in map and route guide publishing during the first decade of the century, established its own cartographic office in 1911, which initially published network-style maps for a predominantly northeastern and midwestern market.[86] By the end of the decade, at least twenty state highway departments had begun issuing maps of their state trunk road system to the general public.[87]

Rand McNally's series of Auto Trails maps, probably the most significant of the early network map issues, began in 1917. By 1921, the maps in this series covered the entire country in twenty-one districts. They were sold directly by the publisher at first, supported by subscription sales and advertisements for roadside businesses whose names and locations were indicated prominently on the map. Gradually, under the guidance of sales manager H. M. Gousha, indirect marketing of the Auto Trails maps was introduced. Following a formula it perfected in the late nineteenth century to sell atlases, Rand McNally published special editions of Auto Trails maps for resale or free distribution as premiums by newspapers, by an assortment of auto-related businesses, and most importantly, by oil companies. Reduced-scale versions of Auto Trails maps, called Junior Auto Trails, were marketed extensively as free maps issued by several oil companies in 1924 and 1925, and by 1926, these oil company editions had effectively replaced the Auto Trails maps as the mainstay of the company's production of separate maps.[88] Rand McNally had also invested heavily in the publication of its famous *Road Atlas* by 1926, which it sold primarily (though not exclusively) under its own name.[89]

H. M. Gousha left Rand McNally in 1926 to form his own company. The new company concentrated at first almost entirely on the oil business, and by the early 1930s had succeeded in prying many of Rand McNally's largest oil company clients from the older firm's grip. Throughout the 1920s and

1930s, a number of other firms specialized in this niche market, either as direct publishers, or, like Gousha and Rand McNally, as publishers of maps issued under the names of corporate clients. These firms included the Gallup Map and Supply Co. of Kansas City and the Mid-West Map Company of Aurora, Missouri. But the warfare between the two well-capitalized Chicago-based firms, Rand McNally and H. M. Gousha, coupled with the Great Depression, drove most of these new companies, as well as older players like Hammond and Clason, from the road map field by the early 1940s. The firms that survived after the Second World War mostly served regional niche markets. The National Survey, for example, emphasized the production of maps for New England hoteliers. Thomas Brothers produced the standard series of street guides for metropolitan Los Angeles; Hagstrom did the same for New York. The third major national producer of free oil company maps, the General Drafting Company, flourished until the 1980s primarily due to its close relationship with the largest of all American oil companies, Standard Oil of New Jersey (Esso, later Exxon) and its subsidiaries.[90]

Whether distributed at a discount, as a perk of auto club membership, or as an apparently free gesture of good will, the network map's appeal as a premium relied on its minimal cost, but it still had to be useful to the motorist. Because motorists actually had to use them to navigate, road maps were free of the deliberate geographical distortions present on free railroad maps.[91] The information about road grades and surface quality that had been essential in early route guides and on early, topographically rich road maps was eliminated from network maps, because such information was no longer essential to highway navigation. Stripped-down network maps remained functional as navigational tools because they could assume that all the roads they showed were essentially passable, and, more importantly, were marked in such a way that motorists could get safely from town to town without constantly referring to landmarks and bulky texts. The 1920 edition of volume 5 of the *Automobile Blue Book* required fifteen and a quarter pages, divided among six logs, to describe the eastbound and westbound passages of the Lincoln Highway through Iowa. Rand McNally's 1920 "Official Auto Trails Map, District 10" required only twenty-four linear inches to navigate a motorist along the same route.[92]

This was possible only because by 1920 the Lincoln Highway was well marked. Verbal instructions or large-scale mappings of major roads could be largely dispensed with only where the routes were named or numbered on maps were adequately marked on the ground. One of the primary stated goals every highway association of the 1910s and early 1920s was the field marking of its route, but this required money and labor to both lay out and maintain the markings. Only the best-capitalized highway associations were able to do this effectively with their own resources. Other routes were marked with the assistance of local motor clubs and even commercial mapmakers. As we have seen, the network maps distributed by Goodrich referred motorists to guideposts placed by the company's "pathfinders" at intersections of "main market highways" indicating the distances and directions to major towns. Rand McNally provided funds to local agencies to assist in the marking effort and went so far as to guarantee that every route indicated on its maps was marked—even undertaking, like Goodrich, to mark some routes on their own.[93]

By such informal and private efforts the rudiments of a national system of marked highways were in place by the early 1920s, but this was an uncoordinated and confusing system indeed. Since the named highway associations were independent of each other and were often rivals for local attention and traffic, there was much duplication of named routes over the best rights-of-way. Many public officials expressed frustration at the associations' inability to make substantial progress on road marking and improvement, and questioned the motives of the road promoters. Richard Weingroff relates that "State Highway Engineer Arthur R. Hirst told a National Road Congress in January 1918 that the trails were established with "a great deal of gusto" and "barrels of paint." Hirst added: "The ordinary trail promoter has seemingly considered that plenty of wind and a few barrels of paint are all that is required to build and maintain a 2000-mile trail."[94]

In an early attempt to regulate the naming and marking of roads, the state legislature of Iowa intervened in 1913 and empowered the state's highway commission to supervise the designation of all routes more than twenty-five miles in length.[95] In the late 1910s New York and New England state highway officials marked their named highways according to a uniform color-coded system. East-west routes were marked by red bands on roadside poles; north-south routes by blue bands; and diagonal routes by yellow bands. In 1918, Wisconsin became the first state to reduce the confusion still further by adopting a numerical system for state trunk highways. The logic of this approach was plain enough, and by 1925, half of the state highway departments had adopted similar systems. In 1924 the American Association of State Highway Officials (AASHO) recommended the establishment of a federal numbering system for interstate highways as well. Despite opposition from many named highway associations, who feared the loss of their "brands," AASHO approved the new system of numbered federal interstate highways in November 1926.[96]

The adoption of highway numbers eliminated the confusion posed by named highways and provided a visible expression of expanding public—that is, state and federal—responsibility for highway construction and maintenance. Cartographers now had a network of automobile pathways endorsed and financed by government at the highest levels, and no longer had to rely on inconsistent and often unreliable informal systems. Moreover, with federal aid for highway improvements, the upgrading of state and federal systems accelerated during the 1920s, and by 1934, 96 percent of the 207,000 miles of federal aid roads was improved.[97] Despite setbacks during the 1930s, the vast majority of federal trunk routes were paved from end to end by 1940. The adoption of uniform route marking, coupled with the substantial road improvements of the 1920s and 1930s, meant that motorized Americans now had nearly unlimited touring options, but it also simplified the cartographers' work. Network maps presented these options to motor tourists simply and effectively at scales that allowed detailed depictions of the extensive state and regional highway systems on maps that were both portable and inexpensive to produce.

The improvements in the American highway system during the 1920s meant that more American motorists could feel comfortable planning transcontinental trips that required daily runs of several hundred miles or more. Early travelers on the Lincoln Highway required one or two months to complete the roughly three-thousand-mile journey from New York to San Francisco; the same journey required only seven or eight days in the mid-1930s. The new pace of travel made many older navigational publications obsolete. The venerable Automobile Blue Books ceased publication in 1926 to be replaced by new cartography that reflected the greater range available to motor tourists. During the 1920s, Clason, George F. Cram, Gallup, Langwith, and Rand McNally introduced road atlases of the United States featuring maps of individual states designed in the network style.[98] The maps that Rand McNally, H. M. Gousha, and General Drafting produced for their commercial clients reflected both the increasing geographical reach of these businesses and the increasing touring ambitions of their customers. Rand McNally published twelve maps of western states for Denver-based Continental Oil (Conoco) in 1926.[99] By 1930, Gousha's Conoco series covered every state in the Union on thirty-one maps,[100] reflecting, but slightly overreaching, Conoco's recent acquisition of smaller firms based in the midwestern and mid-Atlantic states.[101]

The earliest network road maps show that the widest ranging motorists of the late 1920s and early 1930s were still slowed considerably by road conditions. According to the conditions described annually in Rand McNally's road atlas, a 1926 east-to-west cross-country traveler on U.S. Route

40, perhaps the most important transcontinental route, would have ridden a paved road virtually all the way from Atlantic City, New Jersey, to Topeka, Kansas (with the exception of one small unpaved segment each in New Jersey, Ohio, and Missouri). West of Topeka, however, the road was almost entirely unpaved (though "improved" in some stretches) all the way to central California, except in the vicinity of Denver and of Salt Lake City. Five years later, in 1931, the paved highway had reached Salina, Kansas, from the east and Battle Mountain, Nevada, from the west, with few improvements in between. By the mid-1930s, these unpaved stretches were rapidly disappearing. In 1941, only a small, fifteen-mile portion of U.S. Route 40 in western Colorado remained unpaved.[102]

As the navigational task of the road map simplified, its essential design could too. Over the course of the 1920s and 1930s this paradoxically led to the creation of rhetorically more complex maps. The complementary growth in the promotional aspect of automobile cartography also added a new rhetorical dimension to automobile mapping. Road maps took on the task of cajoling motorists to interpret the roadscape and to consume it in specific ways. This was expressed by subtle design changes, such as the addition of signs marking the location of places of interest, and more prominently, by the addition of marginal and cover illustrations, and descriptive text. Pictorial vignettes, inset maps, and separate specialized maps devoted to the promotion of specific tourist itineraries also became increasingly common.

Maps issued by many official state highway departments during the 1920s and early 1930s made highway improvements a centerpiece of their rhetorical strategy. The annual Iowa State Highway Commission maps regularly reported the number of state highway miles that were paved, graveled, graded, or "maintained," rejoicing in 1931 that "Iowa has Stepped Out of the Mud."[103] By the mid-1930s the initial paving of most state and federal highways was substantially complete. Official state maps had adopted a rhetorical strategy that deemphasized road improvements in favor of map features that reached out more directly to potential visitors. Daniel Block has observed that official Iowa maps introduced graphics and text onto map backs that extolled the virtues and variety of the state's landscape, natural resources, and industry.[104] This was a pattern followed by official state maps almost everywhere else, but states that were most likely to benefit from motor tourism were especially quick to convert their road maps into advertisements for the state's attractions. New Mexico's official 1927 road map was a Spartan publication measuring only 43 × 35 cm (17 × 13½ in.) when unfolded, printed in black and white with a few statements

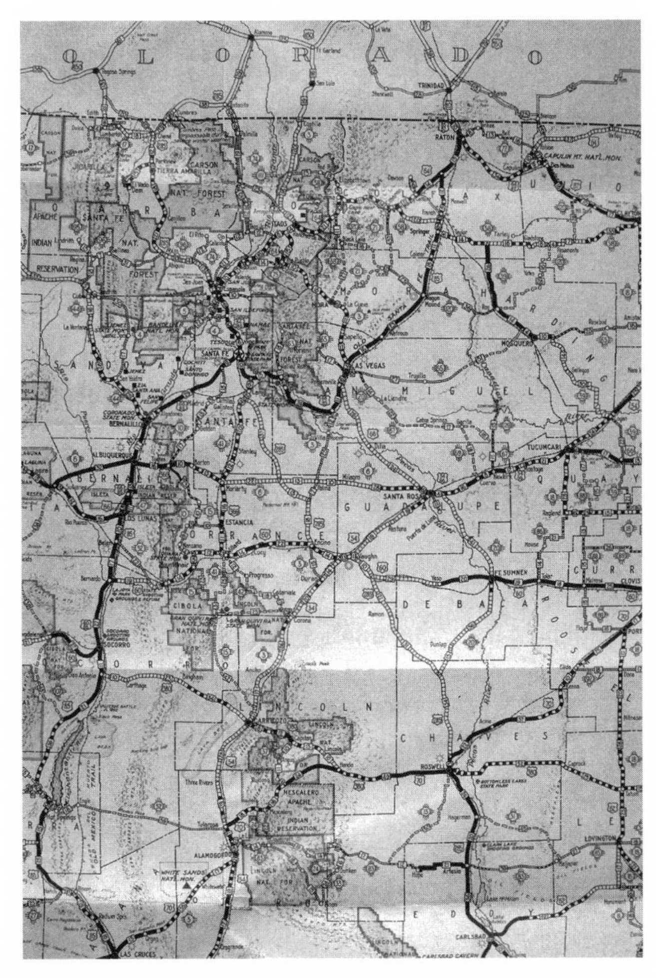

FIGURE 5.11 Detail of "Map of the State of New Mexico" (Santa Fe: New Mexico State Highway Commission, 1927). Rand McNally Collection, courtesy of the Newberry Library.

about highway conditions and regulations (figure 5.11). The 1929 edition was not only larger (65 × 53 cm or 25½ × 21 in.) but had also added relief representation and color. On the back of the map, we find lists of national monuments and forests, Indian pueblos and reservations, other points of interest, and, most significantly, photographs of natural and cultural scenes. By 1936, the outside front and back cover of the map featured an attractive illustration, signed by commercial artist Willard Harold Andrews, of two Indian women in traditional dress with a passing car and a mountain range in the background. Inside, the map was printed in five colors

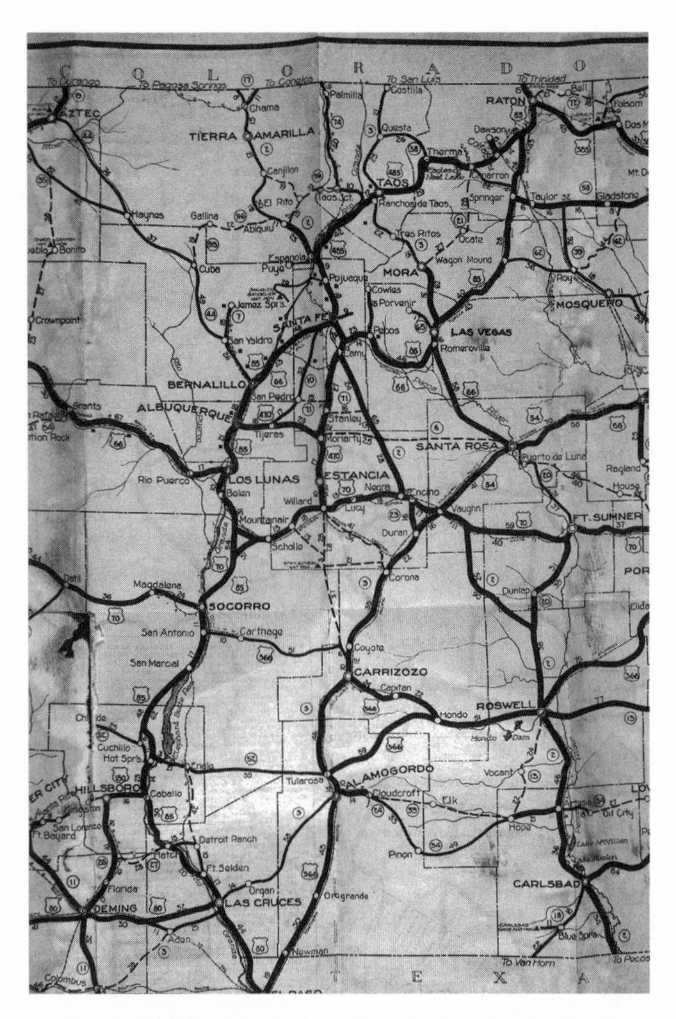

FIGURE 5.12 Detail of "Official Road Map of New Mexico, 'the Sunshine State'" (Santa Fe: State Highway Department, 1936). Rand McNally Collection, courtesy of the Newberry Library.

(figure 5.12), giving new emphasis to Indian reservations (shown in yellow), national forests (green), and mountains (brown hachures).[105] In 1940, on the occasion of Coronado Cuarto Centennial, the state highway department issued in addition to the annual road map a special illustrated map called "Historic Trails through New Mexico, the Land of Enchantment," which marked the routes of five explorers, six "old trails," plus the location of forts, pueblos, and national and state monuments.[106] In keeping with this rhetorical shift from utilitarian emphasis of road improvements to glossy promotional cartography,[107] many state highway agencies followed the lead of the oil companies by contracting with commercial map publishers (including Rand McNally, H. M. Gousha, General Drafting, and The National Survey) to design and print their maps.

These changes in official state highway map design paralleled developments in oil company maps. During the 1930s, gasoline map cover and panel art came to rival that found in advertisements published in glossy magazines. Seeking both to encourage gas consumption generally and to carve out a substantial market share for themselves, oil companies teamed up with map suppliers to develop and staff free travel bureaus that supplied maps, travel information, and travel directions.[108] The maps themselves placed more emphasis on natural features, municipal and county boundaries, state forests, parks, and other points of cultural and historical interest. Oil companies also pressed for the development of new supplementary material, such as pictorial maps, distance tables, touring itineraries, and historic and scenic guides that they hoped would give their maps a competitive edge.

Mapping a National Motorized Space

By 1930, the players who would shape and control the production of automobile road maps until the 1970s were already on the stage. Oil companies, their subsidiaries, and local gas and oil distributors had fully embraced the use of road maps as marketing tools; so had a host of other highway related businesses, including tire and auto parts companies, hotel chains, insurance companies, and independent roadside businesses. The AAA and its local affiliates were publishing network maps, extensive tour books, and route cards that would evolve into the now familiar Triptiks in 1937. Every state in the union published maps of its principal highways for public use on an annual or biennial basis. Much of the private distribution, and some motor club and government output, was handled by commercial cartographic publishers. In addition to the major commercial map publishers, local publishers produced not only highway maps of local and regional interest but also urban street maps intended primarily for motorists. The output of this industry was huge, numbering in the hundreds of millions of maps annually.[109]

Yet, for all this activity the core products were astonishingly similar. Close examination of General Drafting, Gousha, and Rand McNally maps from the 1930s through the 1960s reveals minor differences in drafting style, road classification systems, map symbols, and type faces, but their users likely either did not notice or care, so long as the maps performed the tasks that were required of them. Variation in the content of road maps were accountable primarily to the custom features developed for specific clients or markets. The colors schemes chosen for oil company map linework and lettering, for example, frequently matched the colors used in company logos and service station facades. Conoco and Cities Service maps usually included symbols indicating the location of their service stations.

Ultimately, it was the cover and back panel art that most distinguished one road map from the next. For oil companies in particular the role of these illustrations, indeed of all forms of advertisement, was to foster an image of a particular brand, the products it represented, and the service that customers received at its outlets, in order to encourage continued patronage. A glut of cheap petroleum on the market during the 1920s drove down prices substantially at the very moment that automobile use was expanding rapidly. The resulting fierce competition and low profit margins encouraged the emergence of branded gasolines and motor oils, and the consolidation and geographic expansion of the refiners and gasoline wholesalers that controlled these brands.[110] Free road maps were just one of many premiums and services introduced by gasoline marketers to encourage brand loyalty, but, no doubt because of their genuine utility to motorists, they were the most enduring.

Wherever possible, the backsides and margins of maps were filled with advertisements for specific products. Corporate symbols and slogans, particularly the logographic designs that appeared on gas station signs, facades, and pumps, were ubiquitous (figure 5.13). Subtler persuasion was reserved for map cover art. Advertising historian Richard Ohmann has argued that one of the most important graphic rhetorical strategies developed by the American advertising industry from the late nineteenth century (and, of course, still used today) prompted consumers to imagine "people, places, or occasions to be somehow associated with the product and its use." Among the most potent of these associations, or indexes, are those that establish a correspondence between a product and the values, virtues, social goals, and personal aspirations widely cherished by the advertisement's targeted consumers. Soap is associated with cleanliness and good health, for example; cars with wealth or adventure; soft drinks with youth and energy; food products with thrift, familial responsibility, or refined taste.[111] By the 1920s and 1930s, the kit bag of graphic advertisement had expanded considerably to create what Roland Marchand has called social tableaux. Advertisements—and road maps were no exception—put consumers in idealized social situations demanding the consumption of particular goods or services, or appealed to consumers' desired status or self-image.[112]

In their broadest outlines, the norms, activities, and aspirations indexed by road map cover art were those staked out by automobile advertisements

FIGURE 5.13 Front and back panels, "New York, Amoco, Best Way to Go! from Maine to Florida" (Chicago: Rand McNally for American Oil Co., 1947). Rand McNally Collection, courtesy of the Newberry Library. The back panel's explanation of service station signage for regional American Oil affiliates encourages transcontinental brand loyalty.

and promotional rhetoric from the beginning of the century: freedom, mobility, modernity, wealth and distinction, and a healthy and active lifestyle. But since road maps were associated most particularly with discretionary leisure travel, their social tableaux situated gasoline consumption within automobile tourism. One finds intriguing depictions in road map cover art of the social status, race, and gender of motorists, but all of them, patently, Christina Dando remarks, were "going places."[113]

In the America of road map cover art, most every ride is a joy ride. The destinations of motorists depicted on the covers are often ambiguous, but judging by the smiles on their faces, their relaxed manner, their dress in sport clothing, and the bucolic locales in which they are situated, the implied purpose of their trip is usually recreational. Some motorists are obviously at ease, loading camping equipment into or onto the car or trunk, taking in grand vistas, snapping family photographs, enjoying family picnics, swimming, or playing golf. On other covers, vacation trips or joy rides are merely implied. The motorist asking an attendant for directions or for assistance in reading a map was a common trope. The motoring couple depicted on a 1930 Texaco cover may be lost or slightly disoriented (plate 8),[114] but they do not seem particularly troubled by their predicament. Judging by their dress and the appearance of their car, they are obviously not country bumpkins, but visitors from the metropolis looking to continue the rural joy ride. Road map cover artists were also fond of placing gas stations in bucolic settings, so that even when people were not shown, the relationship between gasoline and a joy ride in the country remained fairly explicit.

The emphasis on touring over practical trips follows an early pattern of automobile promotion. In their first decade of the century as consumer items, automobiles were luxury goods, financially unattainable for anyone below the upper middle class and mechanically unreliable for industry or labor. But thanks to reductions in price, improvements in quality, the stimulus provided by war use of motorized vehicles, and the development of a rudimentary highway network fostered by the automobile interests, the position of the automobile in American social and economic life had changed substantially by the early 1920s. In 1920, 31 percent of American farms had automobiles, and this number would rise to 58 percent by 1930.[115] The first intercity bus service was launched in 1912, and by 1927 there were 650,000 miles of bus lines—about 2.5 times the mileage of the existing rail passenger routes (which nonetheless still accounted for 75 percent of passenger miles in 1930). The American automobile industry produced 25,000 motor trucks in 1914; six years later it produced nearly thirteen times that number (322,000).[116] By 1930, automobiles made at least as important a contribution to the American economy at work as at play. Yet in spite of this, cartographic promotion of automotive consumption continued to stress leisure trips and joy rides. Commercial vehicles are almost entirely absent from map covers of the 1920s and 1930s. Indeed, few cartographic products were developed specifically to suit the specific needs of, say, commercial truckers.

Auto industry executives and the engineers they employed waged vigorous campaigns in the popular and technical press on behalf of the alleged economic benefits of the automobile industry. This could be contentious when it ran against entrenched industries, especially the railroads. With this in mind, Roy D. Chapin of the National Automobile Chamber of Commerce devoted a substantial portion of the 1924 article "The Motor's Part in Transportation" to assessing the economic superiority of automobiles to railroads.[117] Automobile tourism, which (presumably) even railroad executives indulged in, was an easier sell. Political support for creating the infrastructure needed to make America a nation on wheels was best built by sugarcoated rhetoric.

The appeal to the good life in road map cover art was not entirely limited to presenting explicitly recreational trips. Though cars were rapidly entering the lives of middle- and working-class Americans, road map covers—those published by Shell, Sinclair, and Gulf in the 1930s, for example—appealed to the vanity of motorists by dressing them glamorously or by putting them in prosperous settings with a troupe of gasoline attendants to serve them (plate 9). Just as commercial vehicles were largely absent from these tableaux, so the idealized motorist was professional or independently wealthy. One searches the imaginary motoring nation of road map cover art in vain for the trek of the Bonus Army or the migration of Okies. Even in the depths of the Depression, road map cover art depicted the American landscape, to paraphrase Conoco's maps, as a landscape of pleasure.[118]

Looking at the smiling and self-satisfied faces of cover-art citizens, one cannot help but notice the socially limited impression of driving America that they present. Drivers are overwhelmingly male, although a few map covers (their designers perhaps mindful of the automobile's promise of greater mobility for both rural and urban women) showed women behind the wheel—mostly alone, but sometimes in the presence of male companions.[119] These driving women are universally young and glamorous, not matronly, and are apparently driving at some speed, suggesting a certain measure of wantonness, or at least freedom from traditional cares and responsibilities. By the same token virtually all of the motorists in the world of road map cover art during the 1920s and 1930s are young and childless (see plate 8). Children were, of course, taken on car trips, but the prevailing depiction of joyriding appealed to the frivolity, speed, reckless abandon, and romance of the automobile. On the few occasions when children did appear on the covers of road maps before World War II, they were frequently in the presence of their grandfathers, who in contrast to the fast-paced travelers of the younger generation, represented reliability and caution.[120]

The passing of the Second World War brought the automobile still more fully into American life, and road map depictions of family and gender roles changed accordingly. A majority of American households could now afford a car, and entire landscapes were created and old ones modified to suit the

FIGURE 5.14 Front and back panels, "Washington, D.C. and Vicinity Map and Visitor's Guide, Esso" (Convent Station, NJ: General Drafting for Standard Oil of New Jersey, 1957). General Drafting Collection, courtesy of the Newberry Library. © American Map Corporation; reprinted with permission.

car's demand for space and access. The annual summer car trip became an established, and much romanticized, feature of American middle-class life. Drivers remained predominantly male, but males and females alike were no longer romantic vagabonds (figure 5.14). The nuclear family was suddenly everywhere on road map covers: two parents, two children (one male, one female), and sometimes a dog, off to see America.

The faces of these families and other motor tourists remained exclusively white. The belated "discovery" of minority consumers by advertisers occurred too late to influence the covers of oil company road maps. Throughout the era of the free road map, nonwhites were relegated to places in the map covers' roadside scenery, as cultural stereotypes integrated into the social fabric of a particular regional landscape. They were never motorists. The depiction of regional landscapes on the covers of the Cities Service series issued in 1931 is typical: Iowa's landscape is epitomized by farmland, Colorado by mountains, Missouri by a river scene with riverboats, and Alabama by African Americans working in a field of cotton. Owing to their enduring place in American fantasies about the West, Indians also served as local color on map covers well into the 1960s and beyond.[121] These coexisted with white rustic stereotypes, such as midwestern farmers, Maine lobstermen, and (most especially) cowboys, pointing to a perceived customer base that was urban or suburban and middle class as much as it was white.

Another strain in road map cover art celebrated the modernity of motor touring, the enhanced accessibility to the countryside afforded by the metropolitan highway system, and its conquest of natural barriers. City and country were often juxtaposed in the same image (plate 10), suggesting the highway's role in collapsing both the geographical distance and the cultural space between the two.[122] In the middle of a sea of cornfields or grazing cattle, one is never far from a modern highway and gleaming gas stations.[123] The official Vermont state highway maps for 1967 and 1968 offer alternative front covers that make a striking comparison: on one, a photographic image of a tree-lined country lane, resplendent in its fall colors; on the other, the modern superhighway, easing travel throughout the predominantly rural state. Accompanying text explains that a "wonderful thing about Vermont is that it's so accessible. Modern highway systems have brought her borders within easy reach of anyone traveling by car. Once here, the State highway system . . . moves traffic effortlessly from one end of the state to the other. By contrast, Vermont's excellent network of well maintained secondary roads and back-country byways allows the unhurried traveler to explore the quiet countryside at his own pace, in his own way."[124]

Marguerite Schaffer has argued that early promoters of automobile tourism, echoing and amplifying arguments first developed by the railroads, genuinely felt that motor tourism would become a means to making Americans better citizens. Tourism made citizens knowledgeable and appreciative of their country's diverse physical geography and natural resources, enriched by awareness of the regional cultures of their fellow Americans and, through visits to the sites of pivotal events and the reenactment of geographical movements, better-informed in the nation's history.[125]

Many early motorists did indeed feel that a motor tour was a ritual passage into national citizenship. Frederick Van de Water opened his 1927 account of a transcontinental journey with his wife and six-year-old son with the observation that their "pilgrimage" transformed "three originally smug New Yorkers."

> [We] were heated by the sun, powdered by the dust of the Corn Belt and the alkali of the desert, pounded and kneeded by a Ford determined to ignore no irregularity in the highway and at the end of the ordeal [we] were no longer New Yorkers, but Americans, which [we] learned, is something surprisingly and hearteningly different. . . . [My son], when he comes to study geography, will not consider the Rockies as a series of tiny eyelashes spread across a map. He will know that the Mississippi is dirtier than the Hudson and the Missouri dirtier than either and the Columbia wide and swift and a greenish blue. Wyoming will not be to him merely a purplish rectangle requiring to be bounded among the dismal assemblage of Western States. He will recall instead the drab range land and the aromatic smell of the sage and the harsh, bold outline of the mountains against a wide sky, and he will remember that the little square in its northwest corner, uninterestingly labeled "Yellowstone Park," is vivid with bears. If by the time he studies school history, values have been recast and the westward march of the backwoodsmen and later of covered wagons, receives an attention equal to that now expended upon Plymouth Rock he should thrill a little, too, for he has seen much that the pioneers saw.[126]

In the 1930s, map cover art and interior features began overtly to trumpet good citizenship as a motivation for touring, particularly as a means of reliving and learning American history. The covers of Sun Oil (Sunoco) maps published during the 1930s and early 1940s juxtaposed images of automobiles on the open road with a variety of historical scenes and sites accessible by car. The 1941 cover was perhaps the most eloquent of these: A frontiersman armed with rifle and wearing a buckskin coat and coonskin cap pulls aside a stage curtain, revealing a car traveling on a two-lane road that parallels a placid river valley. Above the clouds hover depictions of war memorials, a cannon, and a graveyard, references perhaps to the war sacrifices that Americans would soon have to make.[127] Opening their Sunoco maps, motorists would find, in addition to a traditional road map, a "Historical Pictorial Map" prepared exclusively for Sunoco by Rand McNally, featuring dozens of images of historic sites and events shown in rough relationship to the principal highways of the mapped state or states.

Esso and its mapping partner, General Drafting, pursued a similar strategy from the late 1940s to the early 1960s. Map covers showed individual events and sites appropriate to the state or region featured. The 1952

Arkansas, Louisiana, and Mississippi map featured an image of the signing of the Louisiana Purchase. Covered wagons appeared on the front panel of the map of the central and western United States in the same year. Francis Scott Key and the bombardment of Ft. McHenry appeared on the Delaware, Maryland, Virginia, and West Virginia map; the landing of the Pilgrims on the New England map; the 1775 capture of Ft. Ticonderoga on New York's map; and Lincoln delivering the Gettysburg Address on Pennsylvania's. Esso/General Drafting maps from the same era also put modern tourists, especially families with children, in this historical landscape—visiting memorials, museums, and restored historic buildings and villages. The 1959 New York map shows a couple pausing along the Hudson River to imagine Henry Hudson's first encounter with the Indians of the region. The 1958 Washington, D.C. map features a visit to the Smithsonian Institution by a family of four, the son and daughter dressed appropriately for the occasion in Cub and Girl Scout uniforms. The special map of Williamsburg, Jamestown, and Yorktown published to commemorate the 350th anniversary celebrations at Jamestown shows a modern father and son reverently watching four ghostly images of early settlers piously entering the ruins of Jamestown's Old Church Tower.

The normalized landscapes, social and cultural identities, and historical scenes depicted on road map cover art were given geographical flesh inside. Noting points of interest on network maps—picnic areas and rest stops, parks, and scenic routes provided for the convenience of travelers—programmed the travel experience, privileging a few places and routes as beautiful, significant, or meaningful. Pictorial maps or map supplements had the effect of prescribing experience as well. Their main content was the landmarks, landscapes, and economic and cultural features often found on map covers: a cornstalk stood for Iowa, a cowboy or the Alamo for Texas, skyscrapers for New York or Chicago, steel mills for Pittsburgh. Occasionally these reinforced social and racial stereotypes: watermelon-eating Negroes appear in the southern states; stoic Indians stand in front of New Mexico pueblos; Hillbillies populate the southern Appalachians.[128]

Road maps published by local motor clubs, state governments, and tourism bureaus developed these themes more fully, emphasizing sites, tours, and excursions that conformed with visitor expectations, while glossing over landscapes and features that did not. The Automobile Club of Southern California still publishes a map of the "Indian Country" or Four Corners region of Colorado, New Mexico, Arizona, and Utah that emphasizes historic and current sites associated with native communities in the area.[129] A map published around 1980 by the Allentown-Lehigh County Tourist & Convention Bureau transforms old U.S. Route

22 from Allentown, Pennsylvania toward Harrisburg into the Hex Highway, a chain of roadside attractions united by their Pennsylvania Dutch character.[130]

Other specialized maps and insets encouraged tourists to transform their motor trips into narratives of events and movements critical to national development. A barrage of maps published in the late 1950s promoted a tour of the National Colonial Parkway, a twenty-three-mile long landscaped roadway created in 1930 to link the reconstructed historic sites on Virginia's James Peninsula. A leisurely journey of three or four days along this route would be sufficient to recapitulate 174 years of colonial history from the founding of Jamestown, through the idealized colonial cityscape of Williamsburg, to the end of the American Revolution at Yorktown.[131]

The "Map of the Principal Events in the Life of George Washington" commemorating Washington's bicentennial birthday urged Standard Oil of New Jersey customers to "make this a Washington year. . . . [Y]our family and your guests will be more interested than ever before in visiting the shrines of our country's first President. May we suggest therefore that you will enjoy including at least one of these historic places every time you take an automobile trip." Not technically a road map, the publication was a geographical and pictorial catalogue of places and events associated with Washington in a territory reaching from Georgia to eastern Ohio to western Connecticut. Conveniently, most of the sites indicated on the map lay within Standard of New Jersey's market area.[132]

In the early years of automobile tourism, the comparison between the motoring tourist and the nineteenth-century overland pioneer was often and easily made, partly because the roads were so bad. "The pioneer instinct seems to be in all of us," Henry B. Joy, president of Packard and the Lincoln Highway Association, wrote in 1917, while lamenting nevertheless the poor conditions motoring tourists had to endure. Early motorists liked to think of themselves as latter-day pioneers, and Indians were also invoked, especially by the cover art of the 1920s and 1930s, as emblems of automobile touring itself, as benign keepers of a semiwild landscape now penetrated by cars, or as model trailblazers for the motorists themselves. In one of the most celebrated of early map covers, which appeared on Rand McNally's road atlas from 1926 through 1932 and on some editions of the company's Auto Trails maps, an Indian is seen marking a rock with a directional arrow, unobserved by the white motorists navigating the Lincoln Highway below. The promotion of the Auto Trails maps, the first national series of commercially distributed automobile road maps, drew heavily on this metaphorical equation of pioneer trails to early automobile highways, which Rand McNally called "blazed trails."[133] The authors of many early published travel accounts also chose titles that compared their exploits to the pioneer treks of the previous century.[134]

Poor as they were at the beginning of the twentieth century, the legendary roads associated with the westward expansion of the United States were among the very few true "interstate" highways in existence at the dawn of the automobile era. They were natural focal points for highway improvements, having the additional advantage of a sentimental association with the nation's history that could be exploited by good roads advocates. An enduring cottage industry arose to produce maps and cartographically enhanced guidebooks serving those who wished to retrace the trails more exactly, identifying and interpreting the physical remains of the routes and historical markers on them, and enlarging upon the history of the trails.

The first great western highway, the old National Road from Cumberland, Maryland to Vandalia, Illinois was the focus of much early road publicity and cartography. In the early 1910s, the Daughters of the American Revolution championed the road's repair and improvement as part of a National Old Trails Road linking the National, Boone's Lick Road, the Santa Fe Trail, and other historic western Routes, forming a continuous transcontinental route between Baltimore and Southern California. In the 1910s and 1920s, proponents of the improvement of the National Road for automobile use, including the National Highways Association, published a number of promotional maps and route guides that stressed the road's historic importance and directness, and noted its status as the first federally funded highway as an argument for renewing federal involvement in highway construction. At least one commercial establishment along the route, the George Washington Hotel of Washington, Pennsylvania, published a map in 1925 promoting the National Road and its convenient links to major population centers in the Northeast and Midwest, its critical role in the formation of the federal government, and the amenities of Washington and the George Washington Hotel.[135]

The Oregon Trail, too, was an important target of early preservation and reconstruction efforts, spawning a number of maps and guidebooks for motor tourists bent on following, as near as possible, the wagon ruts of western pioneers. One of the leaders of the movement to find and mark the old trail was Ezra Meeker, who had migrated with his wife to Oregon from Indianapolis in 1852. In 1906 he undertook a well-publicized yearlong eastward journey by "prairie schooner" from his home in Puyallup, Washington, to Indianapolis, placing stone markers along the way.[136] Meeker's cause was eagerly embraced by automobile and highway interests. The Old Oregon Trail Association was founded in 1922 "to permanently mark the road so that its history may be preserved and be a constant reminder to the younger generation of the hardships endured by the pioneer that we may enjoy our present day civilization [and to] correctly and consistently advertise it so that the world may know it for . . . the shortest and quickest route from the East to the Pacific Coast." A map published by the Inland Empire Hotel Association around 1930 identified the westernmost portion of the trail with modern U.S. Route 30.[137]

The Oregon Trail began its route west in conjunction with the Santa Fe and California Trails in what is now metropolitan Kansas City. Promotional maps published in the 1960s by metropolitan chambers of commerce reminded tourists of this important aspect of the region's history by delineating the combined route against a background of modern Kansas City streets (figure 5.15).[138] In like manner, official Nebraska state highway maps from the early 1970s to the mid-1980s featured a special map showing the Platte River Road—the portion of the combined Oregon, California, and Mormon Trails that parallels modern Interstate 80—hoping to capture tourists' attention as they whisked across the state en route to the celebrated destinations of the Central Rockies, the Great Basin, and Northern California.[139]

A number of cartographically rich guidebooks were also published for motorists who wished to study and follow the routes of the pioneer trails more diligently. The depression-era guide to the Oregon Trail prepared under the auspices of the WPA's Federal Writers' Project was one such book.[140] The more recent atlas of the same trail by Gregory Franzwa carefully charted the original route (and its alternatives) on 270 modern topographic and street maps.[141]

The most celebrated of early transcontinental automobile routes commemorated Abraham Lincoln, although the historical association between the slain president and the Lincoln Highway was more symbolic than geographic. The automotive industrialists who honored Lincoln thought it fitting that a route ostensibly binding the country together recognize the leader who held it together in time of war, yet Gettysburg, Pennsylvania, was the only place on the highway explicitly related to Lincoln's life. As the twentieth century progressed, visitation of Lincoln-related sites became a staple of automobile tourism in southern Illinois, Indiana, and central Kentucky. Travelers who followed the map of the roundabout itinerary of the Lincoln Heritage Trail (figure 5.16) could "experience the living presence of this famous American," following the "young Abraham from the National Park at his birthplace to another Lincoln farm on Knob Creek in Kentucky"; through Indiana to New Salem, Illinois, where he chose to become a lawyer; and finally to Springfield, where he rose to national prominence and where his final (and fatal) presidential journey began and ended. Such a journey allowed motor tourists not only to trace the

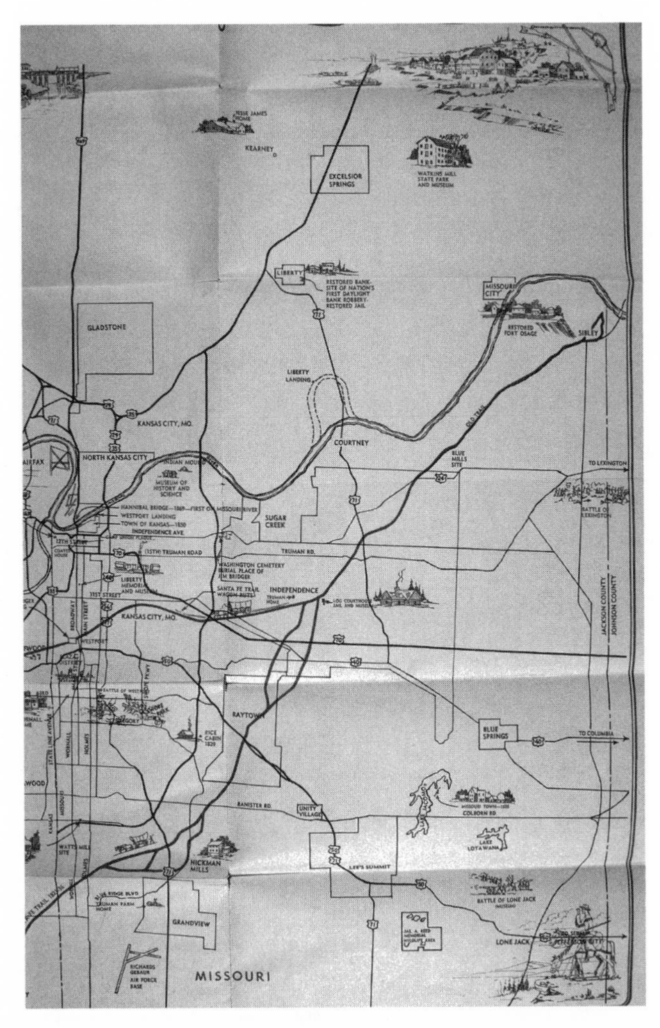

FIGURE 5.15 Detail of "Greater Kansas City Historical Map" (Kansas City: Chamber of Commerce of Greater Kansas City, 1967). Map © Greater Kansas City Chamber of Commerce. Personal collection of the author.

biography of one the nation's most prominent historical figures, it also acted out the heroic national myth his life's career represented: that any American, however humble the circumstances of his birth and rearing, might become successful, even prominent, in the life of his nation.[142]

The pivotal national drama in which Lincoln played the leading role was itself the subject of promotional maps for automobile tourists, reflecting not only the outsized importance of the Civil War in twentieth-century American memory, but also the ubiquity of sites and landmarks associated

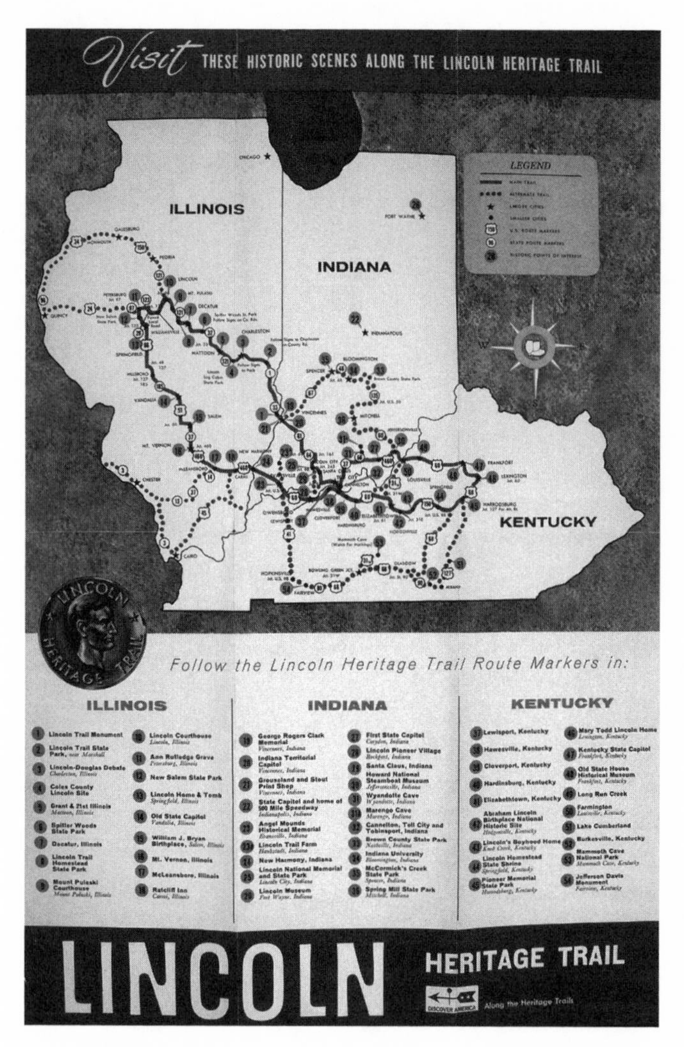

FIGURE 5.16 "Lincoln Heritage Trail" (Champaign, IL: Lincoln Heritage Trail Foundation, 1969). Personal collection of the author.

with the conflict. By virtue of its central importance in the course of the war, its reputation for carnage, and its proximity to major population centers, Gettysburg was a magnet for mourners and curiosity seekers before the guns were silent. Cartographically rich guides for tourists were published in the immediate aftermath of the war,[143] continuing unabated with the arrival of the automobile. Three maps published in a brochure issued by the Gettysburg Chamber of Commerce in the early 1920s emphasize the accessibility of the site to all major eastern and midwestern population centers. Accompanying text compares the apparent geographic centrality of Gettysburg (as portrayed by the map) to its symbolic centrality to the history and political philosophy of the nation: "Gettysburg—the heart of the new America" is the "mecca of millions of tourists, each year, coming from every state in the Union" and the "shrine of all nationalities who cherish a living faith in democratic principles and ideals."[144]

Commercial road mapping exploited the automobile's ability to retrace military movements as well, in some instances assisting motorists who wished to organize trips reenacting entire campaigns. During centennial commemorations, Rand McNally published a map suggesting itineraries linking military sights in each of the three major theaters of war: the Mississippi Valley and Gulf, the mid-South, and mid-Atlantic.[145] Interstate 75 from Chattanooga to Atlanta essentially follows the railroad route that was lifeline and pathway of General Sherman's 1864 campaign for Atlanta. In many places, the battles of the campaign were fought within sight—or even on—the path of the modern roadway. Dave Hunter's 1992 guide to the modern superhighway and its roadside amenities exploited this happenstance, superimposing battle lines on the strip maps that compose the guide, so that modern Northerners bound for Disney World at seventy-five miles per hour could imagine that they were units of Sherman's army outflanking Johnston at Rocky Face, Resaca, and Kennesaw Mountain.[146]

Toward the end of the of the twentieth century even the great automobile routes of the past—the Lincoln Highway, Routes 1 and 40, and especially Route 66—themselves became the subjects of maps recreating past movements and narratives. In their day these highways had been the most modern expression of the American yearning to travel far and fast. Bypassed by the interstates, by the 1980s the popular imagination had transformed these highways into relics of a bygone era, when automobile travel was leisurely and routes engaged the landscapes through which they passed. Above all, U.S. Route 66 had been singled out by virtue of its long prominence in popular culture, as the futile pathway of John Steinbeck's Okies, the swinging subject of Bobby Troupe's popular song, and the title of a 1960s television show. Michael Wallis, one of the leading spokesmen of the Route 66 revival, observed that the route's appeal derives from its mingling of the common man and the common landscapes of mid-twentieth century with the romance and promise of the open road looking West.

> Route 66 is Steinbeck and Will Rogers and Woody Guthrie and Merle Haggard and Dorothea Lange and Mickey Mantle and Jack Kerouac. It's thousands of waitresses, service station attendants, fry cooks, truckers, grease monkeys, hustlers, state cops, wrecker drivers and motel clerks. Route 66 is a soldier thumbing home for Christmas; an Okie family still

> looking for a better life. It's a station wagon filled with kids wanting to know how far it is to Disneyland; a wailing ambulance fleeing a wreck on some lonely curve. It's yesterday, today, and tomorrow. Truly a road of phantoms and dreams, 66 is the romance of traveling the open highway. It's the free road.[147]

Having lost its designation as a federal trunk route in 1985, Route 66 became a martyred road, ironically representing both the restless energy of the southwestern landscape it traversed and the decline of automobile tourism. Dozens of maps, map-saturated guidebooks, and eventually websites dedicated themselves to memorializing the route and its place in American history.[148] Perhaps most notable among these from a cartographic perspective was Tom Snyder's *Route 66 Traveler's Guide and Roadside Companion,* published in 1990, which reproduced the set of forty-two cartographic route cards prepared by the Automobile Club of Southern California depicting the route as it existed around 1930. Snyder overprinted the route of the modern interstate highways that had superceded Route 66 on his reproductions of the circa 1930 maps, calling attention to the partial obliteration and bypassing of the old route, but also helping modern travelers so-inclined to retrace it with the same veneration that they might follow the old National Road and the Oregon Trail.[149]

The national park drives and parkways, such as the Blue Ridge Parkway and Skyline Drive, stand as extreme examples of a broader tendency to program motorists' experience of the national roadscape. In these controlled environments, motorists had little choice but to follow, since the roads offered no possibility of escape (except by foot) for miles a time. These have their maps, too, which clearly mark designated pullouts and scenic views while interpreting the very impressions of surrounding countryside that motorists are meant to take home.[150] This is the type of strangulated and prescriptive map-reading experience that travel writer Bill Bryson must have had in mind when, in *The Lost Continent,* he reflected on maps during a journey through west central Georgia:

> From Warm Springs I went some miles out of my way to take the scenic road into Macon, but there didn't seem to be a whole lot scenic about it. . . . I was beginning to suspect that the scenic route designations on my maps had been applied somewhat at random. I imagined some guy had never been south of Jersey City sitting in an office in New York and saying, "Warm Springs to Macon? Oooh that sounds nice," and then carefully drawing in the orange dotted line that signifies a scenic route, his tongue sticking out of his mouth.[151]

For the most part, we may assume that the programmatic mappings of colonial and revolutionary history, pioneer trails, and Civil War history described above were followed to the letter by a small minority of obsessed motorists, and at the convenience of whoever else heeded them at all. Moreover, the role of such maps in shaping twentieth-century Americans' perceptions of their history—weighed against what they learned in school or from movies, television, and popular literature—was likely limited.

These cartographic narrations of the national space, nevertheless, were the most complete manifestations of the apparently contradictory inclination of promotional road maps—and in this sense of any map, transportation-oriented or otherwise, that enters into popular culture—to program the experience and perception of landscape. Wallis's Route 66 was "the free road," but to the extent that it was singled out as quintessentially American, it has become the *required* road, the road one must travel—as so many Europeans enamored with American popular culture have—in order to see and to understand America. To the extent that Route 66 is *the road* it is not entirely a *free* road. Jack Kerouac realized this when he had Sal Paradise, protagonist of *On the Road*, ponder why his plan to discover America initially went afoul because of his naïve seduction by the route lines on a map:

> I'd been poring over maps of the United States in Paterson for months, even reading names like Platte and Cimarron and so on, and on the roadmap was one long red line called Route 6 that led from the tip of Cape Cod clear to Ely, Nevada, and there dipped down to Los Angeles. I'll just stay on 6 all the way to Ely, I said to myself and confidently started.

Hardly any of this imagined trip turned out as planned, but Paradise had to confess that it wasn't the map's fault. "It was my dream that was screwed up, the stupid hearthside idea that it would be wonderful to follow one great line across America instead of trying various roads and routes."[152]

Conclusion

On one level, the emergence of the national motorized space was the cumulative effect of ever larger numbers of Americans coming in more intimate contact with larger proportions of the national landscape and its regional subcultures across the continent. So we have been concerned here partly with how road maps widened the horizons of average

Americans by making the cross-country pleasure trip thinkable, and how they made it practical. Maps contributed both imaginatively and materially to the construction of a national highway system. They participated in early twentieth century debates over the proper emphasis of federal and state contributions to automobile highway construction. The first informal automobile road mappings, performed by private pathfinders and the semipublic and corporate sponsors, also played a very real role in determining the specific location of major national and regional highways. Most fundamentally, the early road maps published by private concerns, motor clubs, and highway associations appealed to the first generation of motor tourists, making automobile travel seem both practical and desirable.

More broadly, once the national system of automobile highways provided travelers with access to an apparently endless range of touring options, American automobile road maps added to their simple navigational content a second tier of rhetoric that both defined the national character of automobile tourism and privileged specific places, landscapes, and itineraries. Map artwork pointed the way by idealizing the motor tourist and his expectations of the touring experience. Then, maps gave geographical flesh to national touring by mapping out itineraries that ritualized automobile travel, recapitulating historical narratives and stressing certain geographical experiences in motion that were essential to the development of well-informed American citizens.

Seen from the vantage point of a century of transcontinental automobile travel, it is easy to be cynical about the commercialization of patriotic tourism that runs through American road mapping. Early automobile promoters, however, mixed patriotism and profit without a trace of cynicism. Joy's Lincoln Highway was motivated by a genuine love of country and an earnest desire to compress the distances that separated its people through shared motoring experiences. "[W]e are becoming more cosmopolitan," he wrote in 1917. "I have stopped at night at lonely ranches far out in America's vast Western desert at points in Utah and Nevada, and found them crowded with motor tourists from about as many different states as there were motor cars. It is becoming a common sight in cities along the Lincoln Highway and our other main routes of travel to see the license tags of a dozen or twenty different states in a day—Americans out to 'See America First.' "[153] For people like Joy, the necessity of looking homeward for diversion became a patriotic virtue. Seen this way, it becomes easier to reconcile the apparently opposite tendencies of American road maps: to release motoring tourists freely on the landscape, and to control their movement across it.

6

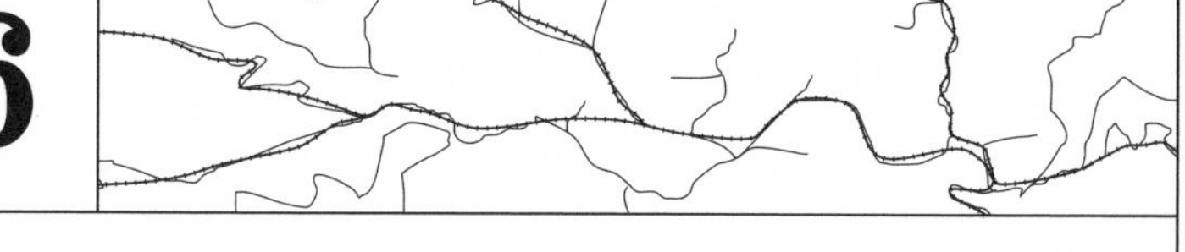

"Up in the Air in More Ways Than One"

THE EMERGENCE OF AERONAUTICAL CHARTS IN THE UNITED STATES

RALPH E. EHRENBERG

Engine-powered air travel is so commonplace today that few people appreciate the complex problems the industry initially needed to resolve. One was simple: easy to read maps designed especially for pilots. "When you are heading for a place at a mile or two a minute it is no trick at all to get lost," noted C. J. Zimmermann, a test pilot for the Aeromarine Plane and Motor Company, following a round trip flight from New York City to Havana in 1920. "In order to distinguish states from each other," he continued, "they are colored pink, blue, yellow and green on ordinary maps. It is a good idea too, but when a pilot who is traveling somewhere at the rate of a hundred miles an hour or so looks over the side of his plane he finds that New York and Connecticut are precisely as like green and brown checkerboards as New Jersey and Pennsylvania and those straight well-defined boundaries are nowhere in sight on the ground. Consequently, the pilot, especially if he is a young pilot, finds that he is up in the air in more ways than one."[1] Indeed he often was! To aid Capt. Zimmerman and his fellow pilots, the aviation industry developed a wide variety of maps and charts for air travel, ranging from aero route maps and air navigational charts for pilots and navigators to airline souvenir maps designed for commercial passengers and the sightseeing trade.[2] The focus of this essay is limited to the early development of the aerial navigational chart in its various forms from its inception in the United States until the end of the 1920s, when major advances in American commercial aviation modified the format and content of the aeronautical chart and introduced the airline souvenir map.[3]

Aviation Charts

Air navigation charts are unique among transportation maps. Their lineage can be traced to both road and railroad maps and to hydrographic charts. Like road and railroad maps, the aviation chart provides a generalized representation of a portion of the earth's surface. Like a hydrographic chart, it serves as a base and spherical frame of reference for computing and plotting navigational information. But the aviation chart is more than a combination of these three maps. Its special nature derives from its purpose as a navigational aid for tracking the movement of airborne vehicles from one point to another high above the earth's surface at fast rates of speed.

The air navigation map reflects the influence of five activities or practices associated with flying, each of which is external to map construction but has an impact on its form and content. These are aerial navigation, the nature of cross-country flying, the design and capabilities of aircraft, the world in which the airman operates, and the infrastructure of the airways system.

AERIAL NAVIGATION

Aerial navigation basically means finding one's way from one point on the earth's surface to another. Pilots who lost their way during aviation's pioneering era simply sought out a farm pasture or a dirt road, landed, and then asked directions to the nearest town or railroad. Sometimes they learned where they were, sometimes they did not. General William "Billy" Mitchell recounts that when he ran out of fuel flying over the Appalachian Mountains and landed about eight miles from a post office to seek information, the mountain people that he encountered did not know of the post office's existence. "Neither did they know the name of the President of the United States," he noted. "[O]r even of the sheriff of their county."[4] More often, however, pilots relied upon one of the three methods of aerial navigation: visual, celestial, or instrument.

The basic technique of visual navigation is known as *pilotage* or *contact piloting,* flying from one place to another guided solely by map, magnetic compass and reference to visible landmarks on the ground. Pilotage is generally combined with *dead reckoning,* another form of visual navigation, which involves estimating one's position from a known point by means of direction and distance. During dead reckoning navigation pilots fly from one checkpoint to the next by compass bearings indicated on a map with periodic visual reference checks of prominent landmarks. Visual navigation was the preferred method of navigating during aviation's formative years, when it was also known as *map and landmark flying*. To quote one pilot from 1924, "Aviation today, to use the figure of speech, is on the ground. Pilots fly by map and landmark. . . . Pilots very seldom fly over clouds and in very few instances do they use any of their navigation instruments."[5] Pilotage and dead reckoning remain important elements of navigation today, though the rapid acceptance of Global Positioning Systems suggests that GPS is becoming the primary navigational system.

Because visual navigation relies so heavily on chart reading and landmark identification, aviation maps generally include only those natural and manmade features easily visible from the sky. Accurate and up-to-date depictions of railroad tracks remain particularly important. Easily seen from high altitudes, they originally provided a link to urban centers for lost or disorientated pilots. Dubbed the "iron compass" by pioneer aviators, railroads were considered more dependable guides than early aerial magnetic compasses, and for this reason many of the first airway routes were designed to follow them.[6] Railroads are also less likely than roads to be concealed by overhanging foliage and were particularly helpful in northern regions during the winter months when highlighted against white fields of snow.[7]

Magnetic compass bearings and compass roses were added to aviation maps to aid pilots using dead reckoning navigation during cross-country flights over unfamiliar territory. Early aerial magnetic compasses, however, were notoriously unreliable, particularly during turbulent weather conditions and aircraft acceleration. Pioneer pilots placed little faith in them.[8] As late as 1929, Commander D. G. Jeffrey of the Royal Navy wrote that the compass "still remains somewhat of a mystery to the average pilot." One pilot that he knew carried a small magnet in his pocket "to ginger up his compass when it stuck."[9]

Celestial aerial navigation is a method of determining position by observation of the sun, moon, planets, or stars. It was developed primarily for navigating large bodies of water but was cumbersome and time-consuming prior to World War II, requiring large charts, a sextant, and an accurate timepiece for computation. A single sighting required nearly fifteen minutes. Although introduced around 1919, celestial aerial navigation was little used until the late 1930s, when Pan-American Airways established its overseas routes and the War Department began developing long-range bombers.[10]

Instrument navigation (also called blind flying) refers to navigating under conditions where neither earth nor sky is visible to the pilot. Instrument navigation uses radio waves to guide pilots. Introduced first in marine navigation on the eve of World War I, technical problems associated with techniques of radio navigation were not fully resolved until the 1930s.

NATURE OF CROSS-COUNTRY FLYING

Flying is basically a series of flight maneuvers that involve take off, climbing, straight and level flying along a predetermined course, descending, and landing. Each of these tasks influenced the development of special types of navigational aids and charts. For flight departures and arrivals, where a pilot's knowledge of his location with relation to the terrain is critical, cartographers created large-scale local area or approach charts (1:250,000) and airport diagrams. For en route flying between two points, where precise estimates of position are not as important, medium-scale route and sectional pilotage charts (1:500,000 and 1:1,000,000) evolved, often accompanied by "chart extenders" or supplemental aids. And for preflight planning, a variety of small-scale pilotage planning charts (1:3,000,000–1:5,000,000) emerged for laying out courses and determining the charts required for en route aerial navigation, approach, and landings. Similar sets of charts evolved for instrument navigation.[11]

AIRCRAFT DESIGN

The structural design of an airplane and its aerodynamic capabilities influence map construction in a variety of ways. During the open cockpit era, which lasted until the late 1920s, aviation maps were formatted as narrow strips to facilitate handling in small spaces and to reduce the risk of being lost in flight (figure 6.1). Vivid examples of maps blown from aircraft or shredded during unexpected gusts of wind or violent down drafts testify to the difficulties. On a cross-country flight from Washington, D.C., to Miami in 1923, Army pilot Lieutenant Irwin Amberg reported, "as I reached the Coosa River and was glancing at my map a blast of air tore it from my hands, which left me in the most unfavorable of predicaments." Maps were also subjected to excessive stress and deterioration from the elements. Another army pilot lost a composite strip map consisting of several state maps when "the poor glue used in preparing them" failed to hold after he flew through a heavy snowstorm over the Southern Rockies. "[F]rom there on I had to rely entirely on my compass," he acknowledged.[12] Such experiences led to the development of special chart holders and map waterproofing schemes.

FIGURE 6.1 Major Reuben H. Fleet following first airmail flight from Bustleton Field, Philadelphia to Washington, D.C., May 15, 1918. During the open cockpit era aviators used different techniques to ensure that their maps were not lost during flight. In this instance, the pilot folded his small-scale map of the United States into a narrow strip along the line of flight, which he attached to his leg with string for easy reference. National Archives and Records Administration, Special Media Archives, RG 342 - FH, Box 1018.

As speed, range, altitude, and maneuverability of aircraft increased, so did the variety of maps and charts used to guide them. A World War I Curtiss JN-4 Jenny biplane, with a cruising speed of 60 miles per hour and a 200 mile range, could fly across an area represented by a typical 1:190,000 to 1:200,000 aviation strip chart (about 20 by 100 miles) in a span of one and a half to two hours—about the average duration of a flight. By the mid-1920s, the increased speed and extended range of aircraft such as the DH-4B De Havilland, which could fly nearly 400 miles at a cruising speed of 105 miles per hour, reduced this time almost by half. When cruising speed and aircraft range more than doubled again in the early 1930s, demand grew for smaller scale en route charts (1:500,000 and 1:1,000,000) to reduce the frequent change of maps required by longer trips.[13]

Airplane capabilities also influence chart content. It was essential, for instance, that early aviation charts include symbols for sites that provided potential landing fields in case of flight emergencies. During an era when forced landings were routine—due to unreliable aircraft engines, low grade aviation fuel, and propellers that often splintered in rainstorms—cross-country flights were planned to ensure a series of emergency landing fields within gliding distance from any point along the route.[14] Pioneer pilots particularly favored automobile and horse racing tracks as emergency landing sites because they were easily located from the air, generally maintained in good order, and situated near urban areas.

"THE AIRMAN'S WORLD"

Flying high above the surface of the earth provides a unique vantage point not available to other map users. Army Captain Harold Hartney described this world in 1920 as an ocean of air that remained to be "explored, surveyed and charted just like the sea so that inexperienced navigators may with ease make their way without loss of time across new territory in peace or war." Brig. General Neil Van Sickle, one of Hartney's successors and the author of a classic work on airmanship, observed that "only in the airman's world can one view horizons encompassing land and sea, mountain and prairie, storm and calm, and even peace and war, at a single glance."[15] This perspective, the ability to actually observe a large part of the earth's surface as one inches along a route map at 5,000 feet and 100 miles an hour, bonds aviator and mapmaker. To airmen, chart symbols representing the visible landscape and invisible radio beams are not mere decorative features, cultural artifacts, or social constructs with political agendas, but lifesaving landmarks, checkpoints and directional finders. The basic cartographic skills of selection and generalization are life and death issues for airmen, and for this reason they have participated from the beginning of aviation in the design, compilation, and evaluation phases of chart construction.

AIRWAYS INFRASTRUCTURE

Airways are aviation's "invisible highways," pathways through the air connecting one airport to another. The earliest airways were little more than air routes that followed major roads, railroads, or other prominent landmarks such as the Amman, Jordan–Baghdad furrow, a 390-mile ditch created in 1922 by the Royal Air Force with a Fordson tractor and plow to guide airmail pilots flying across the Syrian desert on the Egypt to Baghdad route.[16] In the United States, the Post Office Department, Army Air Service, and Navy worked with state and municipal governments, private industry, and civic groups, to initiate a cooperative effort after World War I to establish a national airways system analogous to the nation's railroad and highway systems. The airway system was a permanent structure laid out on the ground to support and guide the flow of air travel. This infrastructure included landing fields, support facilities, and a variety of visual navigational aids that outlined the airway on the ground.

The First Aviation Charts

Despite Orville Wright's epoch flight at Kitty Hawk in 1903, and similar efforts by French aviators shortly thereafter, the airplane was not widely publicized until the Wright brothers provided public flying demonstrations in both the United States and Europe in 1908. These early flights were limited in range, however, and did not require the use of special aviation charts. Preflight planning involved little more than driving one's automobiles along a route beforehand, noting landscape features that would be apparent from the air. This situation changed around 1910 when cross-country races began replacing short exhibition flights. The first transcontinental flight was undertaken a year later by Calbraith Perry Rodgers, who crossed the country in forty-nine days while enduring fifteen major accidents. Flying without a compass, he followed the railroad west from New York City, at an average speed of fifty-five miles per hour. Shortly after Rodgers completed his historic flight, he was killed during an aerial show at Long Beach, California, a not uncommon occurrence. About this same time the Glenn Curtiss Aviation School sponsored an exhibition team that traveled to air shows throughout the United States and Canada.[17]

Longer distance flights increased the need for improved navigational aids, including aviation charts. From the beginning, pilots recognized the charts' unique nature and value. In 1910 George B. Harrison proposed a standardized aerial map designed for both aviators and balloonists "for the reason that aviation is adding another dimension to our knowledge of travel." Writing in the second issue of *The Air-Scout,* a monthly publication of the United States Aeronautical Reserve, Harrison advocated the adoption of existing symbols and standard conventions developed by professional government mapmakers. These included compass roses and "other symbols of use relating to air currents and aerial conditions." Harrison suggested copying these symbols from U.S. Navy Hydrographic Office charts. Similarly, he recommended that the symbols for meteorological data be taken from U.S. Weather Bureau charts, those of power lines and other obstructions from military maps, and topographical contours from U.S. Geological Survey quadrangle maps.[18]

In the same year that Harrison's article appeared in print, Charles C. Turner introduced the cartographic concepts of Prussian balloonist Hermann Moedebeck to American aviators. Moedebeck was the first airman to promote the idea of special maps for aerial navigation, when he prepared a series of aerial navigation maps for Count Ferdinand von Zeppelin's airship lines in 1909. Moedebeck advanced aviation cartography through publications and public forums in continental Europe, and inspired aeronautical clubs to establish map committees. His ideas reached English-speaking audiences through the publication of Turner's *Aerial Navigation of To-Day: A Popular Account of the Evolution of Aeronautics,* which included a chapter devoted to "Special Charts and Landmarks."[19] In an effort to coordinate the work of these new map committees, an International Commission on Aeronautical Maps was held in Brussels in 1911, followed by subsequent meetings in 1912 and 1913. Austria, Belgium, England, France, and Germany were represented. Delegates recommended the creation of an International Aeronautical Map at a scale of 1:200,000, but no agreement was reached with respect to conventional symbols except for high-tension electric overhead wires, which were to be indicated appropriately by "strings of red crosses."[20]

France was the first country to produce a prototype air navigation map that embodied the resolutions adopted by the International Commission on Aeronautical Maps. By 1911, France was the world leader in aviation, claiming some three hundred and fifty licensed pilots, almost twice as many as the rest of the world combined.[21] A prototype aviation map, sponsored by the *L'Aéro—Club de France,* was evaluated during military maneuvers by the French government's Permanent Committee for Aerial Navigation of the Public Works Department. It covered an area one degree of latitude by one degree of longitude at a scale of 1:200,000. Aids to navigation were depicted by special symbols, including enlarged perspective views of prominent landmarks such as the Cathedral of Reims. Primary roads and large towns were highlighted in bright red, woodlands in green. A British commentator noted on reviewing the map that its "general effect . . . on the eye is irritating, owing to the violently contrasted colours used. This, combined with the number of fine lines appearing, is likely to lead to considerable eye-strain if continued study of the map is required." An experimental aero map prepared by the geographical section of the British General Staff "is much superior from this point of view," this reviewer concluded. Despite such criticism, the French model—promoted by Charles Lallemand, president of the French Association for the Advancement of Science—received widespread circulation in the United States when the Smithsonian Institution reprinted Lallemand's article in English.[22] In the United States, the development of aviation charts prior to entry into World War I was limited to the work of a few individuals associated with the Aero Club of America and the Army Signal Corps.

AERO CLUB OF AMERICA

The Aero Club of America, modeled after similar organizations in Germany and England, was formed in New York City in 1905 by members of the Automobile Club of America to advance the science of aeronautics and aerial navigation through conferences, expositions, congresses, and flying contests.[23] The first map published by the Aero Club documented the routes of contestants participating in the 1907 International Gordon Bennett Aeronautic Airship Race, held in St. Louis. Titled "Courses of Balloons in International Race 1907 Showing Positions and Speed for Each Hour," the map depicts the paths of nine air ships carried by air currents as far as New Jersey and Maryland at the average speed of twenty miles per hour (indicated by dots spaced one hour apart). The map was compiled by Williams Welch under the direction of Captain Charles de Forest Chandler, copilot of the airship, *America.* Chandler was the first chief of the army's Aeronautical Division, then part of the Signal Corps. The Signal Corps was the army's primary service corps for conducting scientific work.[24]

Influenced by the work in France and Germany, the Aero Club of America established a short-lived Committee on Aeronautical Maps in 1911 that included Augustus Post, secretary of the Aero Club and a member of the Curtiss Flying Exhibition Team. Post's elaborate and visionary plans for the committee were featured in two *New York Times* articles that appeared in the spring of 1911. They describe the Aero Club's effort to establish and plot a national network of "transcontinental 'highways' through the air." Post also discussed the club's plan to prepare a series of detailed sectional strip maps to guide aviators and balloonists along these routes, but he noted with some concern "that canoeists and 'hoboes' might gobble up the first edition." These maps, according to Post, "will be printed on long strips of parchment" and housed in chart rollers. "As the airman passes over each part of the country he will thus be enabled to identify it without moving in his seat, merely by unrolling the map." Each map was to cover an area about ten miles wide to allow for aviators "being blown off" course.[25]

The Aero Club also sponsored the production of a plaster of paris raised relief map of the western portion of Long Island "designed especially to meet the requirements of aviators" (figure 6.2). Generally considered the first aviation map produced in the United States, this three-dimensional model depicted "the favorable landing places, aerodromes, railroads, roads, high tension wire lines, and other features which are of value to the man in the air." It was produced by the aeronautical department of the *Automobile Blue Book,* which employed a new process that permitted "the rapid reproduction of any section of country in exact miniature." The sponsors

FIGURE 6.2 Aero Club of America, "Aeronautic Map of Western Long Island," 1911. The first map designed specifically for air navigation in the United States was issued in the form of a photograph of a molded plaster of paris raised relief model of the western half of Long Island. Library of Congress, Geography and Map Division, G3802.L6P6 1911. A Vault.

believed that "the aviator of the future will with the aid of a photograph of one of these maps be enabled to find his way in any part of the country in which he desires to fly without the slightest trouble, being able to know at every moment just where he is and where the nearest favorable landing place is located."[26] Although limited to a small region, the Long Island map, according to one contemporary critic, was a "very creditable work for a time when a cross-country aëroplane flight of 25 or 50 miles was still considered an extraordinary event." The map reached a wide audience of flyers when it was reproduced in the November issue of the aviation journal, *Aircraft*.[27]

The Aero Club of America established a second map committee in May 1914 under the chairmanship of Rear Admiral Robert E. Peary, the noted arctic explorer whose early training included work as a mapmaker with the U.S. Coast and Geodetic Survey. Known as the Aeronautical Map and Landing Places Committee, it was largely the creation of Henry Woodhouse, founding editor of two popular magazines devoted to air travel during the second decade: *Flying*, the organ of the Aero Club of America, and *Aerial Age Weekly*. Woodhouse (born Enrico Casalegno in Turin, Italy) worked tirelessly from 1913 to 1920 persuading leading military and political leaders to advance aviation and aerial photography in the United States.[28] His earliest contribution to aviation cartography was a translation of Giovanni Roncagli's report—originally presented in 1913 at the Tenth International

Geographical Congress in Rome—on the need for an international aeronautical map, which appeared in two installments in the 1913 issue of *Flying* magazine.[29]

More ambitious than its predecessor, the Aeronautical Map and Landing Places Committee laid out an extensive and comprehensive program for establishing and mapping a national airway system. This included the promotion of cross-country flying by chart and compass, development of a network of landing fields, the preparation of "an aeronautical map of the United States," and standardization of aeronautical signs and symbols.[30] The committee represented a veritable who's-who of the civilian and military aviation communities, among them Glenn Curtiss, Orville Wright, Cornelius Vanderbilt, and General George P. Scriven, chief of the U.S. Army Signal Corps.[31] Though America's entry into World War I curtailed the committee's goals, it nonetheless provided a pivotal forum for promoting the development of aviation cartography in the United States.

SPERRY AVIATION CHARTS, 1916–1917

Concurrent with these efforts was the work of committee member Lawrence Sperry. A test pilot and aeronautical engineer, Sperry devised and supervised preparation of the first series of prototype air navigation strip charts in the United States. Sperry worked at his father's firm, the Sperry Gyroscope Company (later the Sperry-Rand Corporation), in Brooklyn, New York. From 1914 to 1917, he directed the company's Aeroplane Department, which pioneered the development of several instruments that revolutionized flying, including the automatic gyrostabilizer (a forerunner of the autopilot), the synchronized drift set, and ground speed indicator. Later, Sperry designed the first ground controlled, radio operated aerial torpedo. As a navy pilot during World War I, Sperry made the first night flight in the United States and the first airplane-dirigible aerial hookup.[32]

In response to increasing demands for information concerning flight routes and compass courses, the Sperry's Aeroplane Department produced a series of aeronautical strip charts from late 1916 to early 1917, inspired perhaps by Post's effort five years earlier. Omar B. Whitaker, a Sperry engineer, prepared the charts from U.S. Geological Survey maps at scales ranging from approximately three to eight miles to the inch and printed on map strips ten inches wide. At least three aeronautical maps were published and marketed: (1) "Sperry Aviation Chart [of Long Island . . . Nov. 1916]," printed in color at a scale of one inch to two and two-thirds miles; (2) "Sperry Aviation Chart New York to Newport News, [Virginia] . . . Nov. 23 [19]16," at a scale of one inch to two and two-thirds miles; and (3) an untitled map of the air route between New York and Chicago, at a scale of one inch to eight miles (figure 6.3).[33] While the aeronautical map of Long Island depicted only the location of flying fields, the two route maps were designed specifically for aerial navigation and may be considered the first aeronautical charts published in the United States. In addition to showing the location of wayfinding landmarks such as cities and towns, roads, railroads, and rivers, each chart also included compass roses and compass bearings along the most desirable flight courses. The compass roses, however, were not corrected for magnetic declination, a major failing for an aeronautical chart.

The concept of air navigation strip charts limited to single airway routes originated in Europe. Woodhouse credits his native Italy with this invention but notes that Sperry developed his maps independently of the Italian model. British prototypes could also have influenced Sperry. Lawrence demonstrated the Sperry gyrostablizer to European pilots during a business trip to London and Paris in 1914 and 1915, and it seems quite likely that he saw such maps in England and France. The English engineer Eric Hollocomb Clift prepared a series called Aerodrome to Aerodrome Maps in 1911, and Woodhouse himself published a picture of an airway strip map in 1914.[34]

Sperry aviation charts were formatted for use with the Sperry Map Holder, a mechanical device designed to attach to a cockpit instrument panel. It consisted of two rollers for mounting a strip map and a set of turning knobs for scrolling or advancing the map in flight. "The distinctive feature of these charts," noted Whitaker, "is that, regardless of the number of turns made in any particular flight, the whole chart is presented in a straight strip only ten inches wide. The advantage of such a method is that almost any scale may be used without making the chart cumbersome, since it lends itself readily to a pair of chart rollers, which can be mounted in a case and attached to the instrument cowl, if so desired."[35] The idea for the movable map holder appears also to have been borrowed from Europe, where it was in use as early as 1911.[36]

Lawrence Sperry's aeronautical charts and chart holder were marketed extensively through trade journals, exhibitions, and public forums. They were advertised monthly in Woodhouse's *Aerial Age Weekly* and *Flying* magazines from February through December 1917. "Every aviator appreciates the value of an accurate map when making cross-country flights and the Sperry Aeronautic Maps, and Sperry Map Holder, supply this necessity," one ad proclaimed. The Aeronautic Library of New York City, a Woodhouse marketing entity, sold the maps and map holders, and claimed to have "secured the exclusive agency from the Sperry Gyroscope Company

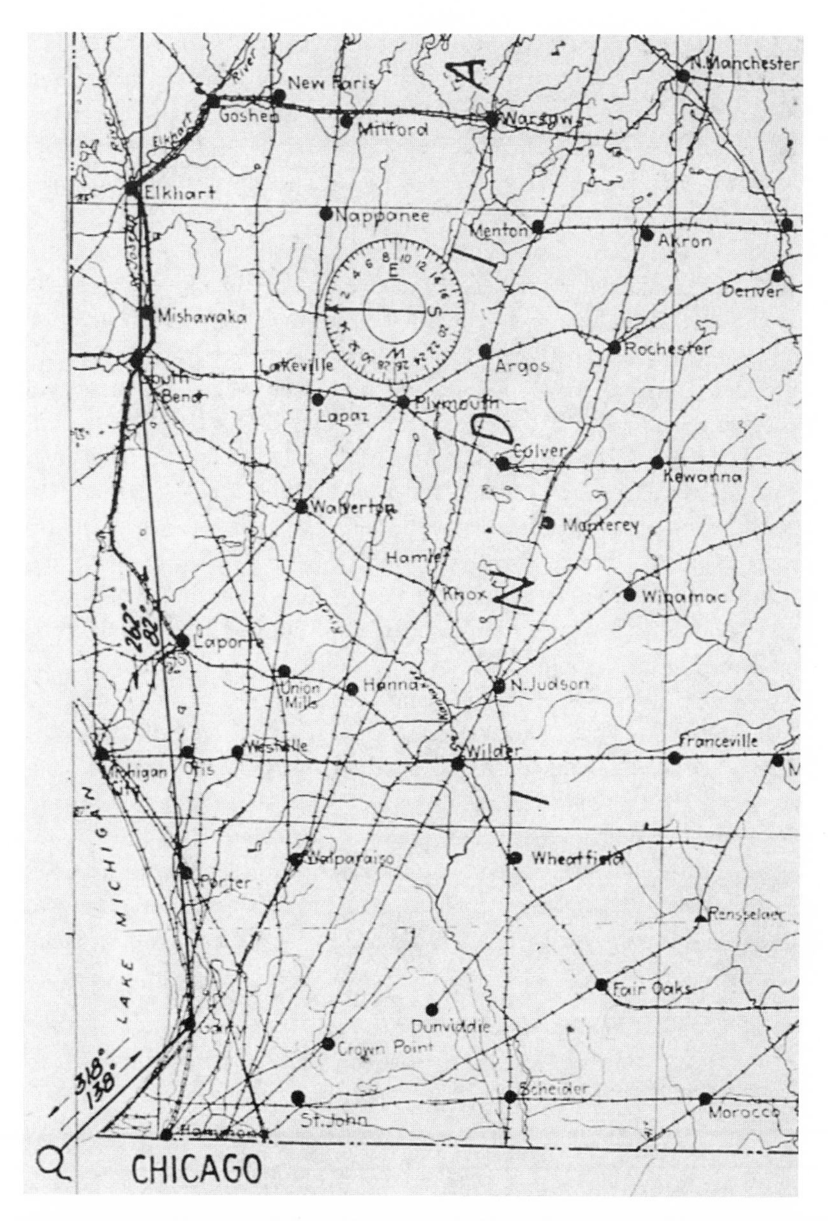

FIGURE 6.3 Detail from untitled air navigation strip map extending from Chicago to New York City prepared by the Sperry Gyroscope Company, 1917. The first aviation maps in the United States were designed and produced commercially by the Sperry Gyroscope Company beginning in 1916 but production ceased with America's entry into World War I. Reproduced from *Flying* (January 1917), 504.

for the sale of the Sperry Aeronautic Maps."[37] In 1917, Sperry exhibited his charts and chart holder at the First Pan-American Aeronautic Exposition, held in New York City's Grand-Central Palace from February 8 to 15. He also presented a paper at the Exposition's professional sessions, Aeronautic Maps for Aviators, one of the first such discourses on aviation charts at a public forum.[38]

Sperry's last cartographic contribution was indirect but no less significant. His aviation charts served as a model for an aeronautical map of the Woodrow Wilson Airway, an outgrowth of the transcontinental route first sponsored by the Aero Club in 1911.[39] Named in honor of the twenty-eighth president, this project was inspired by the Lincoln Highway, a proposed coast-to-coast hard surface road from New York City to San Francisco financed by public subscriptions.[40] Robert A. Bartlett completed the map in 1918 and issued it in book format under the title *The Aero Blue Book and Directory of Aeronautic Organizations.*[41] It consists of seventeen-page-size strip maps depicting similar information found on Sperry's aviation charts and, below the map, a profile or cross section of the region traversed. Bartlett was Admiral Peary's navigator during his expedition to the North Pole and a member of his Aeronautical Map and Landing Places Committee.

Sperry's innovative series of aviation charts did not survive America's entry into World War I. Shortly after war was declared on April 6, 1917, the Sperry Gyroscope Company closed its aviation department for lack of commercial markets. Following navy service, Sperry formed a new company, the Lawrence Sperry Aircraft Corporation, which focused on building airplanes rather than navigation instruments. His untimely death at the age of thirty-one in 1923 during an ill-fated attempt to fly an airplane of his own design across the English Channel ended the Sperry Company's association with aviation maps.

UNITED STATES ARMY PROGRESSIVE MILITARY MAPS

While America's entry into World War I stopped production of commercial air navigation maps in the United States, the war dramatically spurred the development of aviation charts for military purposes. By 1917, Europe could claim more than 7,000 licensed pilots (compared with 300 in the United States) and was producing a variety of aviation charts in large numbers. The primary aviation charts used on the Western Front were two French maps, "Carte Aéronautique de la France du Service Géographique de l'Armée" and "Carte Aéronautique de l'Aero Club de France," each published at a scale of 1:200,000. Italy issued innovative strip charts of airway routes, printed at a scale of 1:250,000. And the

British Hydrographic Office made available several series of aviation charts for coast patrols and antisubmarine warfare.[42] Meanwhile, production of air navigation charts for military purposes in the United States in April of 1917 was "next to nothing," according to a staff officer in the Army Air Service.[43]

Military aviation in the United States began in 1907 with the creation of the Signal Corps' Aeronautical Division, but little progress was made until seven years later when the division was expanded and renamed the aviation section.[44] Two of the army's most brilliant and innovative officers, Lieutenant Colonel George O. Squier and his aide Major William (Billy) Mitchell, led the reorganized aviation section. A highly respected scientist with a Ph.D. in physics, Squier would go on to make major contributions in the fields of radio and cable communications and aerial photography. Mitchell had learned to fly in 1912, and following tours in the Philippines and Alaska establishing telegraph networks for the Signal Corps, he was named chief of the Army Air Service in France during World War I. Both Squier and Mitchell were to become strong advocates of using airplanes and airships for attack on front line troops and targets behind enemy lines, goals requiring new types of maps for long distance flights. This view clashed with the American military doctrine of the day, which considered aircraft as defensive rather than offensive weapons, and would ultimately lead to Mitchell's famous court martial and removal from the army.[45]

As early as 1913, after examining several British aeronautical maps that had been sent to the Signal Corps' aviation school in San Diego, California, army pilots stationed there initiated a program to incorporate aviation data on existing military maps. In a memorandum directed to the Corps of Engineers, the army organization responsible for military mapping, Second Lieutenant Joseph E. Carberry requested that "aeroplane landing fields" be displayed on the army's Progressive Military Map. "I have the honor to recommend," he wrote "that a suitable conventional sign, easily distinguished from any other, be adopted to designate landing places suitable for aeroplanes, that a clear description of what constitutes such landing places be furnished officers on mapping detail and that those officers be instructed to mark (say within every 5-mile square of their territory) two or three of the best landing places. Such a system, checked when possible by officers on aviation duty would be of inestimable value to military pilots in planning and executing cross-country flights."[46]

Captain Arthur S. Cowan, the school's commanding officer, endorsed Carberry's recommendation. Cowan further suggested that two additional symbols be employed, one to indicate "a good landing place" and one for "a possible landing place." Referring to the British maps recently received, Cowan noted that the "adoption of such a sign has already been made in several foreign countries, and has proved of great value to aviators in selecting routes to fly over."[47]

The Progressive Military Map was the army's basic large-scale tactical planning map from about the 1880s to the establishment of the Army Map Service in 1942. Designed to provide topographic coverage of the entire United States and its possessions, work on the map proceeded according to those areas of greatest strategic importance. The base sheets comprising this series were similar to United States Geological Survey topographic quadrangles, but they included additional information of military value (figure 6.4).[48] In response to the requests by Carberry and Cowan, and in

FIGURE 6.4 Detail from U.S. Army, Corps of Engineers, "Progressive Military Map, Farmington, Massachusetts, Quadrangle," Sept. 1918. The army's earliest attempt to aid pilots was to indicate the location of emergency landing fields on its tactical map series. Landing fields were identified by a symbol drawn in the shape of a propeller. The number of blades on the propeller indicated the quality and characteristics of the landing field. Library of Congress, Geography and Map Division.

concurrence with the Signal Corps' aviation section, the Chief of Engineers issued formal instructions for incorporating this data. Later codified as "Aviation Field Notes," these specifications were designed to guide army topographers in the selection and placement of emergency landing sites onto the Progressive Military Map. Accordingly, emergency landing fields were to be depicted when possible near all towns with more than 10,000 inhabitants and within each five square miles of territory. The symbol selected for depicting the location and qualitative characteristics of flying fields was, appropriately, a propeller: four-blades designated a good landing field (defined as "one not less than 500 yds by 800 yards, open country from all sides, level and clear of all obstructions") and two-blades denoted a poor landing site ("rough and rolling with obstacles such as houses, fences, small ravines, etc.").[49]

Aviation Maps during World War I

About the same time that the Sperry Gyroscope Company began issuing its series of Sperry aeronautical maps, Mitchell, then Acting Officer in Charge of the Signal Corps' aviation section, initiated the military's aviation mapping program. On September 6, 1916, he requested that steps be taken by the adjutant general of the army for preparing a set of "essentially aerial navigation maps."[50] These maps were to cover the coastal region of the United States from North Carolina to Boston. Mitchell had already begun to form the ideas that established his reputation on national preparedness, but his views on the role of aircraft as offensive weapons were still evolving. With the war in Europe heating up, he believed that the proper role for aircraft in homeland defense was to provide aerial reconnaissance for harbor and mobile coastal defenses and for attacking enemy submarines and minelayers.[51]

With respect to form and content, Mitchell requested that the maps be "printed on sheets of size convenient to carry in aeroplanes, about ten inches square, each section being made to the scale of ten (10) inches equal to thirty (30) miles." Furthermore, he noted, "on them should be indicated permanent land marks, cities, railroads, principal automobile roads, munitions supply bases and factories; landing places, being indicated on the map as good, bad or indifferent, and the general nature of the terrain indicated by simple conventions, etc." In response, the adjutant general directed the chief of engineers "to take up at the earliest practicable moment the preparation of aerial navigation maps of the United States and its insular possessions."[52] Separate funding was requested for this project in the chief's 1918 budget.[53]

While Mitchell initiated the first significant military effort to prepare maps specifically designed for aviators, his initial request for a flying map of the East Coast was temporarily delayed as the army's aviation mapping program was redirected and expanded following a request from the French premier for an American "flying corps" of 4,500 planes and 5,000 pilots. To meet this challenge, the Army expanded the number of its flight training schools from three to twenty-seven, the majority of which were built in southern states where flying conditions were good throughout the year.[54] At the same time, the aviation section of the Signal Corps requested that the Corps of Engineers produce a series of special purpose aviation maps. "This is urgent," noted the chief signal officer, "in order that pilots may learn to read the same type of map that will be supplied them in Europe. Experience has demonstrated that a pilot trained to use the ordinary geological survey map is extremely liable to find himself in trouble when he tries to use the aeronautical maps now in use in Europe. Entire squadrons have landed behind the German lines, due to misinterpretation of their maps that lead them to believe that they were behind the French lines. The importance of this question is therefore apparent."[55]

On November 15, 1917, the chief signal officer provided the Corps of Engineers with general guidelines for the preparation of a series of aviation strip maps that would provide an integrated network for cross-country training flights.[56] Most significantly, the chief signal officer indicated that the Air Service would provide aerial photographs to aid Army topographers in delineating the woodlands. "It is thought that the Engineer Corps will be able to use these photos to obtain the necessary data to fill in the geological survey maps of the sections of country in the vicinity of the flying schools."[57] The Corps of Engineers responded with an innovative multiphase mapping program that produced the army's first printed specifications and symbol sheets for aerial navigation maps, the first published aerial navigation maps, and the earliest use of aerial photography in the United States for map production.

The Corps of Engineers codified specifications for aerial navigation maps in *War Department Bulletin 64*, issued on December 19, 1918. These specifications defined for the first time the nature of aerial navigation maps, stating that their purpose was to furnish "aviators a route map to be used in flying from one point to another." Accordingly, aerial navigation maps were to be formatted as narrow strip maps drawn at a scale 1:200,000, with each sheet covering an area twenty miles in width. And while its length was not specified, it was expected "to cover the zone between two important terminals, but not too long to be used in the standard map case carried by aviators." These maps were to be printed in five colors, including the four standard colors generally associated with military maps: green (woods), blue (water), black (railroads, prominent buildings, monuments, and all other artificial features), and brown (prominent ridges, hills, and mountains). The color red, normally used to depict confidential information, was chosen for highlighting roads, cities, and towns. The most controversial specification pertained to topography; it ruled out the use of contours.[58]

A separate directive for conventional map signs soon followed. It provided examples of symbols and signs for forty-five different physical and cultural features of interest to aviators, including railroads (four categories), gasoline stations, motor repair shops, water tanks and steel towers, telegraph and telephone lines, and racetracks. Most significantly, it included for the first time a standard symbol for depicting aerial routes: a red line connecting two red circles (representing checkpoints) with compass bearings and distances marked in black numbers and lettering. To eliminate misinterpretation, the symbol for roads changed from a red line to a white line outlined in black.[59]

The production of this series involved a complex process shared by several federal agencies and army field units. The initial base maps, covering strips of land 20 miles wide and approximately 125 miles long, were compiled by topographers from the Corps of Engineers and the Geological Survey's Division of Military Survey (which had been mobilized for the war effort). The finished compilations were then reproduced and lithographed by either the survey's Division of Engraving and Printing or the army's Central Map Reproduction Plant in Washington, D.C. Next, the printed base maps were sent to army topographic field units for review and updating, using ground reconnaissance and aerial photographs. After the Air Service added its air navigational data to the revised base maps, they were ready for publication by either the Geological Survey or the Central Map Reproduction Plant.[60]

Army demobilization in March 1919 ended this program before the base maps were completely field checked or the aviation data added, but at least thirty base maps linking flight training centers were compiled and published.[61] One of the few surviving examples is a map of the San Antonio to Austin, Texas, route. Covering an area twenty miles wide and one hundred miles long, it depicts roads, railroads, watercourses, towns, and cities.

The Corps of Engineer's Southern Department, Fort Sam Houston, Texas, compiled this base map from tracings of existing Geological Survey maps and railroad maps. The army's Central Map Reproduction Plant then lithographed it in black and white at a scale of one inch to three miles (36 × 12 in). Fifty copies of advance sheets, "subject to correction," were furnished the Air Service on January 17, 1918.[62] While the cooperative effort of the various organizations involved was successful from a bureaucratic and administrative perspective, it revealed fundamental differences between map producer and map user that continued to plague the design and production of aviation charts until the Air Service gained full control of the mapping process.

One example of this tension concerned the placement of compass roses, a vital navigational aid to airmen. Air Service specifications called for locating the compass rose with magnetic arrows "within the body of the maps at intervals not to exceed 10 inches." Captain Charles Ruth, the army engineer in charge of the Central Map Reproduction Plant ignored this directive. Instead, he printed two compass roses in the left border along with the map's legend and index. It was "impractical," he informed the chief of engineers in response to Air Service complaints, to place the compass roses within the map, "inasmuch as these magnetic declination indications would obscure some of the topography." Further, Ruth concluded, "where it is possible to follow your instructions without destroying the value of the map they will be complied with."[63] Ruth's inability to recognize that "the value" of an aeronautical chart differed from that of a topographic map reflected an attitude prevalent among army and geological survey mapmakers trained primarily in land surveying or civil engineering. A reserve officer in the Corps of Engineers and a former executive with the *Washington Star* newspaper, Ruth commanded the Central Map Reproduction Plant from shortly after its inception in 1917 to 1920. He is considered the founding father of the Army Map Service.[64]

Among the legacies of this joint civilian-military effort to produce the Army's first aerial navigation map was the application of aerial photography for revising and updating the base maps. This was a critical decision, as the Geological Survey had adequately mapped only 30 percent of the United States by 1917. European aviators first introduced American airmen to the value of aerial photography for mapmaking during the war. The Air Service was further encouraged by the work of the National Advisory Committee for Aeronautics' Airplane Mapping Committee, established March 10, 1917 and chaired by Captain James Bagley.[65] A topographic engineer with the Geological Survey, Bagley was commissioned an officer in the Corps of Engineers along with most of the survey's topographers. From previous experience using panoramic photography for map work in Alaska, Bagley believed that similar results could be obtained with photographs taken from airplanes.

As early as November 1917, Bagley suggested using aerial photographs for the Air Service's aviation map program. Bagley tested his theories during the winter of 1917–18 when he took some of the earliest aerial photographs in the United States during field tests over the nation's capital, using a tri-lens aerial camera of his design for which he became well known. This new type of camera had three lenses—one in the center for taking vertical pictures, and two flanking lenses for overlapping oblique pictures. With the aid of a transformer, the oblique photographs were also converted to vertical images. This resulted in one large vertical photograph, which represented coverage of an area nearly six miles square when taken at an altitude of 10,000 feet.[66]

Aerial photographic surveys were undertaken for eight of the thirty flight training routes, but only three were completed before demobilization. Captain Calvin E. Giffin of the Corps of Engineers conducted these surveys. Another former Geological Survey topographer, Giffin had also worked extensively in Alaska. Assigned to Kelly Field outside of San Antonio, Texas, Giffin used Bagley's tri-lens camera to photograph several routes (San Antonio–Austin, San Antonio–El Paso, San Antonio–Waco) from November 4, 1918 to January 11, 1919.

The routes were initially photographed from an altitude of 7,000 feet, a height that took a photo plane weighted down with camera and film nearly two hours to reach, leaving little time for actual photography. The altitude was later reduced to allow more time for filming. For ground control, Giffin used existing Geological Survey topographic maps when available, or ran compass traverse lines using a motorcycle with sidecar. While none of the nearly 3,500 aerial photographs prepared under Giffin's direction appear to have survived, this project proved invaluable to the development of the aeronautical chart. Giffin was one of the first professional cartographers to work closely with aviators and to actually fly as an observer while he trained Air Service personnel to take aerial photographs.[67] The insights gained during this experience were reflected in several experimental aeronautical charts that Giffin later prepared for the Air Service.

AERIAL NAVIGATION MAP, WASHINGTON–NEW YORK–BOSTON ROUTE

In response to Mitchell's initial request for an aerial navigation map of the East Coast, the Corps of Engineers produced a chart of the route from Washington to Boston. Officers from the corps' regional offices in New York City and Boston conducted the work. Basic data such as road networks and drainage patterns were obtained from Geological Survey quadrangles while significant landmarks, woodlands, and landing sites were plotted from available Progressive Military Maps sheets, field reconnaissance, and aerial photographic surveys.[68] The final base map was printed at the scale of one inch to three miles, and issued sometime between June 1918 and March 6, 1919, when 150 advance copies were forwarded to the Air Service.[69]

One of the map sheets comprising this aerial route map was field tested under actual flight conditions by Captain Charles Ruth, commanding officer of the Army's Central Map Reproduction Plant. Ruth's final report represents one of the earliest examples of flight editing. Flying as an observer in a hydroplane at nearly 3,000 feet, Ruth discovered that "the main roads show much more prominently from the air than they are shown on the map and [that] the secondary roads are practically lost from view." He further noted that "from the air [railroads] do not show near so prominently as the main highways, approximating in color the ground itself." Based on these observations, which generally contradicted the experience of most veteran pilots, Ruth recommended that the prototype be changed to "make the map more in conformity with the view one gets from the air." He suggested that the line weight of the main roads "be slightly increased" and the weight of the secondary roads and railroad lines "be slightly reduced." He concluded that since "it was practically impossible" from the altitude at which he flew to distinguish small features such as water tanks, churches, and windmills, such landmarks should be omitted altogether from aviation maps covering highly developed urban areas, except in the West, "where land marks are few and the map cannot be cluttered with detail."[70]

LOCAL AREA AVIATION MAPS

A variety of local area and planning maps designed primarily for flight training were produced at army flying fields. These differed considerably in content and form but were generally issued as inexpensive blueprints or blueline photostats. They were derived from many sources. When Love Field opened in Dallas, Texas, in late 1917, for instance, the only available maps for cross-country flying were topographic maps based on an 1893 survey.[71] In response to a request from the 2nd Aero Squadron, the Fort Sam Houston Engineering Department produced an aviation map of the San Antonio region from a Post Office Department map with prominent topographic features depicted in brown hachures and roads by single red lines.[72]

The Flying Department at Kelly Field, located southwest of San Antonio, Texas, issued a number of blueprint maps in late 1918 depicting airplane routes, railroads, landing fields, and towns in the vicinity of Ellington, Carruthers, Kelly, and Taylor Fields in Texas and Alabama. Private Joseph Palle drew these maps at various scales under the direction of Second Lieutenant H. C. McGregor.[73] Designed for visual orientation from the air, they depicted only basic information. Standard symbols signify railroads, "shell roads," dirt roads, drainage systems, and towns and cities, and pictorial symbols show oil well derricks. Bearings, courses, and distances are indicated by dashed lines or are overprinted in red.

Student pilots also prepared aviation maps. The commanding officer of Ellington Field, Texas, sent the director of Military Aeronautics a small collection of maps "made from the air" by students during cross-country flights. "These maps," he concluded, "have been checked up and have been found as accurate as ordinary road maps made with a plane table and alidade."[74]

Standard Published Maps Used for Aerial Navigation

During the rapid demobilization of the armed forces following the Armistice, high-level official support for the development of aviation maps diminished. The Air Service inaugurated a cooperative plan with cities and towns to create a national system of municipal landing fields, and sponsored a series of cross-country flights and pathfinding expeditions to promote and advance aviation. At the same time, the Post Office Department inaugurated airmail service between Washington, D.C., and New York. It was quickly extended from New York to Cleveland, then to Chicago in 1919, and to San Francisco the following year, creating in the process the first scheduled airplane service in the United States.[75] By 1925, the number of Air Service, navy, postal, municipal and commercial landing fields had grown nearly tenfold to 643, with an additional 2,817 temporary fields available for emergency landings.[76]

The lack of adequate flying maps throughout the early 1920s remained a major problem. As pioneer airmail pilot Ken McGregor recounted, "Map reading was not required. There were no maps. I got from place to place [by] the seat of my pants [and] the ability to recognize every town, river, railroad, farm, and, yes, outhouse along the route."[77] While a few pilots like McGregor relied strictly on visual navigation, the majority resorted to using some form of published map. These ranged from general purpose and transportation maps to thematic maps, including in one instance a census map.[78] But the majority of published maps used for flying were issued by the major commercial and government mapmakers: Rand McNally, the Post Office Department, and the Geological Survey.

The maps most frequently used by army, navy, and airmail pilots during the leather helmet and goggle era were those produced by Rand McNally, a reflection of that company's dominant position in the map publishing trade. According to veteran airmail pilot Ed Mack, "Navigation in the days of the early air mail was mainly by 'Rand-McNally'—by road and by railroad track."[79] An analysis of map requests submitted to the Air Service by army pilots and aviation centers supports this statement: during a two-year period from 1922 to 1923, 70 percent of the 3,749 maps requested were for Rand McNally maps.[80]

The use of Rand McNally maps was a logical choice. With almost fifty years of experience, Rand McNally was the only contemporary map publisher that provided detailed state maps. The company's Indexed Pocket Map state series included an alphabetical list of potential landmarks such as mountains, rivers, lakes, cities, and towns. Railroad systems were highlighted in black to standout for easy reference. Measuring 21 × 28 inches, these state maps were compact enough when folded to be carried with ease in small cockpits. Rand McNally's popular Auto Trails map series coincided with the army's post–World War I cross-country flying program. First issued as sectional maps in 1917 and then as state maps, this series covered the entire country by 1922.[81] Finally, Rand McNally's were the only maps available on a national basis that depicted terrain features.

Rand McNally maps found support among pilots both for preflight planning and en route navigation. Army Second Lieutenant William McKiernan, Jr., spoke for many aviators in 1919 when he noted, "The maps of the Rand McNally type used by automobilists were found to be very accurate and helpful."[82] Veteran navy pilot Rear Admiral George Van Deurs reminisced that during cross-country flight training in 1923, "We'd have a Rand McNally State map, and we looked at that, picking out railroads and rivers and other landmarks."[83] Reporting on a long distance cross-country flight from Langley Field, Virginia, to Fairfield, Iowa, in 1922, another army pilot noted that "On this trip as on all my other cross-country trips, I drew a straight line on my map from the point of departure to my destination, then flew this course with the aid of my compass and by checking land marks, I consider Rand McNally RR maps the best obtainable at present."[84]

In an effort to make these maps easier to use in the cramped quarters of open cockpits, the Air Service and individual pilots reformatted them as strip maps. Mounted on airplane linen, they were used with government issued aluminum rolling map units that "could be placed in the lap, on the side of the seat or hung on the side of the fuselage."[85] Army Captain

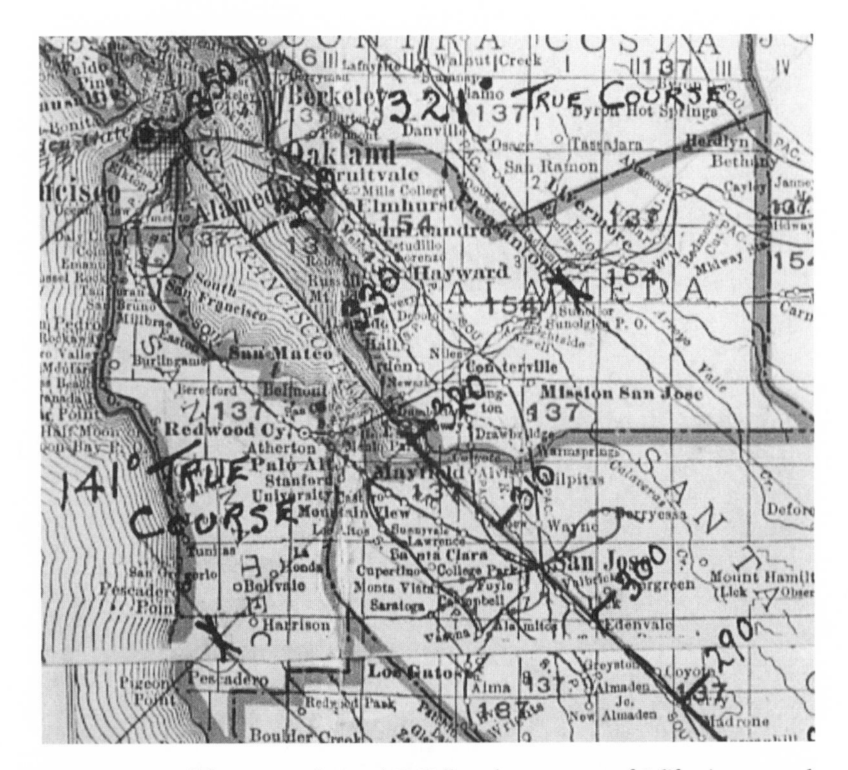

FIGURE 6.5 Detail from composite Rand McNally tack-system map of California annotated with flight course from Los Angeles to San Francisco and return, July 1923. Rand McNally commercial state maps were widely used by aviators for cross-country flying prior to the 1930s because they were inexpensive and readily available. National Archives and Records Administration, Textual Archives, RG 18, E147, box 3, folder 74.

Samuel Skemp related using an eight-inch composite strip map made from Rand McNally "map-tack" state maps of Oklahoma and Texas during a reconnaissance flight in 1922 from Post Field at Fort Sill, Oklahoma, to Dallas. He found it "very satisfactory for the purpose and proved to be very accurate, especially in respect to the shape of the rivers, the lines of railway, and distance of towns from streams"[86] (see also figure 6.5). In another instance, First Lieutenant H. G. Crocker of the Eighth Attack Squadron at Kelly Field, Texas, prepared a strip map measuring 12 × 144 inches from seven different Rand McNally state maps for a twelve-hour flight from the Texas Coast to the Canadian border.[87] On several occasions the Air Service purchased custom produced strip maps directly from Rand McNally, with orders ranging up to 200 maps.[88] As a further aid to pilots, appropriate flight information such as courses, course distances, and compass bearings were added to these reformatted strip maps.

While most pilots found Rand McNally maps helpful, some were critical. Disapproval focused on their small scale, incompatibility of scale from map to map, and excessive detail such as county boundary lines and small towns. Flying as an observer on a cross-country flight of three airplanes in 1918, army pilot Herbert H. Balkam stated that "The map which we used . . . is a very poor map for flying purposes. The map, as it will be noticed, is a Rand McNally map or rather sections from maps of Michigan, Ohio, Indiana and Illinois, all of which are of different scale. . . . Cities and towns all appear alike on these maps and prominent roads and landmarks, such as lakes, etc., are not shown. The map was too small a scale to permit the marking of any fields or roads thereon and it may be said that this map was the only one available upon such short notice as we were given for the flight."[89] To compensate for lack of scale and compatibility across state boundaries, one pilot reproduced a set of blue print strip maps made from Rand McNally maps, all at the same scale. "If [Rand McNally] maps were all made to [the] same scale," he concluded, "I don't believe they could be improved upon."[90]

POST OFFICE "BELT MAPS"

Next in popularity to Rand McNally were the Post Office Department postal route maps, representing nearly twenty percent of the maps requested by army pilots (figure 6.6). Produced by the department's Topographic

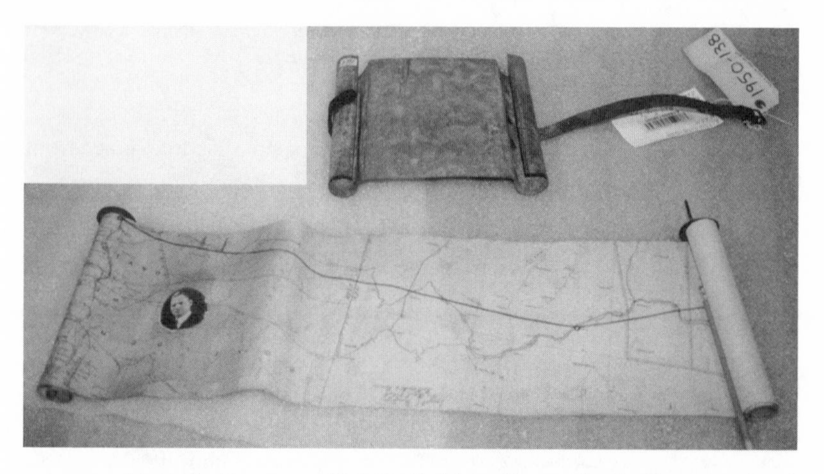

FIGURE 6.6 Post Office Department "Belt Map" with scroll-type galvanized knee-board used by air mail pilot Joseph Mortensen on the Salt Lake to Reno Air Mail Route in 1921. This composite scroll map was made from Post Office Department route maps for Nevada and Utah, cut along the line of flight. The scroll map is eight inches wide and about sixty inches long. Smithsonian Institution, Air and Space Museum, Garber Facility, FMSB24, CAB-49-D.

Division for postmasters and for the Railway Mail Service, postal route maps had several advantages over the Rand McNally maps. They provided the most accessible, up-to-date geographic information for many parts of the country, were readily available to government flyers in large quantities, and were published at scales ideal for visual navigation (1:250,000 and 1:500,000).[91] Commenting on the accuracy of postal maps, one of the Air Service's most experienced pilots reported that on a flight from New York to Omaha he was able to keep his airplane "within a quarter of a mile of our true air line compass course." He further noted that while the postal route maps "do not take in sufficient detail," they were the only available maps that covered all the states through which he passed.[92]

Despite these advantages, postal route maps had several serious drawbacks for aviation: they depicted only the largest towns and only those roads and railroads that carried the mail; showed few prominent landmarks or checkpoints; and their scale varied from state to state. Even more limiting was their total lack of terrain information. Army pilot Harry C. Drayton succinctly summarized their deficiencies in 1919: "In bad weather with low clouds these maps are difficult to follow as they show few land marks, no contours and omit many railway lines."[93]

Another problem was size and format. Designed as large wall maps to aid postal delivery workers in post offices and railroad mail trains, postal route state maps averaged three by four feet and were thus too large for standard cockpits. To overcome this problem the mail service converted its postal route maps to long strip maps, cutting away all unnecessary portions except for the route to be followed. Airmail pilots dubbed them "belt maps." They were designed for use in map roller cases similar to the one sold by Lawrence Sperry from 1916 to 1917. Belt maps generally measured eight inches in width and ranged in length from 40 to 146 inches, often consisting of several separate maps pasted together.[94] The belt map for the aerial mail route from New York to St. Louis via Chicago, for example, consists of six separate state postal maps issued at four different scales. To aid pilots, scale and state boundaries were highlighted with a water-color wash wherever state maps were conjoined.

Major Carl Spaatz, who later served with distinction as commanding general of the U.S. Strategic Air Forces in Europe during World War II, thought that with the addition of the outlines of mountain ranges and peaks, and the inclusion of population figures under the names of towns, the postal route map would have been "an ideal map" for cross-country flying.[95] The Air Service formally "adopted" the postal route map as its first "official" flying map sometime in 1920, an action unpopular with many Army airmen.[96] Though some pilots favored these maps, particularly where mountainous terrain was not a factor, the majority clearly preferred Rand McNally maps.

The comparative usefulness of the two maps as navigation aids was measured under flying conditions during the First Transcontinental Airplane Reliability and Endurance Flight Test between New York and San Francisco, held from October 8–31, 1919.[97] The most famous cross-country flight of that era, it was planned by Brig. General Mitchell as a massive peace time field maneuver to test Air Service staff, equipment, and navigational skills. Sixty-nine aircraft and crews participated, with thirty-three completing one-way transcontinental flights and ten finishing the round trip. Though nine fatalities marred the event, it made important contributions to aviation. Navigation was by "map and compass" over areas that most of the pilots had never flown. Each pilot was provided with postal belt maps that covered the entire route, but twelve crews supplemented their postal route maps with Rand McNally maps.[98] One pilot who used Rand McNally maps to navigate the first leg of his flight from San Francisco to New York and the official postal route maps on his return spoke for a number of the aviators. He reported that while the postal route strip maps were more accurate for the relatively flat region stretching from New York to Cheyenne, Wyoming, the Rand McNally maps "were far superior" in the West "on account of the accuracy with which they showed the mountain peaks and ranges."[99]

Under constant pressure from army aviators, the Air Service finally relented, officially sanctioning Rand McNally maps for use as aerial navigational aids during cross-country flights. On June 29, 1922, the Air Service informed all flying fields that it would furnish each field with ten Rand McNally state maps for each state located within a 500 mile radius of the field, and two sets of state maps for the remaining states. The Air Service nevertheless continued to provide pilots with postal route maps.[100]

U.S. GEOLOGICAL SURVEY MAPS

The Geological Survey produced the third major category of standard published maps used for cross-country flights during this period. These maps represented less than three percent of the maps requested by pilots, but this number is somewhat misleading since the maps were used primarily for planning or reference purposes. The Geological Survey, which traces its origin to the great geographical and geological surveys of the West (1867–79), published two types of maps that held interest for aviators—state base maps (1:500,000 and 1:1,000,000) and topographic quadrangles (1:62,500 and 1:125,000).[101]

The state base maps portrayed principal cities and towns, railroads, county, and Indian reservation boundaries, township lines, and drainage systems, but showed no terrain features or roads. First issued in 1912, the 1:500,000 series provided coverage for virtually the entire United States by 1922.[102] Earlier editions were printed only in black and white but by the early 1920s they were printed in two colors, with drainage shown in blue. One of the great advantages of this map from an aviator's perspective was its uniform scale across state boundaries.[103] Like Rand McNally and postal route maps, the Geological Survey state base maps were converted to strip maps with compass courses and emergency landing fields added as navigation aids.[104]

Topographic quadrangles were designed primarily as base maps for engineering work and geological study. Issued separately as single sheets from their introduction in 1885, they were formatted so they could be conjoined to form a topographic map of a state, region, and ultimately, the entire United States. In addition to depicting relief by contours, these maps also displayed drainage and cultural features. But their large scale and detailed topographic information limited their value as navigational aids, and with less than thirty percent of the country surveyed and mapped by the Geological Survey, the maps were not available for many airways. Pilots used topographic quadrangles primarily for locating and plotting prominent landmarks and for establishing air routes.[105] The chief of Air Service Training and War Plans Division, for example, placed an order with the Geological Survey in 1922 to send separate sets of quadrangle maps to fifteen Air Service flying fields for use in locating prominent landmarks within a 500 mile radius of each base.[106]

Planning Maps for Cross-Country Flights

The immediate postwar period was one of the most creative eras in aviation and aviation cartography. As early as November 1918, Colonel Milton F. Davis of the Office of the Director of Military Aeronautics requested authority to establish five or six "Charting and Mapping Squadrons" to travel throughout the United States "charting the lanes between prominent cities . . . so that any pilot capable of reading an engineer's route book, could fly between any two points in the United States."[107] On June 2, 1919, the Air Mail Service and the Air Service jointly recommended a similar operation focusing on airmail flight routes.[108]

Though these ambitious projects were not implemented, the Air Service's training section pursued the creation of a nationwide system of ground markings or aerial "sign posts" to aid aviators flying by visual navigation. In 1919, Brig. General Mitchell initiated a program to enlist the support of railroad companies to paint the names of towns on railroad station roofs, an effort similar in concept to Rand McNally posting the names of roadways on telephone poles to guide motorists using its road maps. By 1923, the Air Service had formalized the marking of towns with specific instructions for name placement.[109]

At the same time, the Air Service launched a series of pathfinding expeditions to locate, develop, and map aerial routes and municipal landing fields throughout the country at a time when there were no airways, airports, emergency fields, or support facilities.[110] While pathfinding expeditions were basically conceived to promote the Air Service during a period of postwar downsizing, they also were used to collect pertinent information relating to maps and other navigational aids.[111] The Air Service also directed army pilots to prepare cross-country flight reports for its Office of Mapping and Cross Country Flights with the expressed purpose of "compiling useful flying maps of all parts of the country."[112] Some 300 flight reports were submitted in 1918 and 1919, a number of which were accompanied by annotated commercial or manuscript maps depicting flight routes and landing fields. The Air Service and commercial map publishers used the data collected from these flight reports to produce the earliest small-scale preflight planning maps of the United States.

AIR SERVICE AERIAL ROUTE MAP OF THE UNITED STATES

The first official planning map to appear was issued by the Air Service in the spring of 1919 with the title, "U.S. Army Air Service Aerial Routes of the United States Flown and Reported by Direction of Director of Military Aeronautics." It was compiled by First Lieutenant Kenneth C. Leggett, who maintained the master file of flight reports, and drawn by John Locke, an Air Service draftsman. Leggett's map was published at a scale of 1:6,000,000, and measures 18 × 25 inches.[113] It displays the army's network of aerial routes, extending from the Pacific to the Atlantic coast but located primarily in the southern and midwestern tier of states. Two types of routes are depicted, based on a qualitative assessment of a route's terrain with respect to emergency landing sites ("good" terrain is indicated by solid black lines, "poor" terrain by dotted lines), and three classes of landing fields are ranked by availability of services and supplies.

Leggett's map was widely circulated to military and private pilots. "You will find all of the aerial routes at present known to the Air Service very plainly marked," an information officer notified one private pilot preparing for a flight from Forth Worth, Texas, to Detroit in the spring of 1919.[114]

The Air Service issued expanded and enlarged editions of the map in 1920, 1923, and 1924 under different titles and at a larger scale (1:4,000,000).[115] The 1920 map depicted lines of equal magnetic declination, boundaries of "aviation forecast zones," and the location of four types of landing fields. The number of landing fields was expanded on the later maps to seven (government or army, navy or marine, air mail, municipal, commercial, emergency or unimproved, and seaplane), and a proposed airway system was shown in red. The Air Service informed all government and municipal landing fields to make these maps "accessible to pilots or others interested in order that best use may be made of same."[116]

As these maps received wider distribution, private pilots throughout the United States contributed information, helping to fill in areas across the country where the Air Service did not operate. J. G. Rankin, founding pilot of Rankin Aviation Company, submitted an extensive list in early 1924 describing "possible landing fields throughout the Northwest." Rankin reported that he had landed and flown out of most of the fields while accumulating more than 1100 hours of flight time. He also suggested changing a route shown on the 1923 map, which as depicted "leads one over some of the wildest part of Oregon. Mostly mountains and sage brush."[117]

RAND MCNALLY AVIATION MAP OF THE UNITED STATES

Just as earlier generations of private publishers used data obtained from army maps and reports, commercial markets exploited the Air Service files. Using data collected by Leggett's group, Rand McNally published its first aviation map of the United States in April 1923: "Aviation Map of United States Featuring Landing Fields." It was prepared for the National Aeronautic Association of the United States (formerly the Aero Club of America) and the United States Touring Information Bureau, which joined together to promote aviation.[118] It measures 26.5 × 40.5 inches, and was printed on the verso of Rand McNally's popular "Official Auto Trails Map of the United States" at a scale of 1:8,000,000. More than 3,000 landing fields are shown by dot symbols, differentiated according to whether they were improved or unimproved.

Rand McNally complemented its aviation map with one of the first published directories of landing fields. *The Complete Camp Site Guide Including All Airplane Landings in the U.S.A. with Maps Showing Locations of Both* was issued by the United States Information Bureau in Waterloo, Iowa in 1923, as a small handbook (fifty-six pages) to be carried in the cockpit of an airplane. Each airport listing included a brief description of the size and nature of the field as well as its location, prevailing winds, and available facilities such as oil, gas, and hangar.

Army pilots and the Air Service endorsed both Rand McNally products. Reporting on a round-trip transcontinental airways flight from Pope Field, North Carolina, to San Diego, California, in 1923, one Army aviator described the *Camp Site Guide* as "an invaluable aid, especially in the west, where landing fields are generally permanent." "On two occasions," he noted, "when landings were necessitated on account of motor trouble or weather conditions it aided me in locating fields near my course."[119] In another instance, the Air Service provided a copy to the Boston Chamber of Commerce, which used it for planning an airways system in New England.[120]

In 1927, the Standard Oil Company of California published a small handbook listing landing fields for the Far Western states, titled *Airplane Landing Fields of the Pacific Coast.* Each landing site is extensively described. At least four editions were published between 1927 and 1929. The 1929 edition included a set of aerial navigation maps prepared "especially" for Standard Oil by Rand McNally. While Standard Oil advertised that they distributed this free handbook to western pilots "in the interest of their safety," it was also a great promotional device. In addition to airports, the list includes the names of each Standard Oil Company station, which were conveniently painted with the town names on their roofs so that they could easily be identified from the air.[121]

NATIONAL HIGHWAYS ASSOCIATION "HIGHWAY AIRPORT" MAPS

The National Highways Association also used Air Service's landing field planning maps to prepare a series of colorful and decorative maps that promoted its concept for the placement of landing fields along major highways.[122] These "highway airports," as the association designated them, each spanned forty acres and were to be placed at twenty-five-mile intervals. The maps promoting them were printed in the form of large strip maps, one for each of the major national highways. They ranged in scale from 1:3,000,000 to 1:4,000,000.

An example of this genre is the "Map of the National Roosevelt Midland Trail . . . Proposed by the National Roosevelt Midland Trail Association" (figure 6.7). Compiled by John C. Mulford, the association's chief cartographer, it depicts a transcontinental route from Washington and Norfolk, Virginia, to San Francisco and Los Angeles. In addition to an inset of a highway airport, it shows the location of improved and unimproved highway airports, Indian Reservations, national parks and monuments, and

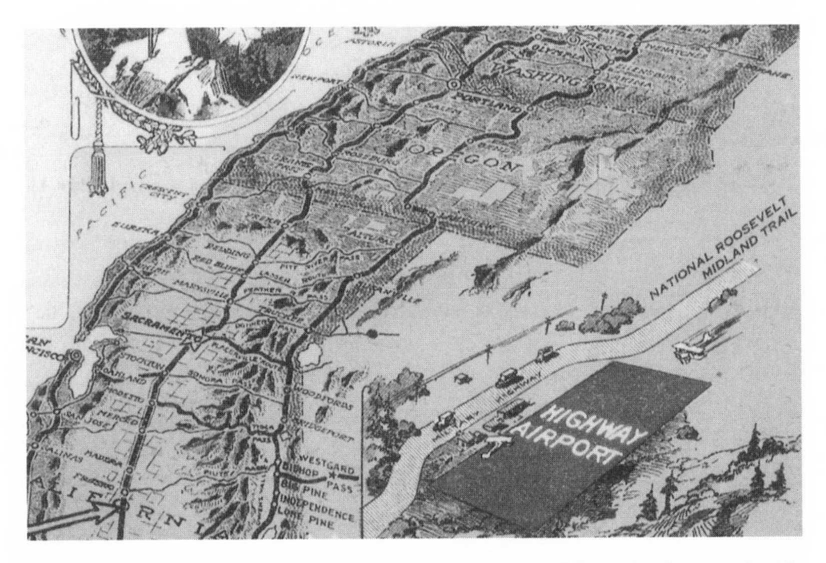

FIGURE 6.7 Detail from National Highways Association, "Map of the National Roosevelt Midland Trail," 1923. Depicts a cartographer's conception of an emergency "Highway Airport," an idea promoted by the National Highways Association during the early 1920s. Library of Congress, Geography and Map Division, US/Roads/1923.

national forests. A cross section or profile along the bottom of the map shows the elevation above sea level of the principal cities along the route. The map measures 16 × 42 inches and was printed by A. Hoen and Company, Baltimore, in 1923. As an indication of its popularity, it was reissued three times over the next two years. Other maps in this series depict the National Old Trails Road/Grand Canyon Route, the Yellowstone Trail, the Custer Battlefield Highway, and Pike's Peak Ocean to Ocean Highway.

Experimental, Temporary, and Emergency Aviation Maps, 1918–1923

As military and airmail pilots increased in number and cross-country flights were extended in range, the limitations of standard published maps as navigational aids became more apparent to pilots and government officials. "There is an immediate and pressing need for suitable maps," the chief of Flying Operations for the United States Aerial Mail Service lamented in 1919. "There are no such maps at present. The nearest approach is the map printed by the Post Office Department. This map has been used extensively, even by the Army. Due to its many confusing and unnecessary features, [however,] it is far from perfect."[123] As a consequence, the Corps of Engineers, the Air Service, and the Air Mail Service set out to develop more useful operational charts and supplemental aids, generally collaborating but sometimes going their separate ways as they competed for scarce resources. Their products ranged widely in form and content, but they helped shape the specifications later adopted for standard published aviation charts.

POST OFFICE DEPARTMENT'S "PILOT'S DIRECTIONS"

Some of the earliest aviation navigational aids were published in the form of narrative route itineraries because they were easier to produce and distribute than maps. A narrative route itinerary provided a unique verbal image of a flight course, a textual ordering of the landscape that guided a pilot from checkpoint to checkpoint almost as effectively as a map, providing that the aviator did not deviate greatly from the course.

The Post Office Department issued the first published narrative route itineraries in February 1921 during the establishment of its transcontinental airmail route from New York to San Francisco. Known as *Pilot's Directions,* these booklets were prepared from information obtained through Post Office Department sponsored pathfinding contests. For a winning route report, contestants received fifty dollars, a substantial sum in 1921. *Pilot's Directions* included information about distances, landmarks, compass courses, and emergency and regular landing fields for a ten-mile wide strip of land along a flight route. An example taken from the Transcontinental Airmail Route is representative:

> [Mile] 174. After leaving Sunbury the next landmark to pick up is Penns Creek, which empties into the Susquehanna 7 miles south of Sunbury. . . .
>
> [Mile] 178. "*New Berlin*—Identified by covered bridge over Penns Creek. . . .
>
> [Mile] 202. The next range of mountains is crossed through the pass at Millheim, a small town. A lone mountain may be seen to the south just across the Pennsylvania tracks. . . .
>
> [Mile] 217. After crossing another mountain range without a pass Bellefonte will be seen against the Bald Eagle Mountain Range.[124]

PHOTOMOSAIC AERIAL ROUTE MAPS

Both the Corps of Engineers and the Air Service were anxious to exploit the new medium of aerial photography in the production of aerial route maps, but major disagreements between these two service units over primary issues of command and control hindered this development somewhat. The demands of the war effort masked these differences, but friction between topographers and aviators intensified following demobilization as each continued to develop and expand their respective aerial photographic programs in the preparation of aviation maps.

The Corps of Engineers, supported by the Geological Survey, believed that aerial photographs must be taken by experienced topographers trained in the mapping process, and as the lead service unit within the Army for surveying and mapmaking, these activities logically fell within its domain. The Air Service, which had been given responsibility by the general staff during the war for all photographic work from aircraft, argued instead that it was essential to train aerial photographers in the techniques of photography rather than mapmaking. This issue was finally resolved in early 1920 when it was agreed that the Air Service would be responsible for taking aerial photographs for the Corps of Engineers and the Geological Survey, and that the Corps of Engineers would be responsible for compiling and printing aerial route maps for the Air Service. Under this agreement, the Air Service was allowed to prepare "such temporary, experimental or emergency maps as may become necessary."[125]

Prior to this accord, the Corps of Engineers had conducted experiments at its Central Map Reproduction Plant to convert aerial photographic mosaics into aerial route maps. The Central Map Reproduction Plant (later renamed the Engineer Reproduction Plant) was the predecessor of the Army Map Service. Located at Washington Barracks (now Fort McNair) in the District of Columbia, it was established in early 1917 as part of the Army's Engineer School. Primarily responsible for the reproduction of maps, its small cadre of officers and civilians also taught military mapping and printing.[126]

One of the surviving examples is an aerial photographic half-tone map depicting the aerial route between Freeport and Amityville, New York (figure 6.8). It was printed at two scales: one inch to one mile and one inch to three miles. Photo-engraved on enameled book paper, the larger scale map measures 9 ½ × 3 ½ inches. While the map was innovative and inexpensive to produce, flight tests proved unsatisfactory. It lacked detail and the photographic paper was too brittle to withstand wear and tear in the cockpit.[127] To solve the latter problem, the Central Map Reproduction Plant planned to test the use of photo-sensitized silk cloth—thus anticipating the cloth aviation escape and evasion maps of World War II—but these tests were apparently not carried out.[128]

As the interagency controversy heated up in 1919, both organizations pushed forward with their own aerial photographic and aviation map programs. "To show . . . that we can prepare this route stuff better than the Air Service," Colonel E. H. Marks of the Corps of Engineers wrote, "we are trying to rush through some of the short routes which are completed."[129]

FIGURE 6.8 U.S. Army, Central Map Reproduction Plant, "Aerial Route Freeport to Amityville, N.Y.," 1918. Army cartographers experimented with a variety of data gathering methods and production media to produce their first aviation maps, including this aerial photo map. National Archives and Records Administration, Textual Archives Division, RG 77, E104, File 061.1 (Eastern Department).

These included aerial photographic mosaics prepared by the corps for several of the aerial route maps begun during the war, including San Diego to Los Angeles; Washington, D.C., to Norfolk, Virginia; Waco to Austin, Texas; Austin to San Antonio; and San Antonio to Yuma, Arizona.[130] In early 1920, Captain Wilburn Henderson photographed the route from Kelly Field to Laredo, Texas, "for the use of air pilots in guiding troop marches between these two points." This mosaic covered 366 square miles.[131]

At the same time, the photographic branch of the Air Service promoted a nationwide program to photograph and prepare aerial photographic mosaic strip maps of major airway routes for training purposes and cross-country flights.[132] Detailed guidelines pertaining specifically to the selection of base maps, plotting control points, flight procedures, and camera types were issued for the preparation of these mosaic maps. Three aerial photographic units at Post Field, Fort Sill, Oklahoma, were directed to prepare mosaics for a twenty-mile-wide airway extending from Fort Sill to Wichita Falls, Texas, and from Fort Sill to Oklahoma City. Only a portion of the mosaic map was completed, but it nevertheless illustrated both the potential and the challenge of this new process. Composed of 4,000 aerial photographs, the mosaic measured 6 × 16 feet, covering some 341 square miles. It required approximately sixty hours of flying time to photograph the route.[133]

Apparently, neither program found strong support in the army during a period of meager funding, and few aviation photographic mosaic maps were completed. Nevertheless, the participating topographic engineers and aerial photographers gained valuable experience that was later put to good use in the preparation of standard aerial navigation maps. These efforts also helped lay the foundation for the use of aerial photography in basic topographic mapping in the federal government. As early as 1921, the Corps of Engineers published two topographic quadrangles (Comanche County, Oklahoma and Santa Monica, California) that were revised and corrected using aerial photographs. At the same time, the Air Service provided the aerial photography for preparation of the Geological Survey's topographic quadrangle of Schoolcraft, Michigan, which was published in 1922 and is generally regarded as the first topographic map prepared directly from aerial photographs.[134] In 1934, an experimental airway chart was successfully compiled from high altitude aerial photographs, but this method of producing aeronautical charts was not fully exploited until World War II.[135]

AIR SERVICE/POST OFFICE BLUELINE STRIP MAPS

During the immediate postwar period, airmail pilots were the only aviators in the United States flying scheduled long-distance flights on a daily basis. In response to their needs, the Air Service's Information Group began producing and testing experimental flight maps for the Air Mail Service that they believed superior to those prepared by army topographers.[136]

The most successful was a set of photoprocessed blueline strip maps initially prepared at the request of Postmaster General Otto Praeger. These special maps were designed for cross-country flying at an altitude of 5,000 to 6,000 feet in accordance with guidelines prepared by James Edgerton, the Post Office Department's chief for Flying Operations. "Simplicity should be the keynote of any map used for flying," Edgerton believed, "[a]ll features not vital should be eliminated." According to Edgerton the most important elements to be depicted on an aviation map included the following, all based on "Aerial Mail experience": (1) railroads: "always prominent and are extensively followed by all pilots"; (2) rivers and lakes: "Water courses of all kinds are vital and can be seen greater distances than any other feature. It is especially important that all shore lines be shown accurately"; (3) towns and cities: particularly important "where near railroad intersections or water courses"; (4) roads: "straight roads connecting cities should be emphasized. They are of aid in checking positions"; (5) vegetation: "peculiarly shaped woods, etc., are good sign-posts." In addition, Edgerton recommended that relief be depicted by contours and that "true North be shown at intervals with variations to true magnetic North at points at which they occur."[137]

The Air Service began furnishing copies to the Post Office Department in late 1919. "These maps," noted Lieutenant Colonel Horace Hickam of the Information Group, "embody all that is necessary for a flier to know in travelling cross-country. Prominent landmarks are made more prominent and non-essentials eliminated."[138] All of the major features mentioned by Edgerton were depicted except contours, as terrain data was simply not available for most of the areas covered. Although no map legends were included, the borders were scaled in miles and numbered at ten-mile intervals for quick reference. Similar in format to the Post Office belt maps, the blueline route maps generally measure eight inches in width and range in length from thirty-three to sixty inches.[139] They were considered the best aerial route maps available for the East Coast, due in part to the numerous watercourses depicted, but a lack of topographic detail and prominent landmarks made them less useful for flying in mountainous regions and during poor weather conditions. Rand McNally state or railroad maps were still considered superior for the Appalachian and interior regions.[140]

By 1923 the major aviation centers along the East Coast and in the Midwest were linked by blueline route maps, which were used by army and navy pilots as well as by those carrying the mail. Based on an examination of existing maps and an analysis of Air Service map procurement reports, at least twenty-two different route maps were compiled by Air Service and Geological Survey cartographers.[141] The Air Service and Post Office retained tracings of the originals from which photoprocessed copies were issued upon request to pilots and flying fields.[142] In addition to providing pilots with better maps, the Air Service and Post Office realized considerable savings because the photoprocessed copies could be supplied to pilots on demand at much lower costs than published maps.[143]

While experimental in design and temporary in nature, the photoprocessed blueline belt maps were an important stepping stone for the Air Service in its development of an aeronautical chart. They demonstrated that simple, precise maps designed according to specifications prepared by aviators were superior to most existing maps. The Air Service also gained the support of an important ally in the Air Mail Service, an organization that represented the broader interests of commercial aviation to a greater extent than any other government agency.

PHOTOPROCESSED MAPS OF LANDING FIELDS

Demand for maps of landing fields increased dramatically after World War I as the Air Mail Service and Air Service began nationwide campaigns to establish an airways system. Both services contributed to the construction of municipal airports throughout the country by providing technical support and donating steel hangars. Other government organizations, municipalities, and flying clubs established emergency landing fields, though many were available only on a seasonal basis, particularly in agricultural regions.

The dramatic increase in the number of landing fields was not matched immediately by the production of adequate maps and diagrams. "It has been one of the great handicaps of commercial aviation that scant graphic information has in the past been available regarding the actual size, shape and approach of our permanent airports," noted a critic in the prestigious journal *Aviation* in 1923. Invoking the name of one of history's great military commanders, he continued, "'A sketch map tells me more in one minute than a verbal report in an hour,' said Napoleon. This is particularly true about landing fields."[144] While this problem was not entirely solved until the late 1920s, pilots and flying organizations developed and tested a variety of approach charts and landing field diagrams to meet growing demand.

The first Air Service landing field maps were prepared during the war. The 272nd Aero Squadron, for example, issued a series of nine blueprint maps of emergency landing fields located in the vicinity of Houston, Texas, dated October 5, 1918. They were drawn "from aerial observation" by Sergeant V. G. Smylie, and depict every feature of interest to pilots landing in a confined space.[145] Following such efforts, the Air Service began maintaining a master file on landing fields, which was used to prepare detailed maps. Pertinent data was obtained through a variety of sources but primarily from questionnaires sent to government, commercial, and municipal flying fields.[146]

By 1922, the Air Service furnished flying fields with preprinted forms containing grid outlines to aid respondents in sketching diagrams of their fields.[147] The submitted sketches were reduced in size and redistributed to Air Service facilities in much the same way that the Air Service/Post Office blueline strip maps were made available.[148] Known as a Standard Sketch and Information Sheet, each photoprocessed landing aid included a detailed map of a landing field and a description of the landing conditions, communication facilities, airplane accommodations, and availability of weather reports. The sketches measured 5 × 3 ½ inches, small enough to be carried in the cockpit or pasted on aerial route maps.[149]

Similarly, the Air Mail Service required local postmasters along airmail routes to submit detailed reports identifying potential or existing emergency landing fields and information concerning their locations and characteristics. Initially, rough sketch maps of the landing fields and of obstructions accompanied these site location reports, but later versions were more detailed and accurate.[150] This information was used by the Air Mail Service to issue large-scale blueprint maps of landing fields.

Other sources for maps of landing fields were the pilots themselves. Airmail pilot Harry Huking submitted a set of blueprint maps of emergency landing fields west of Reno that he had prepared during the course of flying the San Francisco to Reno mail route. These maps were apparently issued by the Post Office Department in loose-bound atlas for each leg of a flight. A personal copy of the Chicago to Cheyenne route atlas belonging to C. F. Egge, general superintendent of the Air Mail Service, contains thirty-six maps of emergency landing fields issued in 1923 and 1924. These maps measure 10 × 13 ½ inches and range in scale from one inch to four hundred feet to one inch to eight hundred feet. They were prepared in accordance with a standard set of conventional signs issued by the Air Mail Service in 1922.[151]

AERIAL ROUTE MAPS WITH LANDING FIELD INSETS

The integration of maps of landing field and aerial route maps was a logical attempt by aviators to reduce the number of maps that had to be carried in open cockpits. Several interesting cases attest to the creativity of the pilots that prepared them. One of the earliest surviving examples of the genre is a blueprint map prepared by Private Joseph Palle for the Flying Department at Kelly Field. The map is titled "Cross Country Landing Fields Showing Their Relations to Kelly Field," it is a small scale local vicinity aeronautical map of the San Antonio region, surrounded by eleven larger scale maps of landing fields. Each inset provided critical information concerning size, condition, and characteristic of the flying field, location of adjacent roads or railroads, distance to the nearest town, magnetic north, and obstructions in the immediate vicinity.[152]

Major Theodore C. Macaulay, commanding officer of Taliaferro Field near Fort Worth, Texas, issued several innovative aviation blueprint en route strip maps that combine maps of aerial routes with inset maps of landing fields (figure 6.9). A veteran pilot, Macaulay flew the first two transcontinental pathfinding flights along the southern border of the United States, specifically to map the location of landing fields. His large-scale map of the route from Taliaferro Field, Texas, to Oklahoma City includes insets with detailed sketches of landing fields and their direction from the closest towns. These sketches were reproduced from aerial photographs taken by George Goddard and depict features of particular interest to airman. Macaulay also devised a set of symbols to indicate permanent landing fields (designated by a circle with a cross), emergency landing fields (shown by the name of a town underlined), line of flight, railroads, and electric lines.[153]

A similar map was prepared and tested by pilots at Langley Field, Virginia, in 1922 during flights from Langley to Bolling Field in

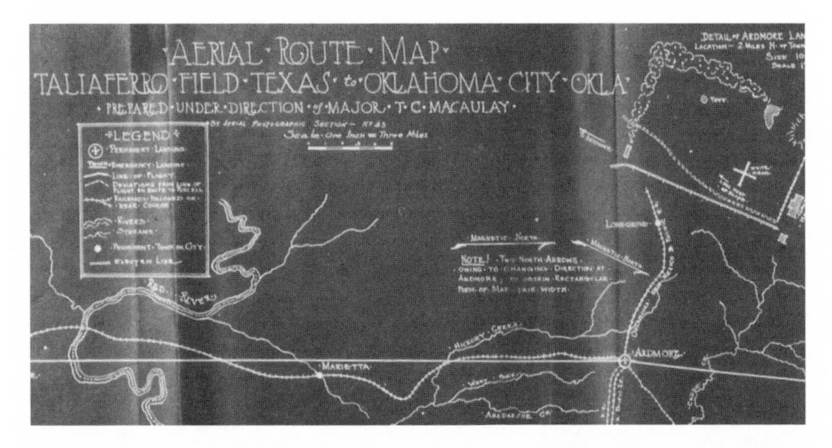

FIGURE 6.9 Detail from experimental blueprint aerial route strip map prepared under the direction of Major Theodore C. Macaulay, commanding officer, Taliaferro Air Field, Fort Worth, Texas, 1918. Macaulay combined large-scale drawings of emergency landing fields with course and route information. National Archives and Records Administration, Textual Archives, RG 18, E143, Box 14.

Washington, D.C. Measuring $5\frac{1}{2} \times 18$ inches, this map was mounted on balloon fabric and covered with transparent material waterproofed with several coats of thin acetate dope to protect it from the elements.[154] "The idea of placing small sketches of landing facilities along the edge of a strip map is a very good one," the executive officer of the Air Service's Training and Operations Branch observed during an evaluation of this map. But he believed that a better course of action was to paste small, updated maps of landing fields on strip maps because of continuing changes to emergency fields, particularly with respect to cultivation.[155]

CORPS OF ENGINEERS EXPERIMENTAL AERIAL ROUTE MAPS

A more conventional aerial route map was compiled in January 1919 by student officers for an aerial mapping course conducted by Captain Charles Ruth at the Central Map Reproduction Plant. Ruth hoped it would become a model for subsequent aerial route maps. "Aerial Route Map Washington [D.C.] to Aberdeen [Maryland]" measures 9×26 inches, and covers an area 20×70 miles at a scale of 1:200,000.[156] Symbols, scale, and lettering are in general accordance with *War Department Bulletin No. 64* and "Aerial Navigation Maps," *Conventional Signs,* U.S. Army Maps *Changes No. 2,* War Department, Washington, December 28, 1918. Basic information was derived from Geological Survey topographic quadrangles and aerial photographs specifically solicited for this project. Ruth's base map was forwarded to the army's director of Military Aeronautics on February 5, 1919 with a request that the Air Service "have placed on this map in red ink the location of such landing fields within the area as have been determined upon, together with [a] line indicating [the] most suitable route for flying, if it is desired that such be placed upon the final maps." It was returned a few days later with an air route inserted, connecting Bolling Field in Washington with Aberdeen Proving Ground, and marking the emergency landing fields at the Laurel and Havre de Grace race tracks in red.[157] Despite the efforts of Ruth and other army topographers, the experimental aerial navigation maps produced by the Corps of Engineers prior to 1920 were deemed unsatisfactory by the Air Service primarily because the scale was too large (1:200,000), elevations were not depicted, and areal coverage was limited.

On December 19, 1919, the Air Service provided the Corps with new specifications for an experimental aerial navigation map.[158] Guided by the new specifications, Captain Giffin, prepared a prototype that was to have a major influence on aeronautical chart design. Two years earlier, Captain Giffin had been one of the first topographers to use aerial photographs to revise maps. "Aerial Route Map Washington, D.C. to Xenia, Ohio" depicts one leg of the army's model airway between Washington, D.C., and Washington, Pennsylvania (figure 6.10).[159] Formatted as two strip maps, each measuring 14 × 38 inches, it was drawn at a scale of 1:500,000

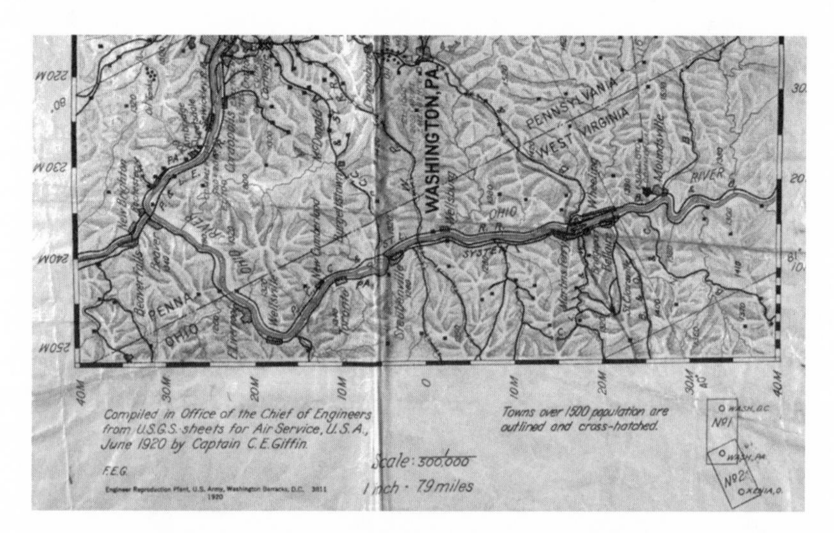

FIGURE 6.10 Detail from experimental aerial route map from Washington, Pennsylvania to Xenia, Ohio, prepared for the army Air Service by Captain Calvin E. Giffin, Corps of Engineers, June 1920. Giffin's major innovation was the inclusion of spot heights and shaded relief. His terrain data was obtained from oblique aerial photographs taken by Lieutenant George W. Goddard, who later made major contributions to the field of aerial photography. National Archives and Records Administration, Textual Archives, RG 18, E143, Box 20.

(1 inch to 8 miles). Each sheet covered an area 80 miles wide and 280 miles long. All towns with a population of over 1500 persons were outlined and crosshatched, with their most prominent features named. Most significantly for many aviators, the elevations for hills and mountains were indicated by spot heights on prominent ridges, and one sheet was further enhanced with shaded relief. The terrain data was obtained from aerial oblique photographs, once again taken by Goddard.[160] Although it contained numerous errors due to its rapid compilation, pilots considered Giffin's prototype the best air navigation map then available.[161]

Published Airway Maps and Charts

The experimental work of the immediate postwar period laid the foundation for a series of air navigation strip maps published by the Air Service (renamed the Army Air Corps in 1926) and the navy. Together, the two service units established a national network of military airways and airports that emerged in 1923.

As early as 1920, the Air Service began drawing up plans for a national airways system. In 1922 it established a model airway between Washington, D.C., and Dayton, Ohio, to develop new methods of conducting scheduled cross-country flights over varied terrain, to test all aspects of an airways infrastructure, and to investigate various aerial navigational aids from route markers and radio communications to aviation charts. Originally extending some 400 miles from Bolling Field in the District of Columbia to McCook Field (now Wright-Patterson Air Force Base) near Dayton, Ohio, the model airway was expanded in 1923 to include airfields in New York and a number of midwestern states.[162]

By the middle of the decade, a national airways network for military and commercial air travel was in operation connecting major urban areas. The national airway system eventually covered some 30,000 miles, comprising three transcontinental east-west routes and four primary north-south routes. Each airway was approximately ten miles in width with intermediate or emergency landing fields spaced at intervals of twenty to thirty miles and major facility airports or terminal hubs every two hundred miles. The airways were outlined on the ground with a variety of artificial landmarks including identification numbers and the names of towns painted on the roofs of railroad stations and the sides of water towers. A series of high-powered rotating beacon lights were spaced approximately ten to fifteen miles apart to guide pilots during night flights. Airports and landing fields were equipped with refueling facilities, weather reporting services, and radio beacons to aid navigation in fog or at night.[163]

During the early 1920s, the navy established a seaplane airway system along the coasts. The first to be laid out was the route between Washington, D.C., and Norfolk, Virginia. Distinctive markers painted on the roofs of lighthouses helped guide naval pilots.[164] With the passage of the Commerce Act in 1926 the responsibility for developing a national airways network shifted from the army and navy to the newly established Commerce Department. Charged with fostering and promoting commercial aviation, the Commerce Department and its mapping agency, the U.S. Coast and Geodetic Survey, assumed responsibility for mapping those airways used most often by civilian airlines.

Map specifications, standardization of symbols, and production procedures for these mapping programs were initially developed by the Air Service in conjunction with the topographic division of the Geological Survey and the Corps of Engineers. The U.S. Board of Surveys and Maps of the federal government further codified symbol usage for aviation maps through the publication of standard symbol sheets. Established in 1919 to coordinate the surveying and mapping work of the federal government, the board was initially comprised of members representing fourteen major government mapping agencies.[165]

ARMY AIR CORPS AIR NAVIGATION STRIP MAPS

A major objective of the Air Service's model airway was the production of "proper and adequate maps for the aviator."[166] Shortly after the formal creation of the model airway, the Air Service's airways section set out to develop an acceptable standard air navigational map for the national airways system. Map tests were carried out under the direction of Captains Burdette Wright and (later Brig. General) St. Clair Streett with much of the developmental work done by Lieutenant J. Parker Van Zandt and test pilot Lewis G. Meister, both with the map section of the Engineering Division at McCook Field.

As a first step, Van Zandt prepared detailed specifications based on a study of English, French, Italian, and American air navigation maps.[167] His specifications called for a scale of 1:200,000 (which conformed to the international standard), hachures or contours for depicting relief, inset diagrams of landing fields in the map margins, spot heights, towns shown in black according to their shape, roads colored white, and aerial routes represented by red lines with distance scale and magnetic compass bearings shown for both directions. Van Zandt's proposed specifications were distributed to some 300 army pilots for review and criticism in February

1922, resulting in seventy-five specific recommendations, including a return to the smaller 1:500,000 scale advocated by Giffin.[168]

In June 1922, Brig. General Mitchell authorized $10,000 for the production of a series of aerial navigation maps based on the new standards, with instructions that they be prepared under contract by either the Rand McNally Company or the Clason Map Company of Chicago.[169] But heavy lobbying by the Geological Survey persuaded the Air Service that the Geological Survey was better prepared to produce the maps quickly.[170] The first experimental map based on the revised specifications was completed by the survey in late 1922. Covering a region of central California that was selected because terrain data were available, the map was prepared under the direction of Acheson Flynn Hassan, the survey's chief cartographer, and flight tested during trials held at Crissy Field, California.[171]

Using the central California map as a model, Hassan constructed a second experimental map, this time in the form of a strip map of the Dayton, Ohio, to Wheeling, West Virginia, route for flight tests by model airway pilots. The map covered an area 80 miles by 220 miles. It was similar in format and coverage to Giffin's map of 1920 but incorporated several new features, including hypsometric tints to depict elevation with principal contours at 500 feet intervals (an idea borrowed by Hassan from the 1:1,000,000 scale International Map of the World) and small-scale maps of landing fields placed in the border opposite their locations on the map. Aerial photographs taken along the route by Goddard were used to determine the outline symbols of cities and towns and to prepare the landing field maps.[172]

Proof sheets were distributed for flight tests on the model airway in January 1923, a practice that was to be followed by the Air Service during the production of all of its later air navigation maps. A major criticism of the first prototype was the use of the color red for designating roads. One of the model airway officers reported that while it was a great improvement over the maps in current use, the "color scheme clashes with the eye and tires the pilot." The roads in particular, he concluded, "are too prominently marked in red, and the map throughout it is too highly marked in red."[173] Another pilot recommended that major towns be blocked in the color red, as in the experimental map produced by Ruth and his students in 1919.[174] As a result of these suggestions, the color white was selected for roads and red for cities on subsequent revisions and new maps. Another major change occurred in 1926 when the border sketches of landing fields were eliminated based on a sample of more than one hundred pilots who claimed

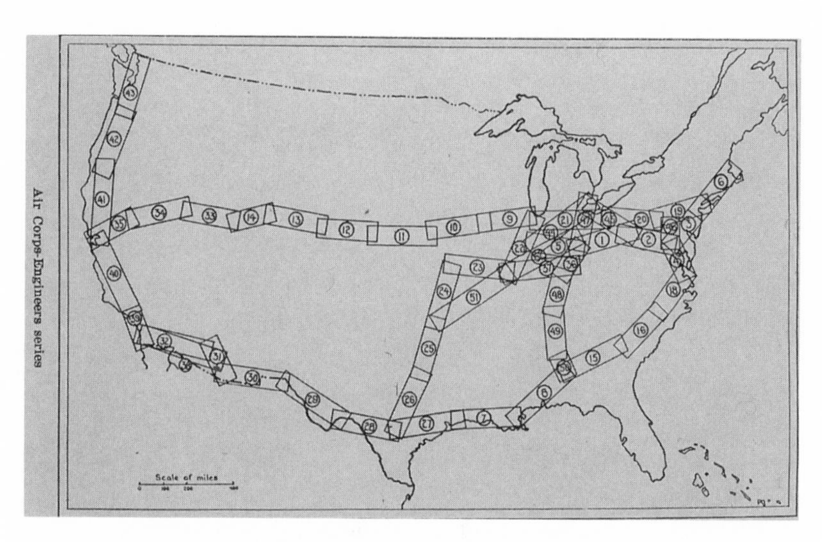

FIGURE 6.11 Index Map showing coverage of the joint Army Air Corps–Corps of Engineers Air Navigation Map Series. Library of Congress, Geography and Map Division, G1201.P6U6 1927, Vol. 1.

never to use them. In many instances, airmen simply cut off the borders, which made the maps easier to fold and manipulate in the cockpit.[175]

Based on the format and content developed during flight testing along the model airway and other sections of the army's airway system, the Air Service and later the Air Corps eventually issued sixty-three air navigation maps in this series from 1923 to 1933, providing complete coverage for the army's airway system (plate 11 and figure 6.11). Of this number, the topographic division of the Geological Survey produced nine maps (including the first four) while the army's Engineering Reproduction Plant prepared the rest.[176] These maps were printed in small press runs of 100 to allow rapid updating, generally on a yearly basis.

AIR CORPS ROUTE INFORMATION AERONAUTICAL BULLETINS

Because the production of published maps and charts was an exceedingly slow process, the Air Service introduced a series of narrative route itineraries in 1925, known as *Aeronautical Bulletins*. Similar in content and form to the Air Mail Service's *Pilot's Directions,* they were conceived as an interim measure to provide aviators with pertinent flight information as quickly as possible for the emerging national airways system.

To facilitate their compilation and production, the airway network was divided into 197 flight segments, which ranged in length from 100 to 200 miles, depending on terrain as well as the fuel capacity and engine endurance of the era's military aircraft. The *Aeronautical Bulletin* covered one flight segment and listed two itineraries, one for each direction of a flight. Bulletins usually contained two to three pages, printed in loose-leaf format for easy handling and updating. Local Air Service facilities were responsible for preparing the route itineraries located within their geographical area. The normal procedure required two officers to survey a route, one serving as pilot and the other as note taker. These terrain reports were then submitted to the airways section for editing and publication. Pilots from the Air Service, the Navy, and the Air Mail Service participated in this process.[177]

The route itineraries contain general instructions for visual navigation augmented with descriptions of prominent landmarks, type of terrain, and landing facilities. Many also included recommendations for appropriate map coverage. For example, Flight Route 68, St. Louis to Kansas City, Missouri, issued in December 1923, instructed the reader to consult Rand McNally's standard map of Missouri. Army itineraries are more succinct than their Post Office counterparts, perhaps reflecting the author's training in writing military reports. Following information about compass course and magnetic declination, Flight Route 68 continues in the same format as the postal itineraries:

> St. Louis to Kansas City, Mo.
>
> Compass course, 283, 271, and 248 true; and magnetic declination, 9 10 east.
>
> Mileage
>
> 0 Leaving St. Louis field at Bridgeton, 18 miles northwest of city, follow between interurban and Wabash Railroad tracks to St. Charles,
>
> 6 crossing Missouri River and passing just south of St. Charles. Pass 4 miles south of big bend of Mississippi River,
>
> 12 and cross Wabash Railroad 3 miles west of Elm Point and 4 miles east of St. Peters. Ground here entirely level; mile-wide landing fields anywhere, but all likely to be wet at most times of the year.
>
> 25 Cross St. Louis & Hannibal Railroad 1 miles north of Enon. Hilly country here; emergency fields to southward. Troy 5 miles north.

A small-scale map of the proposed airways of the United States showed each route segment and served as a quick index to the individual bulletins. Inspired by the automobile blue books discussed in the previous chapter, these narrative route itineraries provide an interesting example of one transportation mode adopting a wayfinding device from another in order to solve an immediate problem.

NAVY AVIATION STRIP CHARTS

The successful development of the Air Service/Air Corps strip maps stimulated production of similar charts by the U.S. Hydrographic Office, the navy's chart maker since the 1830s. The navy's involvement with aviation dated from 1911 when it ordered its first amphibian biplane. During World War I, navy airmen on Curtiss flying boats conducted antisubmarine patrols along the Atlantic and Gulf coasts with the aid of special naval operating charts prepared by the Hydrographic Office.[178] Beginning in 1920 the Hydrographic Office provided navy airman with pertinent aviation information through a monthly newsletter, *Notice to Aviators.* Modeled on the Hydrographic Office's *Notice to Mariners,* it was the first such publication of its kind.

Despite a great increase in seaplane activities after World War I, not until 1925 was a naval aviator assigned to the Hydrographic Office to assist in developing flying charts, which were designed primarily for seaplane coastal navigation. Naval aviators also played a vital role in chart design and testing. The navy's prototype chart, *Baltimore–Washington–Norfolk* (H.O. No. V-234, 1925), was circulated to all naval units for comments and criticism. Nearly two hundred changes were suggested, many of them incorporated in the revised edition (figure 6.12). Between 1925 and 1936, the navy's Hydrographic Office issued sixty aviation charts covering the coastal airways of the United States, Mexico, and the Caribbean at a scale of 1:500,000. Formatted as strip charts, they generally measure 11 × 46 inches.[179]

In addition to standard aviation symbols, special markings pertinent for seaplane navigation were added to show lights and buoys, shallow water, seaplane anchorages, and seaplane courses. The light characteristics of each lighthouse were included for night flying. The navy's experimental chart included photographic insets of lighthouses as well as landing fields within the map border. The lighthouse views were discontinued on later charts, and the half-tone pictures of landing fields were moved to the chart's verso where they were joined by illustrations of "seaplane anchorage chartlets." Chartlets provided detailed information on water depth, tide range, maximum current, bottom characteristics, refueling facilities, communication stations, and the most desirable anchorage for seaplanes.

COAST AND GEODETIC SURVEY AIR NAVIGATION STRIP MAPS

The Coast and Geodetic Survey's mapping program was designed to link the emerging commercial airways system with the existing military network.

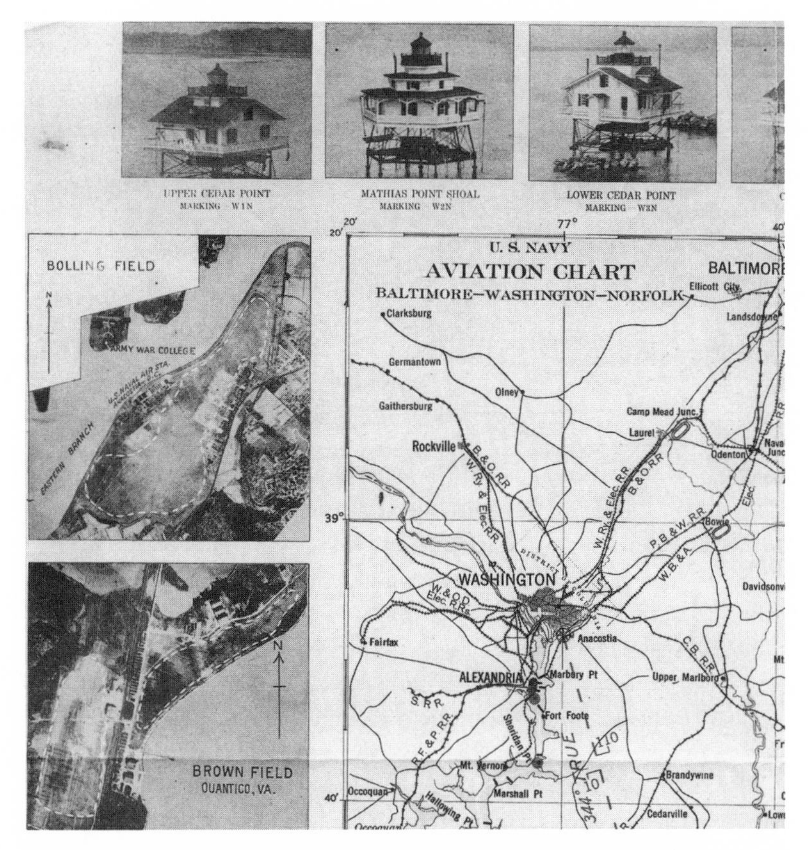

FIGURE 6.12 Detail from U.S. Navy Hydrographic Office, "Aviation Chart Baltimore-Washington-Norfolk," 2nd ed., April 1926. Between 1925 and the early 1940, the Navy issued a series of aviation strip charts designed for seaplane coastal navigation. They covered the coastal airways of the United States, Mexico, and the Caribbean. Author's Collection.

Between 1927 and 1936, the survey's airways map section published forty-three strips maps, each revised several times. These maps measure 11 × 24–48 inches, and covered an area 80 miles wide by 200 to 400 miles in length at a scale of 1:500,000. Like Air Corps strip maps, only significant features appeared, such as railroads and major highways. Terrain was depicted by a combination of gradient tints, contours, and spot heights for prominent towns and conspicuous mountain peaks. One major difference between the survey and the Air Corps airway maps was the depiction of cities and towns; the survey substituted the color yellow for red early in 1927 following testing on the Chicago to Milwaukee map. Subsequently, the color red was reserved for air navigation information such as the line of flight, magnetic courses, and symbols for airports and beacons.

The survey also followed the Air Corps in using flight checks to review all new maps prior to publication and to revise its maps on a timely basis, a system continued to this day. Initially, the survey borrowed Air Corps pilots for field checks but later trained its own cartographers to carry them out. A typical flight check for a new strip map required an average of seven hours of flying time.[180]

AIRPORT MAP DIAGRAMS

In early 1923, the airways section began issuing standard printed map diagrams of landing fields, a practice continued by the Coast and Geodetic Survey. Formatted as small, one- to two-page bulletins (7 ¼ × 4 ¼ inches), the airport map diagrams were designed to be carried in three-ring, loose-leaf binders or in an aviator's logbook. They were published by the U.S. Government Printing Office with the titles *Aeronautical Bulletin State Series* (Air Service/Air Corps) and *Airway Bulletin* (Coast and Geodetic Survey). The first page generally included two sketch maps, one illustrating the airport's position with respect to its immediate surroundings, the other outlining the field. The second page or verso provided standard information on the landing facility, similar to that found on the earlier photoprocessed diagrams. The bulletins were issued in press runs of 1,200 copies, with 1,000 for the Air Service. The Post Office and the Navy purchased the rest.[181]

The data for preparing these maps was obtained by distributing preprinted sketch forms and questionnaires to the towns and flying fields located on the airways. Air Service pilots were also sent out to sketch and photograph landing facilities, particularly those located in towns that did not respond to the questionnaires. Within the first year of this activity, some 200 maps of government, municipal and commercial fields were issued. The navy began issuing similar diagrams for seaplane anchorages in 1926.

Conclusion

By the end of the 1920s, the aviation strip charts and maps issued by the army, navy, and Coast and Geodetic Survey were considered the best maps available for cross-country flying. Together, these three organizations distributed more than 14,000 strip maps in 1929, and 26,000 in 1930.[182] But political, technological, and economic changes in the aviation industry were transforming the form and content of aviation charts. The reassignment of airmail service from the U.S. Post Office Department to private operators supported by large federal subsidies; the development of new and safer airplanes, including multiengine, all metal planes that could carry eight to twelve passengers as well as mail; and the expansion of the national airways infrastructure by the Commerce Department all laid the foundation for a commercial airline industry that crisscrossed the United States by the mid-1930s.

At the same time, federal agencies expanded navigation aids and introduced new ones to mark the nation's emerging airways. By 1932, flashing beacons illuminated more than 18,000 miles of government airways for night flying. In addition to lighted airways, instrument navigation was introduced. Lieutenant James H. Doolittle made the first "blind" takeoff and landing in 1929. A national network of radio range beacons was in place in the United States by the early 1930s, which projected radio beams to aid pilots in maintaining a true course in all kinds of weather. By the mid-1930s commercial airliners were also equipped with radio direction finders and radio compasses appeared.[183] Later that decade, when pressurized cabins began allowing airmen to fly at higher altitudes, "above the weather," navigation gradually shifted from contact piloting or pilotage to instruments.

Coinciding with these technical developments, the number of civilian airplanes purchased for private use mushroomed to nearly eight thousand following Charles Lindbergh's famous flight in 1927. According to a report by the U.S. Federal Board of Surveys and Maps, far more flying in 1928 was done "away from the established airways than on them."[184] In recognition of the standard strip map's limited use for this type of flying, both the Rand McNally Company and the Coast and Geodetic Survey began issuing air navigation maps designed for area or regional flying rather than airway flying, and for instrument as well as visual navigation. Rand McNally's innovative series of Air Trails Maps and the Coast and Geodetic Survey's series of Airway Sectional Charts (figure 6.13) introduced in 1928 and 1930 respectively, dramatically changed the direction of aviation chart design and construction, opening a new chapter in the history of aviation cartography.

Several conclusions can be drawn from this study of early aviation charts in the United States. The first is that the development of the aviation chart was an organizational rather than an individual effort, although individuals played crucial roles, particularly in design. The nature of the process required a coordinated effort from a wide range of specialized technical skills, including surveyors, topographers, aerial photographers, map compilers, map engravers, printers and aviators. The most important contributions were made by the military, particularly the Army Air Service, but commercial map publishers provided critical support, beginning with the Aero Club of America and continuing through the Sperry Gyroscope Company and Rand McNally.

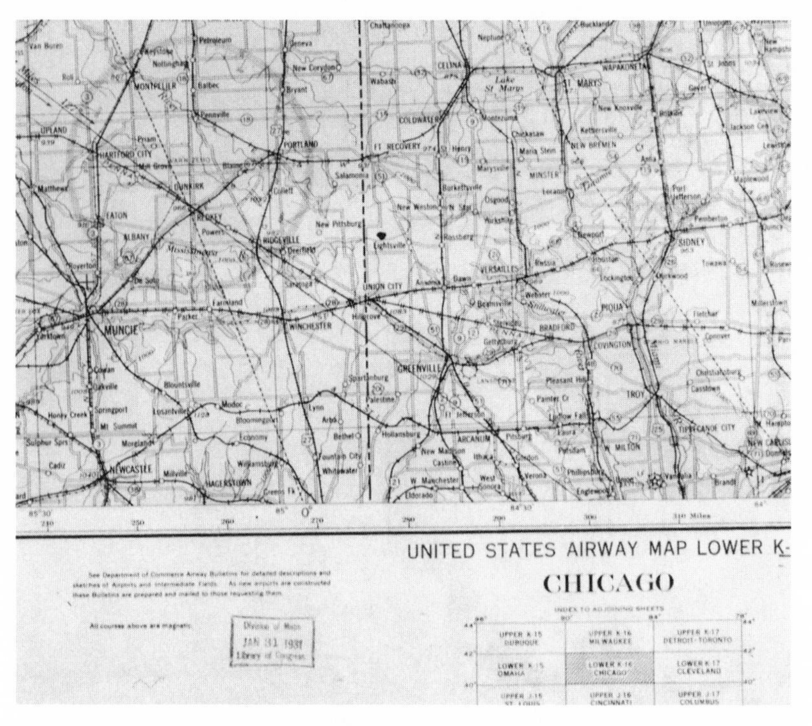

FIGURE 6.13 U.S. Coast and Geodetic Survey, Sectional Airway Map Series, "Chicago, Lower K-16," June 1931. The work conducted by commercial and military mapmakers during aviation's pioneering era laid the foundation for today's sectional airway maps produced by the Coast and Geodetic Survey beginning in 1930. Author's private collection.

Another finding is that from the earliest days of the leather helmet and goggle era, through the initial establishment of a national airways system, airmen played dominant roles in the development of the aviation chart. Aviators considered maps essential wayfinding devices, and to this end, they adapted, designed, and tested a wide variety of maps, map formats, and reproduction processes. From Lawrence Sperry to the flying officers of the Air Service's aviation section, pilots contributed directly to map content by gaining control of the design and editing phases of map production. This was aided by the fact that the Air Service and the Air Mail Service operated independently of the Corps of Engineers and the Geological Survey, the two government agencies nominally responsible for official topographic mapping, and partly because the United States lacked basic topographic map coverage for much of the country.

This study also suggests that the format of the aviation chart was influenced by a combination of technological considerations, airway design concepts, and a general lack of topographic map coverage. It is interesting to note that the only other countries to adopt the strip map as a primary format were Russia, Canada, and Australia, continental size countries with incomplete topographic coverage.

The available information suggests a hitherto unrecognized relationship between the development of the aeronautical chart in the United States and the use of aerial photography for mapmaking. The demand for an aviation chart for pilot training following America's entry into World War I forced the Geological Survey and the Corps of Engineers to set aside their different cultures and inherent distrust of change to adopt this new technology in an effort to meet wartime deadlines. The evidence also hints that the Air Service rather than the Survey or Corps of Engineers was the real agent of change in the advancement of aerial photography as an aid to mapping, in part because of its continuing requirement for accurate and timely aeronautical charts.

Finally, while both the format and content of aeronautical en route charts, approach maps, and airport diagrams have continued to change in response to advances in navigational aids and aircraft, the general principles that evolved during the first two decades of aviation cartography are still reflected today on aeronautical charts used for visual navigation.

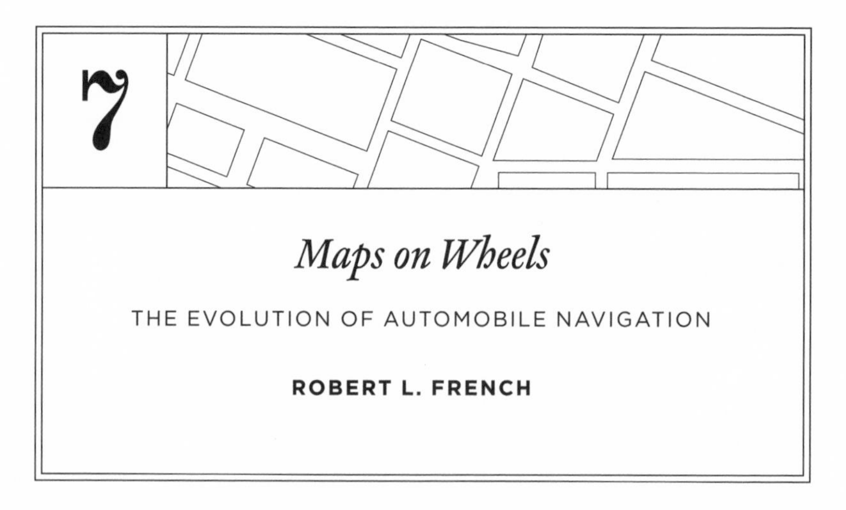

7

Maps on Wheels

THE EVOLUTION OF AUTOMOBILE NAVIGATION

ROBERT L. FRENCH

The term "moving maps" has literal meaning in automobile navigation systems. Around 1910 some of the earliest systems used the automobile's odometer cable to move route instructions printed on a scrolled ribbon or a rotating disk into view under a pointer representing the automobile's location; one such system was sold under the name Live-Map. The scrolls and disks used by these early mechanical systems represented routes to individual destinations in a manner reminiscent of the use by medieval travelers of itineraries for individual routes rather than actual topographical maps as described by Delano-Smith in chapter 2 of this volume.

With the advent of electronic navigation and route guidance systems in the 1970s and 1980s, some systems used an icon to represent the automobile's position on a TV-like screen while a topographical road map stored as a computer file slides underneath. By the 1990s, most electronic systems included on-board software to compute optimal routes to specified destinations. They presented turn-by-turn instructions in the form of symbolic arrows and written commands on the screen and/or by synthesized voice, all without the driver necessarily viewing a topographical map. Automobile navigation systems continue the medieval tradition of travelers using information extracted from topographical maps rather than using the actual maps once an itinerary has been prepared, and are broadly related to the use of road maps in strip format as described by others in this volume.

Compared to paper road maps today's widely available intelligent vehicle navigation systems offer a number of advantages for the driver. Drivers using paper maps must first find the position of both the automobile and the destination on the map, and then choose a particular route to the destination. The driver may occasionally refer to the paper map to confirm adherence to the chosen route and, if necessary, update the route. Once the journey is complete, the driver faces the onerous task of refolding the paper map. In contrast, an intelligent automobile navigation system already knows where it is when first turned on, and the driver may specify a destination by a variety of means such as scrolling through lists of cities and street names, "yellow pages" lists of points of interest (i.e., restaurants, other types of businesses, and public facilities), stored lists of previous destinations, and so on. The system then computes the best route based on the driver's preferences such as shortest time, least or most use of freeways, or avoidance of tolls. The navigation system provides route guidance instructions by synthesized voice and/or screen display of directional arrows in real time (i.e., as approaching the point where an action is required). An updated route is automatically computed by the system if the driver misses a turn or if a detour is necessary.

By the 1990s, automobile navigation and route guidance systems achieved a degree of technological maturity and became a cornerstone for Intelligent Transportation Systems (ITS, called IVHS, or Intelligent Vehicle Highway Systems, until 1994) under development in the United States and other parts of the developed world. The objectives of ITS are to improve transportation efficiency and safety and to minimize traffic congestion and environmental impact through the application of information, communications, and control technologies. In addition to navigation and other vehicular systems such as adaptive cruise control, lane-keeping, and infrared night vision, other aspects of ITS include sensor networks in the roadways of major cities to monitor traffic from a central location and adjust the timing of area-wide traffic signals to improve overall traffic flow. Traffic information thus collected is also displayed on electronic message signs along expressways and major arteries advising motorists on conditions ahead and, when appropriate, to take alternate routes. Although still in the early stages of implementation, traffic information may also be communicated to properly equipped automobile navigation systems for use in calculating or updating routes. A more mature ITS application involving automatic data exchange between automobiles and the infrastructure is electronic toll collection.

Federal Highway Administration (FHWA) research performed in the 1980s confirmed the need for ITS and automobile navigation. One study estimated that in 1987 congestion caused two billion vehicle-hours of delay on urban expressways and cost $16 billion, and predicted that congestion delay will cost $88 billion in 2005 in 1987 dollars.[1] Another FHWA study estimated that poor navigation and route following skills waste almost 7 percent of all distance traveled by noncommercial vehicles and over 12 percent of time spent in such travel, thus contributing to traffic congestion.[2] The study estimated that the annual cost of this excess travel to individuals and to society was $45 billion. Although difficult to quantify in economic terms, heavy environmental costs also result from increased vehicle emissions caused by congestion delay and excess travel. In addition to reducing congestion, vehicles equipped with navigation systems may have a lower risk of accidents because they travel shorter distances and spend less time on the road in performing their missions.[3] Navigation systems further enhance traffic safety by enabling drivers to proceed with confidence to their destinations without hesitating at decision points or having to make last-moment lane changes or abrupt maneuvers to stay on the proper route. However, navigation system controls and displays must be designed with safety in mind to avoid offsetting increases in accident risk.

Today's sophisticated navigation and route guidance systems integrate several technologies to automatically determine vehicle location with sufficient accuracy to identify the road traveled and intersections approached. Figure 7.1 shows a representative system diagram. The distance sensor is usually the odometer that is common to all automobiles. The heading sensor may be any of a variety of means for detecting and measuring direction or changes in direction of travel. The navigation computer uses a process known as dead reckoning to calculate approximate vehicle location relative to a known starting point by tracking distances and directions traveled.

The location sensor gives a measure of absolute location and has routinely been included in automobile navigation systems only since completion of the satellite-based Global Positioning System (GPS) in the 1990s.[4] The map database is used by related software to calculate the optimal route to specified destinations, present real-time route instructions (i.e., just as a turn or other maneuver is approached), and to support map matching, a software process that continuously reconciles the calculated vehicle path with the actual route. Although current systems seldom have a transceiver for data communications with the transportation infrastructure, some are beginning to include a receiver for real-time traffic data

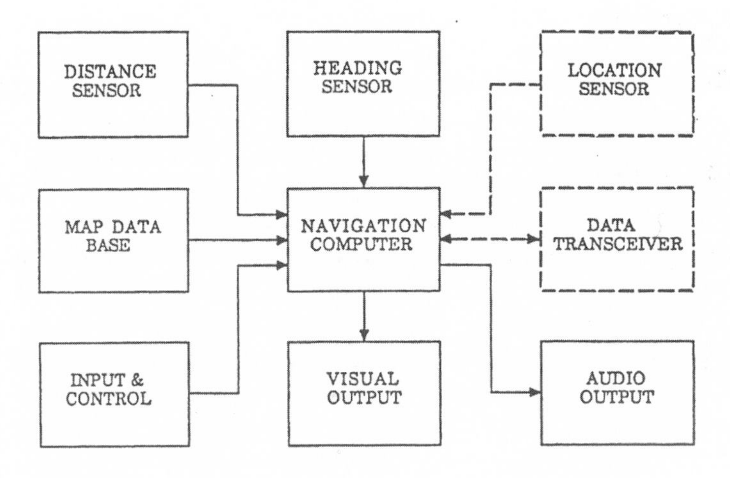

FIGURE 7.1 Typical components and subsystems of vehicle navigation systems. Drawing by the author.

for use in calculating optimal routes. As indicated in figure 7.1, the typical automobile navigation system also requires an effective means of communicating with the driver. This usually takes the form of graphic display and audio communications—primarily synthesized speech emanating from the systems but, increasingly, includes capability to recognize the instructions spoken by the driver as well.

In addition to discussing the various technologies used by intelligent navigation systems in more detail, this essay briefly reviews the historical background of early mechanical and modern electronic automobile navigation before turning to an outline of its development in the major regions involved: America, Europe, and Japan.

Early History

ROAD MARKINGS AND SIGNS

The earliest traveler information systems were in the form of road markings and signs. The Romans marked distances on their roads with marble shafts about two meters high spaced at approximately 1.5-kilometer intervals. Within about 200 km of Rome, these milestones showed the distances to the city. Bronze plaques on the golden milestone erected by Emperor Augustus in 20 BC at a corner of the forum showed the distance to various parts of the empire. Elsewhere, milestone markings were based on distance to the nearest town. A system of signing using either cairns or posts was introduced at French crossroads early in the seventeenth century. Denmark instituted an extensive network of milestones at 1.8-kilometer intervals in 1697. Mathew Simon's 1635 *Directions for English Travillers* notes that directional signs were found in England "in many parts where wayes be doubtful." In 1698, an English law required each parish to place guideposts at its crossroads. Finger posts, often shaped like a pointing hand, and stone mileposts set in the ground became more common from the eighteenth century. Some of these signs included travel times; a forerunner of variable travel times displayed by electronic message signs in today's ITS traveler information systems.

Systematic route marking in Britain's American colonies began with a 1704 Maryland law that required trees along a route to be marked with an elaborate system of notches, letters, and/or colors. In the 1760s, Benjamin Franklin actively promoted the erection of milestones as an aid to postal service. In the United States in the eighteenth century, signposts were often funded by public subscription.[5] But, as Akerman discusses in chapter 5, highway route markings did not become a major objective until the second decade of the twentieth century.

The first attempt at international traffic signs occurred at a meeting of the International League of Tourist Associations held in London in the late 1890s. There the league proposed a system based on arrow signs developed by Italy. Although the matter remained topical among tourist bodies and automobile clubs, formal government action did not occur until 1908 at the initial meeting of the Permanent International Association of Road Congresses (PIARC). After World War II, the United Nations took over sign standardization. More than fifty years later there are still two major systems used in the world, the European and the American, with European symbols, which use language-independent symbols rather than words, gradually becoming the world standard.[6]

ROAD BOOKS AND MAPS

Written navigational instructions were the second type of road navigational tool to be developed. As we have seen in chapter 1, written guides date at least from medieval times. Christopher Colles published the first atlas of American road maps, *A Survey of the Roads of the United States of America,* in 1789. The first major effort toward effective paper road maps was started by cyclists who organized the League of American Wheelmen in 1880. LAW published a number of cyclist road books and maps by 1900 (see chapter 5). The founding of the Automobile Club of America in 1899, followed in

1902 by the American Automobile Association with which it later merged, continued the push for better maps. Nonetheless, the proliferation of the automobile early in the twentieth century created a strong demand for routing information at a time when road signs and accurate maps were still scarce in America. The first guides for helping motorists find their way were descriptive books that included lists of various tours in the region, explicit directions for following particular routes, and limited road maps as inserts.

The tradition of identifying local features such as lodgings, restaurants, hospitals, and tourist attractions began with the earliest road guidebooks. The books also contained other essential tourist information such as the location of garages, gas pumps, and tire repair shops. As Akerman discusses elsewhere in this volume, tire and oil companies recognized the promotional value of road books and maps early, and began distributing them to motorists shortly after 1910 without charge to encourage driving and to advertise products. Road books remained popular until about World War I, when they became so unwieldy (some had 1000 pages) that maps, which had gradually increased in size as well, began to appear separately from books.[7]

THE ODOMETER AND DIFFERENTIAL ODOMETER

Mechanized route guides began to appear in America before 1910, competing with road books and pictorial route guides. All of these pioneering systems used odometers to track distance along a route and automatically display turn-by-turn route guidance. Even today, most automobile navigation systems include an odometer as part of a dead-reckoning subsystem, and some systems use a version known as the differential odometer. The odometer measures distance traveled and takes its name from the Greek words *hodos* (way) and *metron* (measure). The Roman architect Vitruvius wrote the first Western descriptions of the odometer in the first century BC, and the odometer started appearing in China during the late Han Dynasty (25–220 AD).[8]

One early odometer described by Vitruvius recorded distance by periodically dropping stones into receptacles.[9] The odometer was installed on a four-wheeled wagon with each wheel measuring about four feet in diameter, thus corresponding to approximately twelve and a half feet in one complete turn. A drum attached to the inner side of a wheel hub bore a single tooth protruding at a right angle. Another drum mounted on the axle had 400 teeth at equal intervals, including one tooth that protruded further than the others. Each time the wheel completed a revolution, the tooth of the hub-mounted drum pushed the axle-mounted drum ahead one tooth. Thus, the 400-tooth drum made a complete revolution in about one mile. To record distance, the single protruding tooth of the 400-tooth drum advanced a third drum that was mounted horizontally and had a series of holes, each holding a small, round stone. Each time this drum advanced, a stone aligned with a hole in the stationary plate beneath the drum and fell through a chute to a metallic receptacle underneath the chassis. The completion of each mile of travel was thus annunciated by a clanging noise, while the number of stones in the receptacle indicated cumulative travel. An ancient Chinese odometer known as the "drum carriage" used sound effects as well. A gear mechanism driven by the wheels of a cart operated two automatons: one struck a drum at the end of every li (~0.3 miles); the other rang a bell at the end of every ten li.[10]

Mechanical odometers for automobiles use principles similar to those of ancient odometers, rotating numeral-faced cylinders to indicate distance traveled in tenths, units, tens, hundreds, and so on. The electronic odometers that have displaced mechanical ones in most recent automobiles use a computer to add increments of distance to the displayed value for each electrical pulse received from a sensor monitoring the rotation of a wheel or drive shaft. Signals from an electronic odometer, from a transducer attached to the cable of a mechanical odometer, or from one or more wheel rotation sensors, are used as input for dead-reckoning calculations in most modern navigation systems.

Dead reckoning is a process for estimating a vehicle's position by tracking distances and directions traveled from a known starting point. Vehicle heading or changes in heading can be determined by magnetic compass, gyroscopic devices, or by a differential odometer. The differential odometer is essentially a pair of odometers, one for a wheel on each side of a vehicle. When the vehicle turns, the outer wheel travels further than the inner wheel. The change in heading is calculated by dividing the outer wheel's extra distance by the width of the vehicle. This continuous comparison of the difference in travel by the two wheels indicates the occurrence and magnitude of turns.

The differential odometer, invented in China about 2000 years ago (legend dates it even earlier), was the technological basis for the "south-pointing chariot," an automatic direction-keeping mechanism and the world's first vehicular navigation system.[11] A gear train driven by the chariot's outer wheel engaged and rotated a horizontal turntable to exactly offset changes in heading. Thus, a figure with an outstretched arm mounted on the turntable always pointed in its original direction regardless of which way the chariot turned. The obscure principle of operation was periodically

FIGURE 7.2 Working model of south-pointing chariot constructed by H. D. Gardner. Reprinted with permission of the Institute of Navigation.

lost and rediscovered until the south-pointing chariot was finally eclipsed by the twelfth century invention, also in China, of the magnetic compass. Ancient documents recorded use of a south-pointing chariot in the Warring States (475–221 BC), but did not record its design. To settle an argument that such a thing could exist, the mechanical engineer Ma Chuin devised such a chariot operated by a gear mechanism in the third century AD. Because early Chinese literature so thoroughly confused the south-pointing chariot with the magnetic compass, British engineers and historians did not establish that the south-pointing chariot had nothing to do with magnetism until early in the twentieth century.[12] Several working models have demonstrated the south-pointing chariot's principal of operation. The model shown in figure 7.2 was constructed by H. D. Garner.[13]

MECHANICAL GUIDANCE SYSTEMS

Dozens of devices for automobile route guidance based on the odometer were patented by 1920.[14] These pioneering devices incorporated route map information in various forms including sequential instructions printed on a turntable, punched in a rotating disk, and printed on a moving tape, all driven by an odometer shaft in synchronization with distance traveled along a route. The mechanical route guides thus automatically provided explicit route instructions at decision points along the way, an approach now common in electronic systems. One of the earliest guidance devices, the route indicator patented in 1909 by Lindenthaler and Protz,[15] contained

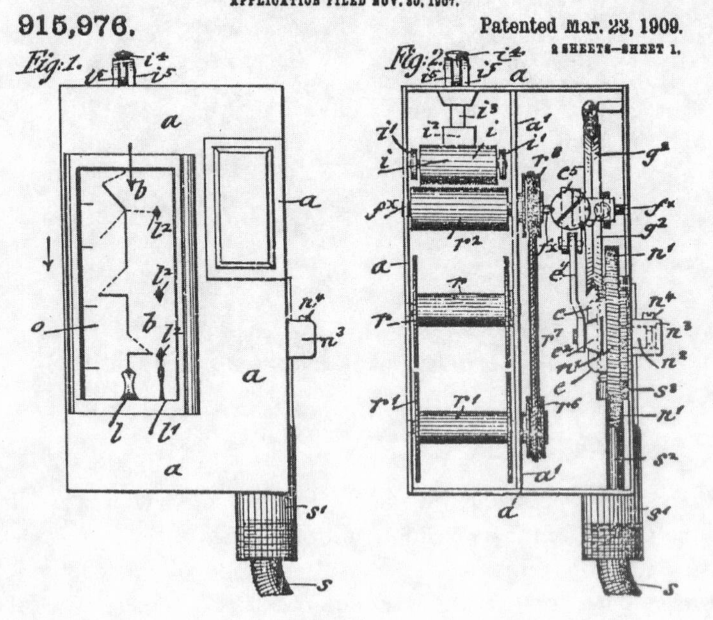

FIGURE 7.3 Diagram of Lindenthaler & Protz route indicator. Lindenthaler, F. J., and Protz, J., "Route-Indicator for Automobiles," U.S. Patent No. 915,976 (1909).

a ribbon printed with graphic route instructions that was wound by a roller mechanism driven by an odometer cable. The route-indicating ribbon was marked at one edge with mile notations and at the center with heavy lines of different lengths, indicating the direction and distances of individual route segments (figure 7.3). As the graphic turn indication reached the lower edge of the glass cover, the driver was cued to turn in the direction shown.

Several mechanical guidance devices were described in the 1911 annual automobile issue of *Scientific American:*[16] "All tourists by automobile know the difficulties and annoyances of finding and keeping on the best routes to their objective. It has been the object of a number of inventors to overcome these difficulties and simplify the unwelcome task of finding the route by producing mechanical devices for attachment to the car that would serve to guide the motorist unerringly over any chosen route. Three such instruments have been perfected and put in the market."

One device the article described, the Jones Live-Map, consisted of a turntable slowly rotated by a gear train connected to one of the vehicle wheels by a flexible shaft. Paper disks for individual routes had a scale of miles printed around their perimeter as shown to the left of the steering wheel in figure 7.4. These interchangeable paper disks, available at some garages, were mounted on the turntable beneath a glass cover. Detailed route instructions keyed to specific distances from the beginning of a route come into view under a pointer when they are to be executed. An advertisement for the Jones Live-Map claimed, "You take all the puzzling corners and forks with never a pause. You never stop to inquire."

The somewhat more sophisticated Chadwick Road Guide was introduced in 1910.[17] Like the Jones Live-Map, the Chadwick Road Guide rotated a calibrated disk in synchronization with distance traveled. This disk, though, consisted of metal punched with holes spaced to coincide with decision points along the route represented. An array of spring-loaded pins behind the slowly rotating disk was normally depressed, but when a punched hole traversed a pin, the pin released and activated a signal arm

FIGURE 7.4 Jones Live-Map. Photograph by the author. Courtesy of the Newberry Library.

bearing a color-coded symbol indicating the action to be taken. Simultaneously, a bell sounded to draw the driver's attention to the symbol. The symbols represented the following instructions:

Continue straight ahead
Crooked road ahead
Bad road, water crossing, or bridge ahead
Turn sharply to the left
Bear to the left
Danger, go slow
Speed trap, go slow
Railroad crossing ahead
Bear to the right
Turn sharply to the right

According to an advertisement, "The Chadwick Automatic Road Guide is a dashboard instrument which will guide you over any highway to your destination, instructing you where to turn and which direction. You will be warned upon your approach to rough roads, railroad tracks, speed traps.... A warning signal will alert you before all turns or danger points."

J. B. Rhodes invented another automatic route guide in 1910, an odometer-like device whose cylinders were printed with directional signals as well as numerals.[18] Whereas the Jones and Chadwick devices required a specially printed paper disk or punched metal disk for each route, Rhodes' Route Indicator was programmable. Following a key given in a master route book, the user positioned and initialized alternating cylinders printed with direction signals and numerals in such a way that the proper route instruction (e.g., "straight ahead, 4.9;" "right, 1.7;" "left, 3.5;" etc.) would automatically appear under a sliding index at the necessary time. A bell sounded automatically to alert the driver to each new route instruction.

EARLY ELECTRONIC SYSTEMS

The U.S. Army Corps of Engineers developed the vehicular odograph (figure 7.5), one of the earliest dead-reckoning navigation systems to incorporate electronic elements, for jeeps and other military vehicles during World War II.[19] The major components of the vehicular odograph included a magnetic compass whose needle position was read by a photocell. The compass output drove a servomechanism to rotate a mechanical shaft

FIGURE 7.5 Vehicular odograph. Photograph courtesy of the author.

corresponding to the vehicle's heading. The heading shaft and an odometer shaft were coupled to a mechanical computer that transformed distance increments and associated directions into x, y components and drove a stylus to automatically plot the vehicle's course on a map of corresponding scale. The vehicular odograph exhibited a wide range of accuracy depending upon terrain, vehicle jolting, operator skills, and so forth, but typically accumulated one mile of error for each 50 to 150 miles driven. Some 5,000 units were built, most of which were installed in jeeps as shown in figure 7.5.

Another pioneering early development was the Electronic Route Guidance System (ERGS), a vehicle navigation system researched in the late 1960s by the U.S. Bureau of Public Roads (now the Federal Highway Administration) as a means of controlling and distributing the flow of traffic.[20] ERGS uses strategically located short-range transmitters known as proximity beacons to broadcast their location to passing vehicles. Thus, reception by a vehicle of the signal from a short-range proximity beacon identifies the receiving vehicle's location. With the ERGS concept (see figure 7.6), equipped vehicles were to interact via two-way communications with the roadside beacons to obtain traffic-responsive routing instructions at decision points in the road network. An in-vehicle console permits the driver to enter a selected destination code. The code is transmitted when triggered by a roadside beacon while approaching key intersections. The roadside unit immediately analyzes routing to the destination and transmits the next route instruction for display on the vehicle's console.

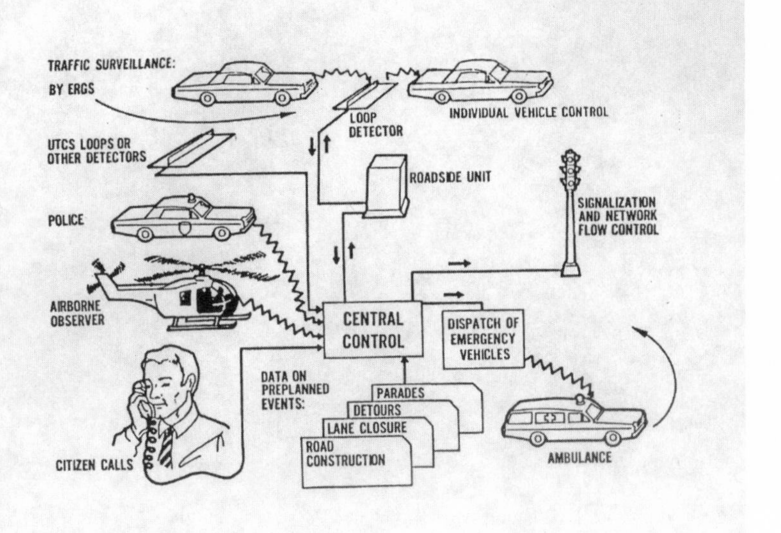

FIGURE 7.6 ERGS dynamic route guidance. Author's private collection.

Although technically sound, ERGS required expensive roadside infrastructure, and the development effort was terminated by congressional mandate in 1970 following limited testing of various subsystems. However, similar approaches were carried through further stages of development in Japan and Germany during the 1970s, and ITS system architectures now emerging throughout the world function comparably.

The Automatic Route Control System (ARCS), developed in 1970–71 by Command Systems Corporation,[21] was the first autonomous route guidance system to use an on-board digital computer with digitized maps and map-matching software in conjunction with a dead-reckoning subsystem.[22] ARCS based its digital maps on the concept that networks of roads and streets may be modeled by the coordinates of points (called nodes) representing intersections, and vectors (straight lines with directionality) representing roads connecting the points, and that particular routes may be represented as unique sequences of mathematical vectors. ARCS included an electronic differential odometer for dead reckoning and used map-matching software to correlate the apparent vehicle route with the actual route map stored on a digital tape cartridge. Prerecorded audio route guidance instructions were automatically issued upon approaching each turn. A second version issued operator prompts (figure 7.7) and real-time route guidance instructions in the form of simplified graphics (figure 7.8) on a plasma display panel, which foreshadowed concepts commonly used in modern automobile route guidance displays.[23]

The *Fort Worth Star Telegram* sponsored the development of ARCS for use in automating home delivery of the daily newspaper. This application required high accuracy (±15 feet) to unambiguously signal which houses should receive newspapers from a delivery van driven through home delivery routes (figure 7.9). Operators wearing headsets were signaled by a special tone to toss a rolled newspaper as passing each subscriber's house.

FIGURE 7.7 ARCS display panel (center), digital tape cartridge (left), and control panel (right). Photograph courtesy of the author.

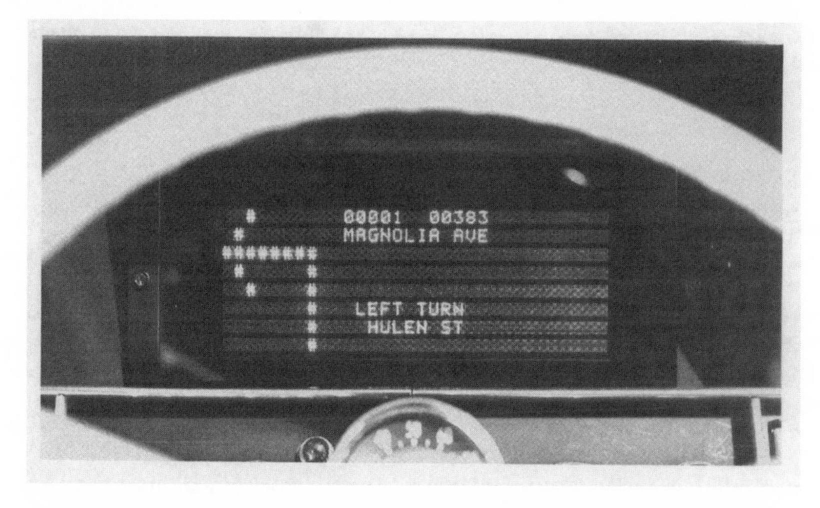

FIGURE 7.8 ARCS route guidance instruction. Photograph courtesy of the author.

FIGURE 7.9 ARCS newspaper delivery van. Photograph courtesy of the author.

The route guidance function was added almost as an afterthought to help drivers follow the proper route. ARCS's overall functionality as a route guidance and newspaper delivery control system was successfully field tested in daily use over a one-year period by the *Fort Worth Star Telegram*.[24] However, the newspaper decided to discontinue the effort to automate delivery due to objections, including sabotage to the ARCS vehicle by labor interests. One automated delivery vehicle with a crew of three could displace approximately twenty conventional delivery workers.

DIGITAL MAPS AND MAP MATCHING

Suitable digital maps were not available when ARCS began. The ARCS project thus required development of special digital maps for the delivery routes (or itineraries) selected for the *Fort Worth Star Telegram* newspaper delivery field test. These digital route maps were generated by operating the ARCS equipment in a data acquisition mode to automatically record distances between turns and the direction and magnitude of turns during preliminary drives over the test routes. An operator simultaneously noted the address number for each house on each side of the street, and pressed a switch for ARCS to record the distance of each house from the last turn. A centrally located IBM 1130 computer processed the recorded data into a map file stored on a digital tape cartridge, with subscriber status added for each house on the newspaper route. Figure 7.10a is the map of a typical route segment and figure 7.10b shows the corresponding ARCS mathematical model used for map matching and route guidance and for signaling the delivery of newspapers.

Contemporaneous with the specialized digital maps developed for ARCS, the U.S. Bureau of the Census promulgated the GBF/DIME

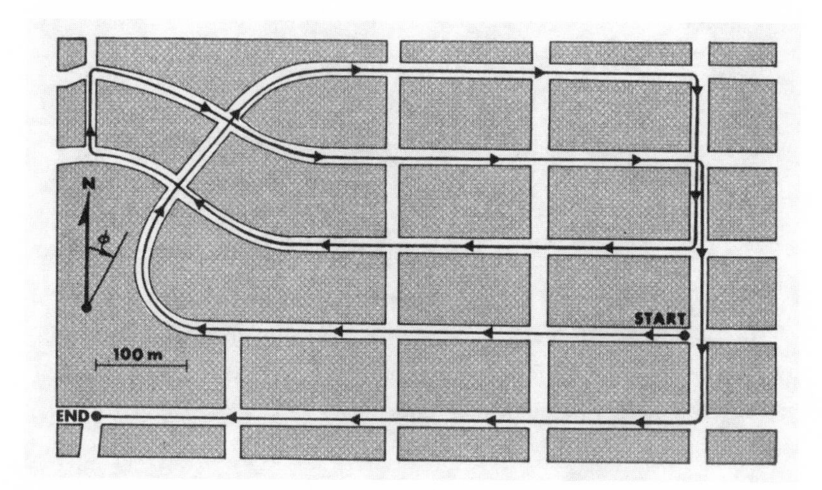

FIGURE 7.10a Typical newspaper delivery route segment. Map drawn by the author.

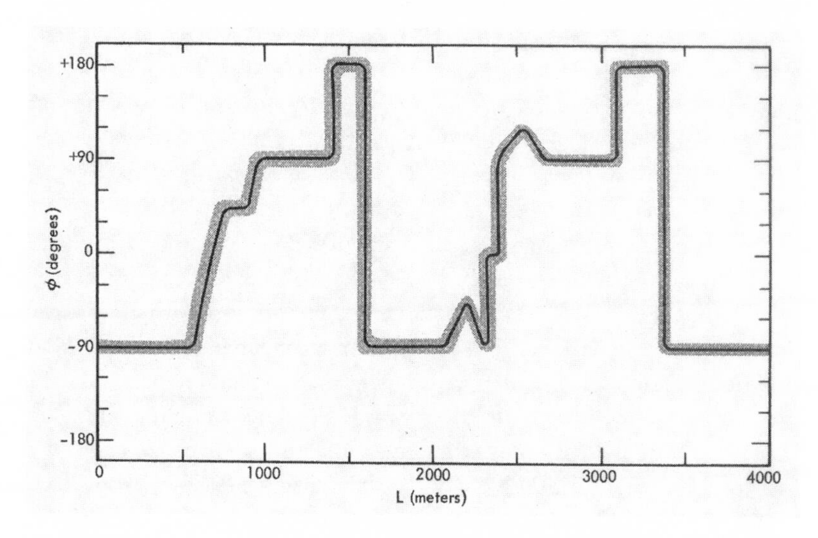

FIGURE 7.10b ARCS mathematical model of route segment. Chart drawn by the author.

FIGURE 7.11 Nodes and street segments of GBF/DIME map.

(Geographic Base File/Dual Independent Map Encoding) map system for processing census data from major population centers.[25] In a manner that was mathematically equivalent to that used by ARCS, the GBF/DIME represents geometrical features such as roadways and boundaries in abstract form with a minimum of data. By considering each road or street as a series of vectors and each intersection as a node, a map may be viewed as a set of interrelated nodes, lines, and enclosed areas as illustrated by figure 7.11. Nodes are identified by their coordinates (e.g., latitude and longitude). A straight road segment connecting two nodes is represented by a single vector. Curved road segments are approximated by a series of vectors connecting intermediate "shape points." The coordinates of node points may be taken from paper maps, aerial photographs, satellite images, and other sources.

The classic approach for digitizing maps uses special computer workstations, which record the coordinates of a given point when the cross hair of an instrument is placed over the point and a button pressed. This process has been automated to varying degrees. In some cases, the printed map is digitally scanned to obtain a raster image (i.e., a TV-like image) that is then converted to vector form by software. Today, some digital map data is acquired by driving the roadways in a GPS-equipped vehicle operating in a recording mode reminiscent of that first used by ARCS in 1970.

The digital map databases currently used with automobile navigation systems evolved from the same concepts used in ARCS and GBF/DIME, and are produced by commercial firms such as NAVTEQ (www.navteq.com) and TeleAtlas (www.teleatlas.com). In addition to defining road location and geometry, these digital maps identify the names, road classification, and address ranges for each segment of a road or street. Map databases used with systems that give turn-by-turn route guidance also require traffic attributes such as turn restrictions and delineation of one-way streets. "Yellow pages" information giving the locations and descriptions of service stations, garages, parking facilities, public buildings, hotels, restaurants, tourist attractions, and other frequently used directory information is commonly included in map databases for automobile navigation.

Virtually all modern vehicle navigation and route guidance systems use map-matching algorithms, the software process invented for ARCS to recognize a vehicle's location by matching the pattern of its apparent path (as estimated by dead reckoning and/or GPS) with the road patterns of digital maps stored in computer memory.[26] The distance along road segments is measured and any accumulated distance error is automatically removed at each node upon confirmation that an expected change in heading actually occurs within acceptable margins. If the vehicle's measured path appears to be slightly displaced but parallel to the roadway as defined by the digital map, this is recognized as position error and is automatically corrected. If the vehicle clearly departs from the defined road network, (e.g., into a parking lot), the algorithm continually compares the vehicle's estimated coordinates with those of the road links that enclose the vehicle's off-road location in order to recognize where the vehicle returns to the road network. Like dead-reckoning and GPS location estimates, digital maps used in vehicle navigation systems may also have some degree of error. The map matching process reconciles these errors as well to maintain an accurate indication of the vehicle's location relative to the road network as defined by the map database.

Automobile Navigation Comes of Age

Most major technologies required for modern automobile navigation systems were already established when the microprocessor emerged in the 1970s to support their integration and enhancement by computer software. These technologies subsequently underwent extensive refinement, and a variety of system architectures had been explored by the time practical systems reached the market in the late 1980s. Among other enhancements of the 1980s was the development of color displays for digital maps and of CD-ROMs for digital map storage. One of the most significant improvements in the 1990s was the addition of GPS satellite positioning receivers, which quickly became a common feature of automobile navigation systems as the full constellation of twenty-four GPS satellites approached completion in 1993.[27]

Although satellite positioning for naval vessels had been around since 1964, the first such system, Transit, deployed only five satellites and provided neither sufficient accuracy nor continuity for use in automobile navigation. The current GPS system comprises twenty-four satellites plus a few spares, enabling a receiver at any point on earth to estimate its position to within five meters if signals from at least four satellites are received simultaneously. Unfortunately, the high frequency GPS radio signals (~1,500 MHZ) are often blocked by buildings, tunnels, foliage, and such. Thus, the most sophisticated automobile navigation systems typically include dead reckoning sensors and map matching software to compensate when GPS signals are not available.

The synergies among GPS, dead reckoning, and map matching for tracking vehicle location in automobile navigation systems may be illustrated by the analogy of balancing a stool. A person could balance on a one-legged stool for a short period of time, and on a two-legged stool for a longer period. Add a third leg to the stool and it becomes so stable that a person could balance indefinitely. Likewise, a navigation system based on dead reckoning alone (as was the case for some early systems) can track vehicle location until the errors inherent to all distance and heading sensors accumulate to the point of becoming useless. Add map matching to remove errors every time the vehicle makes a turn and the system will maintain sufficient location accuracy as long as the vehicle remains on an accurately mapped road network. However, eventually dead reckoning with map matching will fail, which requires that the vehicle location be reset by a tedious manual process. However, add a GPS receiver as a third "leg" and the navigation system becomes very robust and will recover on its own in the event of GPS signal blockage or map error.[28]

Development of ITS started in Europe, Japan, and the United States during the 1980s and 1990s, spurring development of automobile navigation systems and influencing system architecture. Although there was considerable exchange of information and technology among the European, Japanese, and American ITS programs, some distinctions among the systems developed in different geographic regions are worth noting. The following regional summaries are drawn largely from an analysis and comparison of world ITS progress performed in response to questions about international competitiveness raised by the U.S. Congress.[29]

Beginning in 1973, Japan's Ministry of International Trade and Industry (MITI) sponsored CACS (Comprehensive Automobile Traffic Control), a seven-billion yen, six-year ITS route guidance research project.[30] Much like the earlier Electronic Route Guidance System (ERGS) research in the United States (see figure 7.6), CACS used short-range proximity beacons in the form of inductive loop antennas buried in the roadway to communicate route instructions to equipped vehicles as they passed over the buried antennas. However, unlike ERGS, which was tested at only two intersections, full-scale CACS infrastructure was established in a twenty-eight-square-kilometer area in southwestern Tokyo for trials involving a fleet of 330 test vehicles. Similar to ERGS, the test vehicles were equipped with a console for entering a destination code and displaying route guidance.

The CACS operational trial and related computer modeling confirmed the efficacy of dynamic route guidance to improve traffic management, and led to MITI's establishment in 1979 of JSK (Association of Electronic Technology for Automobile Traffic and Driving), a nonprofit membership foundation that was created to popularize the CACS results and to expedite the widespread introduction of in-vehicle route guidance and information systems. Shortly after the establishment of JSK, Honda, Nissan, and Toyota each introduced the first generation of navigation systems for their automobiles to the Japanese market in 1981. The Nissan and Toyota systems used odometer and electronic compass inputs in dead-reckoning subsystems for tracking vehicle location relative to starting and destination positions.[31] Indicator lights and LEDs (light emitting diodes) displayed distance and directions to destinations in these systems. The Honda navigation system, known as the Electro Gyro-Cator, was the only first generation system to include a form of map display.[32] Dead reckoning based on odometer signals and a gyroscopic device for sensing heading changes calculated the approximate route driven by the automobile. The route was graphically traced on a CRT screen behind a transparent map overlay of appropriate scale to show present location and direction of travel, as illustrated in figure 7.12. Although none of these early systems succeeded on the market, their development involved considerable research on the characteristics and limitations of different techniques and sensors used for dead reckoning. They set patterns for the refined dead reckoning subsystems used in the more sophisticated navigation systems that began to reach the market in Japan in the late 1980s.

In the mid-1980s, the Ministry of Construction (MC) and the National Police Agency (NPA) of Japan initiated parallel and somewhat competitive

FIGURE 7.12 Honda Electro Gyro-Cator. Reproduced from K. Tagami, T. Takahashi, and F. Takahashi, "'Electro Gyro-Cator,' New Inertial Navigation System for Use in Automobiles." Reprinted with permission from SAE Paper 830659 © 1983 SAE International.

ITS field tests involving data communication between navigation-equipped vehicles and the traffic management infrastructure. The RACS (Road Automobile Communication System) project of the MC, and the AMTICS (Advanced Mobile Traffic Information Communications System) project of the NPA differed mainly in the communication links tested and in the jurisdictions of the sponsors (the MC manages expressway traffic whereas the NPA manages surface street traffic). Navigation systems from twelve different Japanese automobile manufacturers and electronic firms were entered in the RACS and AMTICS field tests. Many of these systems included features such as color CRT map displays and CD-ROM digital map storage in addition to the underlying dead-reckoning subsystem. Some were prototypes or adaptations of navigation systems that had already reached the market as optional factory-installed automobile equipment.

The Toyota Electro-Multivision, which was developed by Nippondenso, is a leading example that illustrates the evolution of automobile navigation systems in Japan during the late 1980s and early 1990s.[33] When introduced in 1987, the original version of the Electro-Multivision was the first factory-installed system to include digital maps stored on CD-ROM for display on a color CRT. The digital maps for the original version were of limited detail, and the navigation function gave only approximate position based on dead reckoning alone. Nonetheless, this pioneering system included limited "yellow pages" information that gave the location of facilities likely to be of interest to motorists. The reference atlas structure of the first Electro-Multivision was also innovative. An initial display showed a color map of Japan with sixteen superimposed rectangles. Touching a particular rectangle causes the map area it encompasses to zoom and fill the entire screen, once again with grid lines superimposed to form sixteen rectangles. Thus a few touches of the screen would take the viewer from an overview of the entire country down to major roads and landmarks in some section of a major city. The vehicle position and destination appeared as icons on the map display. Improved digital maps along with map-matching software and a GPS receiver were added in subsequent versions of the system. In 1991, a routing feature was added to calculate a suggested route and highlight it on the color LCD (liquid crystal display) map. A later version added synthesized voice route guidance instructions.[34]

Nissan's competing Multi AV system, which became available on top-line Nissan automobiles in 1989, included a navigation subsystem developed by Sumitomo Electric. The original Multi AV was the world's first factory-installed navigation system to feature map matching. A later version was the first production automobile navigation system to include a microwave beacon receiver for information transmitted by microwave beacons installed in limited areas under a pilot program by the Ministry of Construction.[35] A subsequent version included a gyroscopic device in lieu of the differential odometer used for dead reckoning in earlier versions. Figure 7.13 shows the typical layout of Multi AV components installed in an automobile. Like virtually all factory-installed navigation systems in Japan, the original Multi AV was bundled with comprehensive video and audio entertainment features.

EUROPE

An early European example of automobile navigation was ALI (Autofahrer Leit-und Informationsystem), a route guidance system that was tested on the German autobahn in the late 1970s.[36] Bosch and Volkswagen jointly developed ALI, which used inductive loops to communicate with in-vehicle equipment in an approach similar to that of the earlier ERGS project in the United States (see figure 7.6) and CACS project of Japan. Development of autonomous navigation and route guidance systems (i.e., capable of operating without ITS infrastructure) got underway in Europe during the 1980s. In the Netherlands, Philips developed CARIN (Car Information and Navigation System) in 1984.[37] CARIN, which used dead reckoning with map matching, was the first route guidance system in the world to use

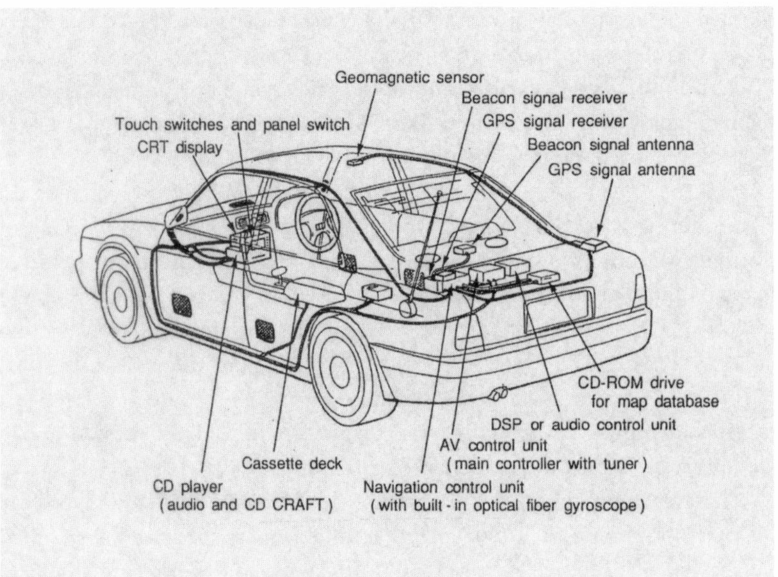

FIGURE 7.13 Nissan Multi-AV Installation. Reproduced from T. Hirata, T. Hara, and A. Kishore, "The Development of a New Multi-AV System Incorporating an On-Board Navigation Function." Reprinted with permission from SAE Paper 930455 © 1993 SAE International.

digital maps stored on CD-ROM and the first to use a color CRT for map display. The CRT could also present real-time turn-by-turn route guidance in the form of simple graphics augmented by synthesized voice (see figure 7.14). Later versions of CARIN were equipped with communication links for use in European ITS operational field tests. BMW in Germany began offering a version incorporating a GPS receiver as a factory option in 1994.

Bosch, an early developer of autonomous navigation systems in Germany, started with a dead-reckoning/map-matching route guidance system called EVA that was developed for tests and demonstrations in 1983. EVA generated optimum routes and presented turn-by-turn route guidance by LCD and synthesized voice.[38] Rather than developing EVA into a product for the market, Bosch subsequently worked under a licensing arrangement with Etak, Inc. of the United States to introduce TravelPilot, a new version of the Etak Navigator, in Germany in 1989.[39] The Etak Navigator/Bosch Travelpilot, shown in figure 7.15, is described below. Bosch subsequently phased out the Travelpilot and collaborated with Mercedes-Benz to develop Auto Pilot, a more advanced route guidance system with color LCD and GPS receiver sold as a factory option.

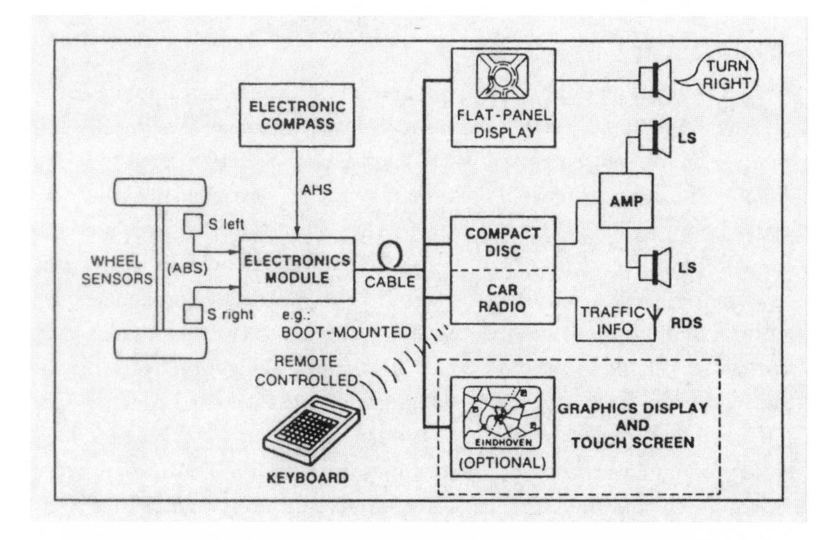

FIGURE 7.14 Philips CARIN Configuration. Reproduced from Martin L. G. Thoone, Leon M. H. E Driessen, Cees A.C.M. Hermus, and Kees van der Valk, "The Car Information and Navigation System CARIN and the Use of Compact Disc Interactive." Reprinted with permission from SAE Paper Series 870139 © 1987 SAE International.

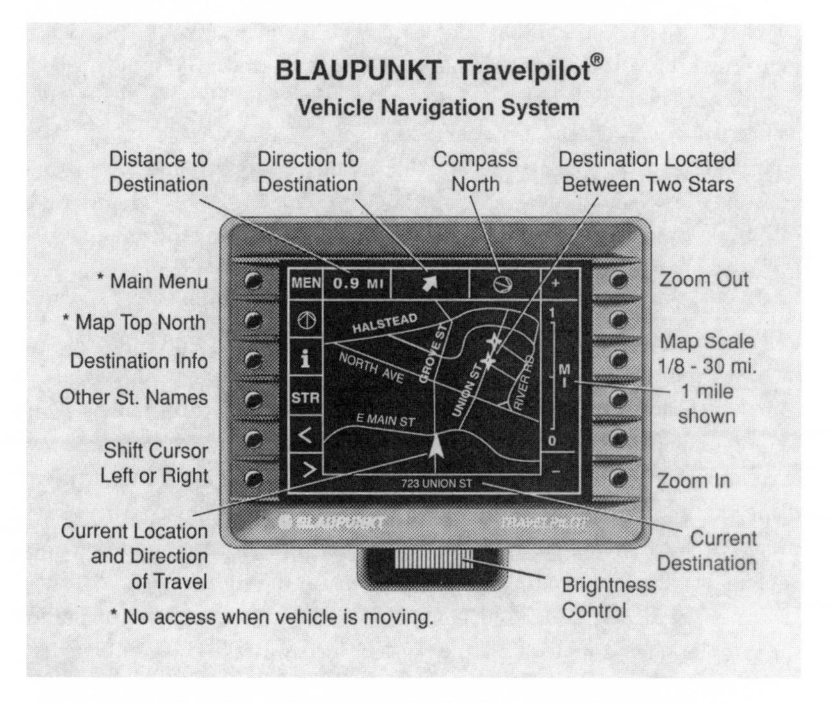

FIGURE 7.15 Bosch/Etak TravelPilot. Reproduced with permission from Blaupunkt USA.

The U.S. Bureau of Public Roads pioneered the role of automobile route guidance in ITS with the Electronic Route Guidance System (ERGS) project, and the U.S. Department of Defense developed the satellite-based Global Positioning System (GPS), both of which are described in earlier sections. Although ERGS development was discontinued by political mandate, the concept was carried forward in the 1970s in Japan and Germany, thus encouraging foreign industry to pursue ITS-related developments more aggressively than the United States. Another early U.S. development that, in retrospect, helped pave the way for automobile navigation development worldwide was the GBF/DIME (Geographic Base File/Dual Independent Map Encoding) project of the U.S. Bureau of the Census, also described above. Although the accuracy and completeness of GBF\DIME maps were inadequate for automobile navigation, the resulting map encoding techniques are now widely used for road map databases in automobile navigation and route guidance systems as well as in traffic management centers and fleet dispatch offices.

The first commercially available automobile navigation system in the United States was the Etak Navigator introduced in California in the mid-1980s.[40] This system included digitized road maps, dead reckoning with map matching, and a monochromatic CRT for map display. It used an electronic compass and differential odometer for dead reckoning. The equivalent of two printed city street maps was digitized and stored on 3.5-Mb tape cassettes for map matching and display purposes. Although sales were modest, the highly publicized Etak Navigator drew widespread attention to the concept of an electronic map display with icons showing current location and destination.

Etak, Inc. and Bosch jointly designed the Travelpilot (figure 7.15), essentially a second-generation of the Navigator, and introduced it in Germany in 1989 and in the United States two years later.[41] A major enhancement of the Travelpilot was its use of CD-ROM storage for digitized maps. The 640-Mb capacity of the CD-ROM permits the entire map of some countries to be stored on a single disc. Like the original Etak version, the Travelpilot displays a road map of the area around the vehicle as illustrated by figure 7.15. The arrowhead icon below the center of the screen indicates the vehicle location and heading. The vertical bar at the right edge of the map indicates the display scale, which can be zoomed in to one eighth of a mile for complete street detail or zoomed out to thirty miles to show only major highways. The map is normally oriented such that the direction in which the vehicle is heading points straight up on the display, thus allowing the driver to easily relate the map display to the view outside.

A menu accessible through the MEN button permits use of a scrolling approach to enter destinations by street address, intersection, and so on when parked. Travelpilot then displays the location of the destination as a flashing star on the map. As illustrated in figure 7.15, two flashing stars bracket the destination. In this case, the stars mark the block whose address range includes the street number of the destination. A line of information across the top of the map indicates the straight-line distance and points the direction from the vehicle's current location to the destination. Travelpilot can store up to 100 input destinations for future use.

Submenus provide several methods for resetting the vehicle's position on the map if the Travelpilot gets off-track. The frequency with which the system requires reinitializing depends on dead reckoning errors and the completeness and accuracy of the map data for the area being driven. When operating in an area like greater Los Angeles, dead reckoning with map matching typically fails once in a thousand miles. In typical locations, Travelpilot claims to have infinitesimal error relative to the map.

The Oldsmobile Guidestar, announced in 1994, was the first navigation and route guidance system offered as a dealer-installed option from an automobile manufacturer in the United States. The Guidestar was supplied by Zexel USA Corp. and is an adaptation of Zexel's NAVMATE system that had been under development for several years specifically for the U.S. market.[42] Zexel subsequently licensed NAVMATE to Seimens in Germany and eventually sold its interest in NAVMATE to Rockwell International, which, in turn, sold its automobile navigation business unit to Magellan, a subsidiary of Orbital Sciences. Orbital Sciences recently sold the Magellan subsidiary to Thales Navigation. Called PathMaster by Rockwell and Magellan, the system integrates a GPS receiver with dead reckoning and map matching. The dead-reckoning process uses gyroscopic and odometer inputs. The map database includes the locations of emergency services, restaurants, major retail stores, schools, office buildings, tourist attractions, and so forth. Destinations are entered as specific street addresses or road intersections, or by selecting categories and scrolling through the points of interest included in the database. Routing preferences (e.g., avoiding expressways) can also be specified. The navigation computer then calculates the route and highlights it on a four-inch color LCD. Once underway, the distance to and direction of each turn is displayed graphically on the screen as illustrated by figure 7.16, and a voice prompt advises the driver as each turn is approached. The driver may switch back and forth between

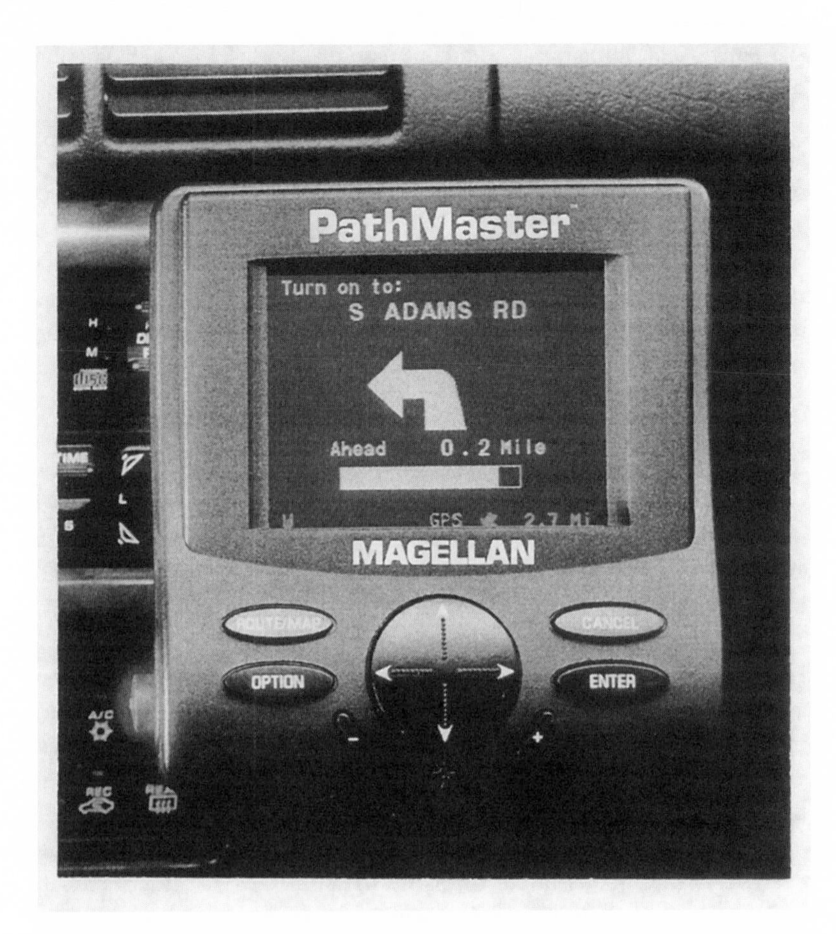

FIGURE 7.16 Magellan PathMaster.

displaying the map and displaying the graphic route guidance instructions. A version of the Magellan system called NeverLost installed in Hertz rental cars has introduced the ease and confidence of intelligent automobile navigation to over four million drivers.

Recent Directions

The foregoing account of the evolution of intelligent automobile navigation systems focuses on systems designed for permanent installation in automobiles. More recent systems for permanent installation have improved functionality at lower cost, and are increasingly integrated with entertainment features and location-related ITS functions such as real-time (i.e., current) traffic information and emergency notification, and are joined by a plethora of aftermarket and portable navigation systems. One manufacturer has even integrated a navigation display screen and controls in a replacement rearview mirror.

Significant advances are being made in realizing the long-time but elusive goal of automobile navigation systems to take into account real-time traffic conditions in calculating and adjusting routes. The problem has been due mostly to the lack of a widely available data link that is viable for the wireless communication of traffic data to automobile navigation systems. RDS-TMC (Radio Data System–Traffic Message Channel), which is used extensively in Europe, multiplexes inaudible traffic data on FM radio broadcasts. Japan's Vehicle Information and Communication System (VICS), which uses a combination of proximity beacons and multiplex broadcasting from FM radio stations, provides traffic data to over nine million navigation-equipped automobiles in that country. The patchwork traffic data communication approaches used in Europe and Japan are less well suited for the United States with its greater geographical area. Proximity beacons are not used at all for this purpose while RDS-TMC sees only limited use. For example, the Clear Channel Radio network of FM radio stations uses RDS-TMC to communicate traffic data aggregated by TeleAtlas for use by AudioVox navigation systems. However, with the advent of satellite radio a seamless nationwide means of delivering traffic data to navigation systems is finally available. Both XM (www.xmradio.com) and Sirius Satellite Radio (www.sirius.com) now offer traffic data aggregated by NAVTEQ for a growing list of major metropolitan areas.

Navigation and selected ITS functions may be provided by a new generation of personal computers for installation and use in automobiles. The innovations gained momentum in 1998 when Microsoft Corporation and allied companies announced the AutoPC.[43] The AutoPC is a complete information and entertainment system that fits in the same space in the automobile dashboard occupied by a radio or CD player, and includes many features in addition to a navigation capability. While different manufacturers offer additional features, the base system includes speaker-independent speech recognition for about 500 words and responds to voice commands for the majority of its functions. Speech synthesis enables the AutoPC to communicate information to the driver, including traffic information, route guidance instructions, and e-mail. Turn-by-turn route instructions may also be displayed graphically if preferred. AutoPC includes an AM/FM stereo tuner that stores up to twenty preset radio stations and a CD player that has all the features of a regular CD system. Numerous automobile manufacturers and aftermarket electronics suppliers have allied with

Microsoft to produce and market AutoPC; Clarion became the first company to offer an AutoPC product in 1998.

Although not a navigation system in the usual sense, General Motors' popular OnStar is by far the most widely deployed emergency notification system. OnStar, which includes a GPS receiver integrated with a dedicated cellular communication link, automatically notifies a twenty-four-hour OnStar call center of the automobile's location in the event an airbag deploys or special sensors detect a crash. An advisor at the call center will attempt to contact the driver and dispatch appropriate aid. The driver may also press a button on the in-vehicle unit to request roadside assistance in case of a breakdown or to use a wide variety of "concierge" services, including verbal driving directions to a particular destination. Other available OnStar services include remote engine diagnostics, remote door unlock, stolen vehicle tracking, and remote horn and light signaling in case the driver needs assistance to locate the vehicle in a parking lot.

Factory installed and aftermarket automobile navigation systems are facing increased competition from less-expensive portable navigation systems that require no installation. Portable systems employ a variety of platforms including notebook computers, portable data assistants (PDAs), and even cell phones. GPS receivers may be plugged in and navigation software and map databases downloaded to notebook computers and PDAs. Many newer cell phones have GPS built in to meet FCC requirements for extended 911 services, and some providers now offer navigation services based on the cell phone's knowledge of where it is located. Nextel, for example, offers a service called TeleNav that, similar to the OnStar service, provides route guidance instructions using a map database residing on a network server rather than the phone itself.

All navigation systems are becoming more accurate and reliable as a result of GPS receiver improvements. Many GPS receivers for automobile navigation now have up to twelve channels rather than the minimum of four channels used by most early receivers. This allow simultaneous signal reception from all of GPS's twenty-four satellites that may be in view, thus increasing the probability of receiving sufficient signals when driving roads shadowed by foliage, buildings and other obstacles. Improved receiver designs are also more capable of receiving signals attenuated by obstacles. Some GPS receivers are also capable of acquiring signals from the Federal Aviation Administration's Wide Area Augmentation System (WAAS), which is based on GPS reception at a nationwide network of ground stations at known locations in North America and broadcasts GPS error correction signals that assure position accuracy better than ten feet 95 percent of the time. These improved technologies are particularly important for portable navigation systems that typically do not augment GPS with dead reckoning.

Conclusion

Most drivers with a navigation system usually have at least one yarn to tell about missing roads, wrong turns or conflicting instructions even though the system guides them perfectly to their destinations the vast majority of the time. Such failures are typically due to incomplete or out of date digital maps. Although map manufacturers release new versions every three months, the system or car manufacturer may only offer them once a year. In spite of lingering limitations, automobile navigation systems are already commonplace in Japan where the majority of drivers use them. The penetration rate in Europe is also higher than in the United States, where they are available as factory equipment on less than one-half of automobile models sold. However, a recent inquiry to Click & Clack of *Car Talk,* the popular National Public Radio program featuring Tom and Ray Magliozzi illustrates growing consumer awareness and interest in automobile navigation systems. The inquirer, seeking advice on whether a navigation system could be added to her car, which was not designed to accommodate a factory system, was steered to several aftermarket systems. In another example of growing awareness and interest in the United States, *Consumer Reports* started publishing occasional reports on automobile navigation systems in 1999.

Just as gasoline is a necessary commodity for automobiles, digital map databases are a vital commodity for automobile navigation systems. In addition to providing a basis for route guidance, digitized maps are used in modern automobile navigation systems for map matching, a powerful software process that augments GPS and dead reckoning to maintain precise knowledge of vehicle location relative to the roadway system. The map databases required by these systems continue to evolve to cover expanded geographical areas and to include additional attributes such as road elevation. In addition to basic road location, geometry, number of lanes, category (e.g., residential, arterial, or freeway), and traffic regulations, state-of-the-art digital map databases for automobile navigation systems also incorporate extensive listings of businesses and other points of interest, all keyed to map location.

Encoded road maps have played vital roles in intelligent automobile navigation systems for 100 years. Early in the twentieth century, encoded maps for mechanical route guides started a trend that still holds: the encoding or storage of road map information on entertainment media. Examples from around 1910 include the phonograph-like Jones Live-Map and the music box-like Chadwick Road Guide. The trend was continued with the ARCS digital tape cartridge in the early 1970s, and in the mid-1980s by the Etak Navigator, which stored maps on digital tape cassettes. Also in the 1980s, both Philips and Toyota adapted compact disks for digital map storage, an approach that became commonplace in the 1990s. Today, many contemporary automobile navigation systems use DVD storage for digitized road map databases, thus continuing the practice of using entertainment media for map storage. Is iPOD next?

In addition to their use in intelligent automobile navigation systems, which in the coming decades are likely to become as ubiquitous as automobile radios, digital road maps are also finding many other uses. One of the latest new uses is for tracking the real-time location of aircraft relative to the road networks of the United States and Canada. Although digital road maps are obviously not used for navigation aboard aircraft, Flight Explorer, a leading global provider of Internet-based real-time flight tracking and weather information services uses digital road maps to provide clients with information about where planes are located, while in flight, relative to the earth.

Digital road maps are one of the more useful resources available on the Internet. For example, MapQuest (www.mapquest.com) will display a map showing the location of a particular address or provide driving directions from one address to another based on the same map database used in most automobile navigation systems. Both Google (www.maps.google.com) and Yahoo (http://maps.yahoo.com) recently added real-time traffic information to its maps for major cities and offer a similar service based on digital road maps.

Though today's in-vehicle navigation systems and online road navigation cartography are impressive technological achievements, we have seen that they have ancient ancestors. In many respects they are the descendents of the verbal itineraries Catherine Delano-Smith describes in the second chapter of this volume, and in this sense—technology aside—they precede the paper road maps many observers now expect them to replace. Yet, although digital maps offer strong advantages for finding the location of specific destinations and for calculating best routes to them, paper road maps are still more effective for uses like studying the layout and planning a tour of a wider region such as a state. Paper maps also have the advantage of being useful anywhere at anytime without the use of special equipment such as a navigation system or a computer. Though an increasingly large proportion of drivers and map consumers are now familiar and even adept with the use of digital navigational tools and cartography, it is unlikely that digital maps will ever fully displace the use of conventional paper maps.

Notes

Chapter One

1. On the ambiguities in the use of portolan charts see Tony Campbell, "Portolan Charts from the Late Thirteenth Century to 1500," in J. B. Harley and David Woodward, ed., *The History of Cartography,* vol. 1, *Cartography in Prehistoric, Ancient, and Medieval Europe and the Mediterranean* (Chicago: University of Chicago Press, 1987), 438–45.

2. See O. A. W. Dilke, "Itineraries and Geographical Maps in the Early and Late Roman Empires," in J. B. Harley and David Woodward, ed., *The History of Cartography,* vol. 1, 234–42.

3. For a straightforward review of navigational terminology and its meaning in a twentieth-century context, see Richard R. Hobbs, *Marine Navigation 1: Piloting,* 2nd ed. (Annapolis, MD: Naval Institute Press, 1981), 1–3.

4. For a summary history of this class of travel map, see Alan M. MacEachren and Gregory B. Johnson, "The Evolution, Application and Implications of Strip Format Travel Maps," *Cartographic Journal* 24 (1987): 147–58.

5. Obvious exceptions are remotely controlled and robotic vehicles, such as unmanned spacecraft, pilotless aircraft, and automated light rail transport.

6. See Neil McKendrick, John Brewer, and J. H. Plumb, *The Birth of a Consumer Society: The Commercialization of Eighteenth-Century England* (Bloomington: Indiana University Press, 1982); Roy Porter, *English Society in the Eighteenth Century,* rev. ed. (London: Penguin, 1990), 214–50; and John Brewer and Roy Porter, ed., *Consumption and the World of Goods* (London and New York: Routledge, 1993). Though speaking to the rise of reading, leisure, travel, and consumption of the arts, general histories of early modern consumption have not dealt directly with the consumption of maps. These issues, however, are addressed by Mary Pedley, *Commerce and Cartography: Making and Marketing Maps in Eighteenth-Century France and England* (Chicago: University of Chicago Press, 2005).

7. On the eighteenth-century development of atlases and maps designed specifically for children, see Mireille Pastoureau, "French School Atlases: Sixteenth to Eighteenth Centuries," in *Images of the World: The Atlas through History*, ed. John A. Wolter and Ronald E. Grim (Washington: Library of Congress, 1997), 125–32. In the United States, the development of school atlases and geographies for American children was seen as essential to the development to American citizenship; see Jeffrey C. Patton and Nancy B. Ryckman, "American School Atlases and Geographies 1784–1900," *Cartographic Perspectives* 33 (Spring 1999); and Martin Bruckner, "Lessons in Geography: Maps, Spellers, and Other Grammars of Nationalism in the Early Republic," *American Quarterly* 51, no. 2 (1999): 311–43.

8. See E. A. Reitan, "Expanding Horizons: Maps in the *Gentleman's Magazine*, 1731–1754," *Imago Mundi* 37 (1985): 54–62; and Mark Monmonier, *Maps with the News: the Development of American Journalistic Cartography* (Chicago: University of Chicago Press, 1989), esp. pp. 25–68.

9. See Porter, *English Society in the Eighteenth Century*, 227–29; and Catherine Delano-Smith and Roger J. P. Kain, *English Maps: A History* (Toronto: University of Toronto Press, 1999), 175–78.

10. James E. Vance, Jr., *Capturing the Horizon: The Historical Geography of Transportation since the Sixteenth Century* (Baltimore: Johns Hopkins University Press, 1990), 1–13.

11. Exceptional examples of planned highway systems predating the nineteenth century include the system of Roman military highways and the highly centralized French national route system begun in the eighteenth century. For a comprehensive review of the general history of road making, see M. G. Lay, *The Ways of the World: A History of the World's Roads and of the Vehicles That Used Them* (New Brunswick, NJ: Rutgers University Press, 1992).

12. The long pedigree of distinguished works on this subject properly begins with A. E. Nordenskiold, *Periplus: An Essay on the Early History of Charts and Sailing-Directions* (Stockholm: P. A. Norstedt & Soner, 1897); the essential work on this subject remains E. G. R. Taylor, *The Haven-Finding Art: A History of Navigation from Odysseus to Captain Cook* (London: Hollis & Carter, 1956).

13. Notable among these are Derek Howse and Michael Sanderson, *The Sea Chart: An Historical Survey Based on the Collections in the National Maritime Museum* (Newton Abbot, Devon: David & Charles, 1973); Michel Mollat du Jourdin and Monique de La Ronciere, *Sea Charts of the Early Explorers: 13th to 17th Century* (New York: Thames & Hudson, 1984); and Robert Putman, *Early Sea Charts* (New York: Abbeville Press, 1983).

14. For example, see Leo Bagrow and R. A. Skelton, *History of Cartography*, 2nd ed. (Chicago: Precedent, 1985); and Lloyd Arnold Brown, *The Story of Maps* (Boston: Little, Brown, 1949).

15. See, for example, Kenneth Nebenzahl, *Atlas of Columbus and the Great Discoveries* (Chicago: Rand McNally, 1990).

16. Of this small number, three concern the work of a single individual, Erhard Etzlaub.

17. Andrew M. Modelski, *Railroad Maps of North America: The First Hundred Years* (Washington, D.C.: Library of Congress, 1984).

18. Walter W. Ristow, *Aviation Cartography: A Historico-Bibliographic Study of Aeronautical Charts* (Washington, D.C.: Library of Congress Map Division, 1960). Monte

Duane Wright's *Most Probable Position: A History of Aerial Navigation* (Lawrence: University of Kansas Press, 1972) makes only a few brief mentions of the use of charts in early air navigation in an international context.

19. T. R. Nicholson, *Wheels on the Road: Maps of Britain for the Cyclist and Motorist 1870–1940* (Norwich, England: Geo Books, 1983); and Douglas A. Yorke, Jr., John Margolies, and Eric Baker, *Hitting the Road: The Art of the American Road Map* (San Francisco: Chronicle Books, 1996).

20. Michael J. Blakemore and J. B. Harley, *Concepts in the History of Cartography: A Review and Perspective*, Cartographica monograph no. 26 (Toronto: University of Toronto Press, 1980); Mark Monmonier, "Where Should Map History End?" *Mercator's World* 6, 3 (May/June 2001): 50–51.

21. Norman J. W. Thrower, *Maps and Civilization: Cartography and Culture in Society* (Chicago: University of Chicago Press, 1996).

22. A. G. Hodgkiss, *Understanding Maps: A Systematic History of Their Use and Development* (Folkestone, Kent: Dawson, 1981).

23. Delano-Smith and Roger J. P. Kain, *English Maps*, 142–78.

24. Walter W. Ristow, *American Maps and Mapmakers: Commercial Cartography in the Nineteenth Century* (Detroit: Wayne State University Press, 1985).

25. David Buisseret, ed., *From Sea Charts to Satellite Images: Interpreting North American History through Maps* (Chicago: University of Chicago Press, 1990).

26. Walter W. Ristow, "A Half-Century of Oil-Company Road Maps," *Surveying and Mapping* 24, 4 (1964): 617–37.

27. George H. Douglas, *All Aboard! The Railroad in American Life* (New York: Paragon House, 1992); John F. Stover, *American Railroads*, 2nd ed. (Chicago: University of Chicago Press, 1997); and James E. Vance, *The North American Railroad: Its Origin, Evolution, and Geography* (Baltimore: Johns Hopkins University Press, 1995).

28. James J. Flink, *The Automobile Age* (Cambridge, MA: MIT Press, 1988); and John B. Rae, *The Road and the Car in American Life* (Cambridge, MA: MIT Press, 1981).

29. J. Valerie Fifer, *American Progress: the Growth of Transport, Tourist, and Information Industries in the Nineteenth-Century West Seen through the Life and Times of George A. Crofutt, Pioneer Publicist of the Transcontinental Age* (Chester, CT: Globe Pequot, 1988).

30. Jan Palmowski, "Travels with Baedeker: The Guidebook and the Middle Classes in Victorian and Edwardian Britain," in Rudy Koshar, ed., *Histories of Leisure* (Oxford and New York: Berg, 2002), pp. 105–30; and Stephen L. Harp, "The Michelin Red Guides: Social Differentiation in Early Twentieth-Century French Tourism," in Rudy Koshar, ed., *Histories of Leisure* (Oxford and New York: Berg, 2002), pp. 191–214.

31. Stephen P. Hanna and Vincent J. Del Casino, eds., *Mapping Tourism* (Minneapolis: University of Minnesota Press, 2003).

32. Jordana Dym, "More Calculated to Mislead than Inform: The Cartography of Travel Writers in Central America, 1821–1950," *Journal of Historical Geography* (forthcoming).

33. Cramer's work was first published as the *Ohio Navigator* apparently in 1801, covering only the Ohio River. The 1804 edition was the first to include the Mississippi navigation; the first edition with maps was published in 1806. The last recorded edition was 1824. Cumings's derivative work first appeared in 1822. For a bibliographical summary of these and early American river pilot books, see Karl Yost, *The Ohio and*

Mississippi Navigator of Zadok Cramer, Third and Fourth Editions (Morrison, Illinois: Karl Yost, 1987).

34. This activity was primarily conducted by the U.S. Coast and Geodetic Survey, of the U.S. Lakes Survey (established 1841), the Mississippi River Commission (established 1879), and the Army Corps of Engineers. In the later nineteenth and twentieth centuries, excursion companies and ferries operating on the Great Lakes and the Hudson also supported promotional maps roughly comparable to that made for railroad and automobile travel. In response to the proliferation of gasoline powered motorboats in the mid-twentieth century commercial publishers such as Rand McNally and H. M. Gousha published series of small-scale cruising charts for distribution by oil companies.

35. Ken Garland, *Mr. Beck's Underground Map* (Harrow Weald, Middlesex: Capital Transport, 1994).

36. David Buisseret, ed., *Monarchs, Ministers, and Maps* (Chicago: University of Chicago Press, 1992).

Chapter Two

The generosity of many people underpinned the original lecture and subsequent drafts. I should like to thank in particular Peter Barber, Tony Campbell, Paul Harvey, Donald Hodson, Roger Kain, Paul Laxton, and Mary Pedley. My interest in maps for travel started in a modest way as a "brown bag" lunch-time talk I gave whilst a Fellow in the Hermon Dunlap Smith Center at the Newberry Library in April 1990: I am indebted to Kenneth Nebenzahl for this opportunity to explore the theme further.

1. Only land travel is considered here. On the use of rutters and charts in an English context, see Catherine Delano-Smith and Roger J. P. Kain, *English Maps: A History* (London: British Library Publications; Toronto: Toronto University Press, 1999), chapter 5, "Maps and Travel," esp. 145–48 and 153–59.

2. Norbert Ohler, *The Medieval Traveller* (Woodbridge: Boydell Press, 1989), 240. First published as *Reisen im Mittelalter* (Munich: Artemis Verlag, 1986).

3. Ohler, *Medieval Traveller* (note 2), 240.

4. The phrase is cited by Geoffrey Parker, *The Army of Flanders and the Spanish Road 1567–1659* (Cambridge: Cambridge University Press, 1972), 50. Parker attributes it to Fernand Braudel but does not identify the source, which is presumably Braudel's *The Mediterranean and the Mediterranean World in the Age of Philip II* (first published in Paris, 1949 as *La Méditerranéen et le Monde Méditerranéen à l'Epoque de Philippe II*, English translation of the second revised edition, London: Collins, 1972, 2 vols.). See especially vol. 1, 276 ff.

5. In regions of openfield cultivation, villagers used well-defined if not always physically distinct routes across the fields, formed from a sequence of single strips set aside to provide a way into each field. It was also possible to cross the field by means of headlands when these were not under growing crops. Green lanes and funnel-shaped "outgangs" served to link the village with its fields, pastures and common grazings. For headlands as routeways, see Stephen Upex, "Leicestershire Headlands" *Current Archaeology*, 149, vol. 13, no. 5 (1996): 191–93. The only village to retain the key features of the formerly extensive openfield farming system in England is Laxton, Nottinghamshire.

6. According to the medieval lawyer Hugh Bracton, a market that set up less than six and two-thirds of a mile [approx. 10.74 km] from another (a distance calculated to allow reasonable time in which to reach the place, do business, and return home within the day) was a "wrongful nuisance": S. E. Thorne, ed., *Bracton, De Legibus et Consuetudinibus Angliae* (Cambridge, MA: Belknap Press for Seldon Society, 1968), 3, p. 198. The majority of the 1200 or so settlements in England and Wales granted market charters between 1232 and 1350 were in fact no more than ten kilometers apart.

7. Data from Marjory Chibnall, ed., *Select Documents of the English Lands of the Abbey of Bec* (London, Offices of the Royal Historical Society, 1951), 41–42. Highclere is the nearest today; Hungerford and Andover are twelve miles away. I am grateful to Paul Harvey for drawing my attention to this material.

8. Duncan's travels cost the manor 6 shillings: Mark Page, ed., *The Pipe Roll of the Bishopric of Winchester 1301–2* (Winchester: Hampshire County Council, 1996) Hampshire Record Series, vol. 14, p. 215.

9. See table of speeds in Ohler, *The Medieval Traveller* (note 2), 107. Debra Birch, *Pilgrimage to Rome in the Middle Ages: Continuity and Change* (Woodbridge: The Boydell Press, 1998), 59–60, discusses this aspect in connection with pilgrim travel.

10. J. E. A. Jolliffe, "The Chamber and Castle Treasures under King John," in Hunt, R. W., W. A. Pantin, and R. W. Southern, eds., *Studies in Medieval History Presented to Frederick Maurice Powicke* (Oxford: Clarenson Press, 1948), 117–42, esp. 119.

11. Ohler, *The Medieval Traveller* (note 2); p.28 reproduces a thirteenth-century French manuscript illumination of a four-wheeled wagon (traveling-cart) with one passenger and the driver. In August 1390, Henry Earl of Derby was transported by cart from Rixhöft, Pomerania, where had landed, to Putzig, where he bought a horse and saddle: Lucy Toulmin Smith, ed., *Expeditions to Prussia and the Holy Land made by Henry Earl of Derby (afterwards King Henry IV) in the years 1390–91 and 1392–93 being the accounts kept by his Treasurer [Richard Kyngersoton] during two years* (London: The Camden Society, n.s. 52, 1894), xxvii, xxix, and xxxvi. See also R. A. Donkin, "Changes in the Early Middle Ages," in H. C. Darby, ed., *A New Historical Geography of England* (Cambridge: Cambridge University Press, 1973), 75–135, esp. 119.

12. J. J. Jusserand, *English Wayfaring Life in the Middle Ages,* translated by Lucy Toulmin Smith (first English edition London: T. F. Unwin, 1889; another London: Ernest Benn, 1950). First published in French in 1884 as *Les Anglais au Moyen Age: La vie nomade et les routes d'Angleterre au XIe siécle.*

13. For Edward I, see Henry Gough, *Itinerary of King Edward the First throughout his Reign AD 1272–1307* (Paisley: Alexander Gardner, 1900, 2 vols.), with maps of his eight expeditions into Scotland. Edward spent less than six years of his reign abroad and whilst in England moved on average almost two hundred times a year, thus yielding upward of five thousand moves for the period 1274–1307. See also Brian Paul Hindle, "Roads and Tracks," in Leonard Cantor, ed., *The English Medieval Landscape* (London: Croom Helm, 1982): 193–217.

14. Raymond Irwin, *The English Library* (London: George Allen and Unwin, 1966), 176.

15. See Toulmin Smith, *Expeditions to Prussia* (note 13).

16. J. H. R. Moorman, *Church Life in England in the Thirteenth Century* (Cambridge: Cambridge University Press, 1945), 175 and 185–96.

17. Ibid., 175.

18. Anne Kolb, "Transport and Communication in the Roman State: The *cursus publicus*," in Colin Adams and Ray Laurence, eds., *Travel and Geography in the Roman Empire* (London and New York: Routledge, 2001), 95–105.

19. Charles A. J. Armstrong, "Some Examples of the Distribution and Speed of News in England at the Time of the Wars of the Roses," in R. W. Hunt, W. A. Pantin, and W. W. Southern, eds. *Studies in Medieval History Presented to Frederick Maurice Powicke* (Oxford: Clarendon Press, 1948), 429–54.

20. Ohler, *Medieval Traveller* (note 2), 65.

21. Ohler, *Medieval Traveller* (note 2), 65. The Roman Church produced an unbelievable quantity of letters and paperwork. Robert Markus has analyzed the destinations of 970 surviving copies of letters sent by Pope Gregory in the sixth century. Over half (51.5 percent) were addressed to places in southern Italy, 7.2 percent to northern Italy, 9.8 percent to Gaul, 7.7 percent to Constantinople, 7.2 percent to Ravenna, and the rest to Illyricum and the Balkans, North Africa, Dalmatia, Antioch, Jerusalem, Alexandria, Spain, and Britain: Robert A. Markus, *Gregory the Great and His World* (Cambridge: Cambridge University Press, 1997), appendix, 209.

22. Ohler, *Medieval Traveller* (note 2), 66.

23. Antonia Gransden, "Antiquarian Studies in Fifteenth-Century England" *Antiquaries Journal* 60 (1980): 75–97, reprinted in Antonia Gransden, *Legends, Traditions and History in Medieval England* (London and Rio Grande: Hambledon Press, 1992), pp. 299–327, esp. 323. The itinerary Harding suggested first to Henry VI and then to Edward IV, the same as one Henry IV had taken in 1400, ran from Berwick to Ross by way of Edinburgh. It is printed in Henry Ellis, ed., *The Chronicle of John Harding with Richard de Grafton's Continuations* (Rolls Series, 1812, 2 vols), 422–29 (original verse form), and 414–20, n. 12 (revised prose version). One version of the map, with the itinerary, is in the British Library, Lansdowne MS. 204, f.226^{v}: see P. D. A. Harvey, *Medieval Maps* (London: The British Library, 1991), fig. 55.

24. E. T. Hamy, *Le Livre de Description des Pays de Gilles Le Bouvier, dit Berry* (Paris: 1908; Receuil de Voyages et de Documents pour servir à l'histoire de la géographie depuis le XIIIe jusqu'à la fin du XVIe siècle, no. 22); Charles Potvin, ed., *Gilbert de Lannoy, Oeuvres* (Louvain, 1878), 51. On Lannoy in context, see Hilda F. M. Prescott, *Jerusalem Journey: Pilgrimage to the Holy Land in the Fifteenth Century* (London: Eyre and Spottiswood, 1954), 23–25.

25. Entry in the household book of Henry VIII, printed in J. S. Brewer, ed., *Letters and Papers, Foreign and Domestic, of the Reign of Henry VIII* (London, 1864), vol. 2, part 2, 1456.

26. The phrase is Fernand Braudel's: *Civilization and Capitalism, 15th to 18th Century*, vol. 2, *The Wheels of Commerce* (first published as *Les Jeux de l'Echange*, Paris: Armand Colin, 1979; English translation, London: Collins, 1982).

27. Eugene Weber calls agricultural trade an "involuntary" trade: *Peasants into Frenchmen. The Modernisation of Rural France 1870–1916* (London: Chatto and Windus, 1979), 197.

28. Wool was the single most important English export up to the mid-fourteenth century, after which the development of trading monopolies (the various staples) stimulated the development of an English weaving industry. Between the autumn of 1338 and the winter of 1339/40, over 40,000 sacks of wool were shipped from

England to, mostly, the Netherlands: E. M. Carus Wilson and Olive Coleman *England's Export Trade 1275–1547* (Oxford: Clarendon Press, 1963), 45–46.

29. Lecari's travels are described by Robert-Henri Bautier, "Recherches sur les routes de l'Europe médiévale" in *Bulletin Philogique et Historique* 1 (1960): 99–143, esp. 142. The relevant document is in the Archivio di Stato di Genova, *Materie politiche*. Aigues-Mortes was founded by St. Louis as point of embarkation for the crusades of 1248 and 1270. It was France's only Mediterranean port in the thirteenth century, when Provence at the time was a separate political unit.

30. Lecari is recorded as being in Mende and in Montbrison in 1292. The motive behind the deviation was primarily political, relating to Philippe-le-Bel's desire to control certain bishoprics and territories along the central route.

31. Robert S. Lopez and Irving W. Raymond *Medieval Trade in the Mediterranean World. Illustrative Documents. Translated with Notes* (London: Oxford University Press, 1955), document 131. See also *Francesco Balducci Pegolotti. La Pratica della Mercatura*, Allan Evens, ed. (Cambridge, MA: Medieval Academy of Arts, 24, 1936), 256–58.

32. Lopez and Raymond, *Medieval Trade* (note 31), document 132.

33. Parker, *Spanish Road* (note 4), p. 50. The following description of Spain's military road is taken from Parker. See also his "Maps and Ministers: The Spanish Hapsburgs," in David Buisseret, ed., *Monarchs, Ministers and Maps: The Emergence of Cartography as a Tool of Government in Early Modern Europe* (Chicago and London: University of Chicago Press, 1992), 124–52, and also his *Empire, War and Faith in Early Modern Europe* (London: Allen Lane, 2002), 96–121 and 320–29 (notes).

34. The Spanish Road is so-named on Sully's copy of Fernando de Lannoy's map of Franche-Comte. Lannoy's map had been printed in 1567 but was suppressed by Philip II on grounds of security. The French minister added the line of the road (in red) and its name to his 1606 copy. See Parker, *The Spanish Road* (note 4), plate 5 (monochrome). After the loss of the duchy of Savoy to France, the Spanish corridor was confined in mid-Alps to a narrow valley and a single bridge (*Pont de Grésin*) over the Upper Rhone, uncomfortably near the French frontier and in 1601 an eastern alternative was negotiated that took the Spanish army through the Tyrol.

35. Parker, "Maps and Ministers" (note 33), 140.

36. James Westfield Thompson, *The Medieval Library* (Chicago, Chicago University Press, 1939), 108.

37. Thompson, *Medieval Library* (note 36), 109.

38. Bede, *A History of the English Church and People* (preface).

39. Veronica Ortenberg, "Archbishop Sigeric's journey to Rome in 900," *Anglo-Saxon England*, 19 (1990): 197–246. Sigeric's journey was recorded in the Anglo-Saxon chronicle.

40. Antonia Gransden, "Realistic observation in twelfth-century England" *Speculum* xlvii (1972): 29–51, reprinted in Gransden, *Legends* (note 23), 175–97, esp. 180.

41. Ibid.

42. It is not clear whether Gerald himself provided the sketch maps of the British Isles found in extant manuscripts of the *Topografia Hiberniae*, e.g., British Library, Add MS 33,991, fol. 26r, reproduced in G. R. Crone, *Early Maps of the British Isles, AD 1000–AD 1579* (London: Royal Geographical Society, 1961), p. 14, and Delano-Smith and Kain, *English Maps* (note 1), 15. For his map of Europe, see Thomas O'Loughlin, "The Map of Europe (N.L.I. 700): A Window into the World of Giraldus Cambrensis," *Imago Mundi*, 51 (1999): 24–39, plate 2.

43. Susanne Lewis, *The Art of Matthew Paris in the Chronica Majora* (Berkeley: University of California Press, 1987 and also Aldershot, Scolar Press with Corpus Christi College, Cambridge, 1987), 4.

44. Gransden, "Antiquarian studies" (note 23), pp. 324–35, and John H. Harvey, ed., *William Worcestre, Itineraries* (Oxford: Clarendon Press, 1969). The manuscript of William Worcestre's *Itinerarium* is in Cambridge, Corpus Christi College, MS 210.

45. Jean Gimpel, *The Cathedral Builders* (first published as *Les Bâtisseurs de Cathédrals* (Paris: Editions de Seuil, 1964; English translation, London: Michael Russell Ltd. 1983), 65–66.

46. Philippe Braunstein, "Leggende 'Welsche' e itinerari Slesiani: la prospezione mineraria nel Quatrocento," *Quaderni Storici* 70 (1989): 25–50. I am grateful to Marco Tizzoni for making this and other papers available to me.

47. C. Cucini Tizzoni and M. Tizzoni, "Li Pertij Maestri – l'emigrazione de maestranze siderurgiche bergamesche della Val Brembana in Italia e in Europa (secoli XVI–XVII)," *Bergomum* 3 (1993): 79–178.

48. Gervasone left Locarno, on foot, in the morning of June 21, 1662 and reached Introbio in Val Sassina in the evening of the June 23: Marco Tizzoni, *Momenti dell'Attività Mineraria e Metallurgica in Valsesia* (Milan: Club Alpino Italiano, Sezione di Varallo Sesia, 1988, Monografie del Comitato Scientifico, n.s.1), esp. 17–18.

49. Bautier, "Routes de l'Europe" (note 29), 115–17.

50. Bautier, "Routes de la France" (note 29) points out that this formerly important north-south artery had lost all importance by the sixteenth century.

51. In 1434 King Henry VI granted licenses permitting 2,433 pilgrims to go to Santiago di Compostella (the shrine of St. James) alone: Christopher Hibbert, *The Grand Tour* (London: Thames Methuen, 1987), 13. William Wey's advice to pilgrims, written on his return in 1478, was one of the earliest books printed in London (*Informacion for Pylgrymes unto the Holy Londe*, published by Wynkyn de Worde in 1495). On maps in the context of medieval pilgrimage to the Holy Land, see Catherine Delano-Smith, "The Intelligent Pilgrim; Maps and Medieval Pilgrimage to the Holy Land," in Rosamund Allen, ed., *Eastward Bound: Travel and Travellers 1050–1500* (Manchester: Manchester University Press, 2004), 107–30.

52. The sixteenth century saw a rise in the number of people traveling for pleasure. By the end of the century the fashion for touring Europe was well established. In 1772 it was noted that "Where one Englishman travelled in the reign of the first two Georges, ten now go on a grand tour. Indeed, to such a pitch is the spirit of travelling come in the kingdom, there is scarce a citizen of large fortune but takes a flying view of France, Italy and Germany": cited by Hibbert, *The Grand Tour* (note 51), 39. A number of these travelers wrote about their experiences and Hibbert's bibliography contains over a hundred contemporary travel accounts for the period 1600–1900.

53. Franciscus Petrarch, *De suo in Montem Ventosum ascensu*, Epipstola 1, *Le Familiari, Libro Quartus*. The letter is addressed to Franciscus Dionisius. Petrarch lived at the foot of Mont Ventoux (1910 metres, 6271 feet) by the Fountain of Vaucluse. Although he wrote a guide, *Itinerarium ad sepulchrum domini nostri Yesu Christie*, he never made the journey to the Holy Land.

54. Hamy, *Description des Pays* (note 24), appendix 5.

55. De Caumont left his home on February 17, 1417 for Santiago di Compostella and was on his way to the Holy Land on February 27, 1418. He reached home finally

on April 14, 1420. For his itinerary, see Hamy, *Description des Pays* (note 24), 217–37 (appendix 3).

56. Fynes Moryson, *Containing His Ten Years Travel through the Twelve Dominions of Germany, Bohemia, Switzerland, Netherland, Denmark, Poland, Italy, Turkey, France, England, Scotland and Ireland* (London, 1617).

57. Moryson, *Ten Years Travel* (note 56), part 3, 12 (printed erroneously as 10).

58. Moryson, *Ten Years Travel* (note 56), part 3, 17.

59. Thomas Coryate, *Coryate's Crudities, hastily gobbled up in five moneths travells in France, Savoy, Italy, Rhetia commonly called the Grisons country, Helvetia alias Switzerland, some parts of Germany and the Netherlands* (London, 1602), Epistle to the Reader.

60. For examples of medieval pilgrim narratives, see the publications of the Palestine Pilgrims Text Society of the 1880s, collected into ten volumes (London, 1897); John Wilkinson, *Jerusalem Pilgrims before the Crusades* (Warminster, Aris & Phillips, 1997); and John Wilkinson with Joyce Hill and W. F. Ryan, *Jerusalem Pilgrimage 1099–1185* (London: The Hakluyt Society, 1988).

61. Birch, *Pilgrimage to Rome* (note 9), says something about practical matters but is silent on other aspects of a pilgrim's preparations. For maps a pilgrim could have studied, see Delano-Smith "The Intelligent Pilgrim" (note 51).

62. Felix Fabri, *The Wanderings of Felix Fabri*, translated by A. Stewart (London, Palestine Pilgrim Text Society, 1892–97), vols. 7–10, 48. See also Conrad Dieter Hassler, ed., *Fratris Felicis Fabri Evagatorium in Terrae Sanctae, Arabiae et Egyptae Peregrinationem*, 2 vols. (Stuttgart, 1843–49); and Prestcott, *Jerusalem Journey* (note 24).

63. Fabri, *Wanderings* (note 62), vol. 1, 50.

64. Hassler, *Fabri Evagatorium* (note 62), 3 (the dedicatory "Epistola" is omitted from Aubrey Stewart's translation).

65. Fabri, *Wanderings* (note 62), vol. 1, 135.

66. Moryson, *Ten years travel* (note 56), part 3, 14.

67. Thomas Frangenburg, "Chorographies of Florence: The Use of City Views and City Plans in the Sixteenth Century" *Imago Mundi*, 46 (1994): 41–64, esp. 48. There had been guides to single towns from the ninth century, especially for Rome: Gavin de Beer, "The Development of the Guide-book until the Early Nineteenth Century," *Journal of the British Archaeological Association*, Third Series, 15 (1952): 35–46.

68. Turler's book was reissued in Nuremberg in 1591 as *De arte peregrinandi*: Justin Stagl, "The Methodising of Travel in the 16th Century: A Tale of Three Cities," *History and Anthropology*, 4 (1990), 303–38.

69. Turler, *The Travailer* (note 68), 5.

70. See Stagl, "The Methodising of Travel" (note 68). Frangenburg, "Chrorographies of Florence" (note 67), 49–52. Germany and England were the two countries of Europe most uncomfortable with what their inhabitants saw as their cultural barbarity. Anxious to catch up with what they saw as the sophistication of French and Italian culture, more Germans and English traveled about Europe for nonessential reasons after the sixteenth century than Frenchmen or Italians.

71. Theodore Zwinger, *Methodus Apodemica* (Basle, 1577), 47.

72. One exception is the reference *per forestam* (in the Ardennes) in the late medieval Rome itinerary from Titchfield and *per nemora* (the Black Forest, near Würtzburg) in one of the Bruges itineraries. For the itineraries in question, see text and note 88 below and Hamy (note 24) respectively.

73. See text below and figure 2.8. Lombardy's political instability seems to have been a chronic problem for travelers. In 1176, John, bishop of Norwich avoided the land route to Rome because of the threat or war in the region, choosing to travel by sea from Genoa: Birch, *Pilgrimage to Rome* (note 9), 51, citing Ralph de Diceto, *Ymagines Historium*, edited by W. Stubbs, *Rolls Series* 68, 2 vols. (London, 1876), vol. 1, 416. A similar emphasis on political boundaries rather than physical features in the fourth century has been noted by E. D. Hunt, "Holy Land Itineraries: Mapping the Bible in Late Roman Palestine," in Richard Talbert and Kai Brodersen, eds., *Space in the Roman World. Its Perception and Presentation* (Münster: LIT, 2004), 97–110, who comments that the pilgrim "Egeria... was meticulous in noting the crossing of provincial boundaries," referring to examples from her account of her travels in the Holy Land.

74. Weber, *Peasants into Frenchmen* (note 27), 217.

75. Ohler, *The Medieval Traveller* (note 2), 54.

76. Fernand Braudel, *Civilization and Capitalism. 15th–18th Century, Volume 1. The Structures of Everyday Life* (London: Collins, 1981; first published as *Les Structures du Quotidien: Le Possible et L'Impossible*, Paris: Armand Colin, 1979), 41.

77. Weber, *Peasants into Frenchmen* (note 27), 195, chapter heading.

78. Jusserand, *English Wayfaring Life* (note 12), 41.

79. H. G. Fordham, *Studies in Carto-Bibliography, British and French, and in the Bibliography of Itineraries and Road-Books* (Oxford: Clarendon Press, 1914), 26.

80. Werner Elias, "Road Maps for Europe's Early Post-Routes 1630–1780," *Map Collector* 16 (1981): 30–34. Similar comments are made by Alan M. MacEachren and Gregory B. Johnson, "The Evolution, Application and Implications of Strip Format Travel Maps," *Cartographic Journal* 24 (1987): 147–58. For a more balanced appraisal, see G. H. Martin "Road Travel in the Middle Ages. Some Journeys by the Warden and Fellows of Merton College, Oxford, 1315–1470," *Journal of Transport History* (1976), new series, 3:159–78, who comments wryly that: "it is not only candidates in public examinations who see England as a trackless waste" after the departure of the Romans and the modern times. Brian P. Hindle, "Seasonal Variations in Travel in Medieval England," *Journal of Transport History* (1977), new series, vol. 4: 170–78, also finds no evidence that even the difficulties imposed by British winter weather prevented those that had to move from doing so, without, moreover, evidence of undue distress.

81. Information on both words comes from the Oxford English Dictionary.

82. Monique Gilles-Guibert, "Noms des routes et des chemins dans le Midi de la France au Moyen Age," *Bulletin Philologique et Historique*, 1 (1960): 1–39, esp. 10.

83. Such rights were formally sanctioned, either locally (in the case of open fields, for example, by a manorial court) or nationally (by the law of the realm, as in the case of a "king's road" or public highway). Similar ordinances governed the state of roads. In 1258, the Statute of Winchester was drawn up for the widening of "highways leading from one market town to another," and to ensure that vegetation was cleared back on both sides of the road to a distance of 200 feet to remove cover for robbers and thieves: Statutes of the Realm, 13° Ed. I.

84. H. B. Wheatley and E. W. Ashbee, eds. *William Smith. The Particular Description of England and Wales, 1585* (London, 1879), 69.

85. See plates 5 to 8 in Brian Paul Hindle, *Medieval Roads* (Princes Risborough: Shire Publications, 1982). See also Brian Paul Hindle, *Roads, Tracks and Their Interpretation* (London: Batsford, 1993); and B. P. Hindle, "Roads and Tracks," in Leonard Cantor, ed., *The Medieval English Landscape* (London: Croom Helm, 1982), 193–217.

86. On *tabellaria,* see Benet Salway, "Travel, *Itineraria* and *Tabellaria,*" in Colin Adams and Ray Laurence, *Travel and Geography in the Roman Empire* (London and New York: Routledge, 2001), 22–66, and ibid., "The Nature and Genesis of the Peutinger Map," *Imago Mundi: The International Journal for the History of Cartography,* 57, 2 (2005): 119–35. Other essays in Adams and Laurence's book contain many details that are pertinent to the present essay.

87. Salway, "Travel, *Itineraria* and *Tabellaria*" (note 86), 54–58 and ibid., "The Nature and Genesis of the Peutinger Map" (see note 86), 130.

88. British Library, Add. MS 70,507, fols. 74–75v. See Bruce Dicken, "Premonstratensian Itineraries from a Titchfield Abbey MS at Welbeck," *Proceedings of the Leeds Philosophical and Literary Society (Literary and Historical Section)* 5 (1938): 349–61, and J. T. Munby, *People and Fields in Medieval Portchester* (Portsmouth Record Series, in preparation). Munby describes the compiler of the book as "an obsessive and indefatigable gatherer of information": I am grateful to J. Munby for allowing me sight of the relevant pages of his draft text, and to Paul Harvey for help with transliteration and translation. For examples of similar information in Greek and Roman itineraries, see Salway, "Travel, *Itineraria* and *Tabellaria*" (note 86), esp. 22–28.

89. Respectively *Et est in medio [it]ine[er]e* and *Et ibidem rediment equum.*

90. Public Record Office, SP 12/125 f. 46. Headed *A breef Shew of the Scituacion of the severall howses named in her Majestys jests w[i]th the nombre of myles betwene every of them,* the route is drawn on a single sheet of paper by William Bowles that was attached to a report on houses thought to be suitable for Her Majestey's overnight stops made: see Zilla Dovey, *An Elizabethan Progress* (Stroud: Alan Sutton, 1996), 17–19. In addition to the map, Bowles provided four plans of two buildings or groups of buildings. Nothing suggests the plan was carried on the journey or that it was heavily consulted, or that Bowles had made use of prepublication copies of the relevant sheets of Christopher Saxton's county maps. A similar type of sketch was made in planning Anne of Cleves's journey to England in 1539: British Library, Cottom MS, Augustus I.ii, 63.

91. J. H. Harvey, *William Worcestre, Itineraries* (Oxford: Clarendon Press, 1969), xiv, points to documents from the 1230s relating to payments made for official travel.

92. British Library, Royal MS 12.D.xi. The letter, addressed to a law student in Salisbury, Hampshire, and sent from Orleans, France, dates from the late thirteenth century.

93. British Library. Royal MS. 12.E.25, fol. 183v.

94. British Library, Sloane MS. 683. fol 42.

95. British Library, Royal MS 17.c.38, fol 2v. What remains of the route goes from Vils in Swabia to Innsbrook, over the Brenner Pass to Venice and then on to Ferrara, Bologna, and Florence.

96. British Library. Egerton 1900. fol. 151v. The itineraries in Maffre's book describe routes between Prague and London and Edinburgh, via Germany and Flanders; Edinburgh to France, Spain (Santiago de Compostella, Portugal and Italy; Venice to Jerusalem and Mt. Sinai; Constantinople to Denmark, Sweden and Norway. Over sixty places are named. Distances are indicated in (German) miles.

97. British Library, MS Add.62540, fol. 3. St. Quentin is called by its ancient name, *Augusta Viromanduorum.* Laurence's map occupies fols. 3v and 4, and there are further (English) itineraries on the verso of fol. 4 that are continued on the following folio. A note records that "Ld. Burleigh carried this map always about him."

98. The folio is in Burghley's "Little Atlas": British Library. Add. MS, 62540, fol. 3.

99. British Library, Royal MS 18.D.3, fols. 4v and f.5.

100. British Library, Add. MS.61342. I owe this example to Peter Barber.

101. British Library, MS Add. 70,507, fols 33r–45. See note 88 above.

102. Universiteitsbibliotheek, Ghent, Belgium, now MS 13, ff. 54–60v. The volume measures 50.5 × 38 cm. I am grateful to Dr Martine De Reu, Keeper of Manuscripts and Rare Books, University of Ghent, for her description of the physical condition of the manuscript. The itineraries are published in Hamy, *Description des Pays* (note 24), 157–216 (appendix 4). Hamy suggests the volume was made at the end of the fifteenth century for Raphael of Marcatellis (1437–1508), one of the illegitimate sons of abbot Philippe-le-Bel of St. Bavon, Ghent. The manuscript is in Latin but the spelling of some place names suggest that the Flemish scribe had difficulty in transcribing them from sources of different dates and different places of origin. Several of the itineraries give alternative routes. Merchants had a choice between, for example, a direct route to Nuremburg (via Prague, on the Krakow route) and an indirect route (by way of Frankfurt-on-Main), and there were also three different ways to Lübeck, identified respectively as "direct," "indirect," and "another way." There was a similar need for reference material in maritime commerce. About 1489, a miscellany of travel-related information (including a portolan or written description of the coasts of the Mediterranean) was copied out, together with thirty-five charts of the Mediterranean compiled by various Venetians but here presented in uniform style on prepared pages (with simple but attractive corner decoration), and the whole bound between hard covers to form an attractive folio for ready consultation (British Library, Egereton MS 73). I am grateful to Peter Barber for drawing my attention to this volume.

103. British Library, Cotton MS, Tiberius. B.V., f.23v. The account includes a list of the churches and other places Sigeric visited while in Rome. For a reproduction of part of the itinerary, see Harvey, *Medieval Maps* (note 23), 9.

104. The term "expanded itinerary" is de Beer's, "Development of the guide-book" (note 67), 37, n.8. The comment on Worcestre comes from Harvey, *William Worcestre* (note 91), xiv, who also notes that Worcestre was following the normal practice of copying his itineraries from another source.

105. Corpus Christi College, Cambridge, MS 407. The manuscript is catalogued under the title *Libri Secreti Secretorum*, another work in the volume. I am grateful to Paul Harvey for drawing this manuscript to my attention.

106. William Wey's account of his 1458 voyage to the Holy Land also contains a Greek vocabulary "for the use of future pilgrims." His map of Palestine shows no roads or routes but there is a table of distances that was to be used to work out the order of travel: see E. Duff, ed., *The Itineraries of William Wey (Fellow of Eton College) to Jerusalem AD 1458 and AD 1462 and to Saint James of Compostella AD 1456* (London, J. B. Nichols & Sons for the Roxburghe Club, 1857). The map is reproduced in Kenneth Nebanzahl, *Maps of the Bible Lands. Images of Terra Sancta through Two Millennia* (London: Times Books, 1986), plate 17. For Leland's diary, see *The Itinerary of John Leland the Antiquary*, edited by Thomas Hearne, 3 vols. (Oxford, 1711).

107. British Library, 21.a.10, fols lxxxii–lxxxiii. Augmented in c. 1521, reprinted in London in 1811 with the title *The Customs of London. Otherwise called Arnold's Chronicle*.

108. Donald Hodson, "The Early Printed Road Books and Itineraries of England and Wales," Ph.D thesis, University of Exeter, November 2000, vol. 1 of 2, 73. I am indebted to Donald Hodson for making a copy of this thesis available to me.

109. William Middleton, *A chronycle of yeares from the beynnynge of the worlde* (London, 1543): Hodson, Early Printed Road Books) (note 108), 73–74 (British Library microfilm, Mic. B57/1795).

110. The routes to Holyhead (via Chester) and Haverfordwest (via Bristol), ports for Ireland, were subsided out of political necessity. In 1494, under Henry VII, all Irish legislation came under English control and in 1540 Henry VIII assumed the title King of Ireland: Philip Beale, *A History of the Post in England from the Romans to the Stuarts* (Aldershot: Ashgate, 1998), 181–82.

111. Royal Geographical Society, 262.A.1. Donald Hodson has pointed out (personal communication) that the *Guide*, with its associated *Discours*, appears to have been copied from a manuscript source, possibly British Library, Royal MS. 16.E.36, which contains a description of England written in French and dated January 2, 1571. The manuscript is addressed to Monseigneur de Saulves (Simon de Fizes, Baron de Sauves, conseiller du roy en son conseil privé, secretaire d'estat et des finances).

112. Estienne's book is generally acknowledged as the earliest known printed French road-book. However, an undated example in the British Library (C.55.a.9), appears (Hodson, personal communication) not to be one of the three (dated) examples referred to by Herbert George Fordham, "An Itinerary of the Sixteenth Century: *La Guide des Chemins d'Angleterre*," paper read at a meeting of the Cambridge Antiquarian Society, December 6, 1909 (Cambridge: J. Webb, 1910) or in "Notes on British and Irish Itineraries and Road-Books," paper read to the Geographical Section of the British Association for the Advancement of Science, Dundee, September 1912 (Hertford: Stephen Austin & Sons, 1912).

113. See Hodson, "Early Printed Road Books" (note 108), for details.

114. The text is a translated and expanded excerpt from Justus Lipsius's *Epistola de peregrinatione Italicae* (1592).

115. The author is given only as "A Lover of His Countrymen." I am indebted to David Webb for pointing me in the direction of Speed as the source of the maps in *The English Traveller's Companion*. The set is incomplete in both the British Library's copy of both the *Companion* and the *Theatre*, but the two publications between them have all five maps. The arterial roads represented all start from London and include: the North-Road (to Berewick), the North-West-Road (to Holyhead), the West-road (to Bristol), the Western-road (to Lands End) and, together on one map, the southeast, south and southwest roads.

116. Swall's pocket book publication included Ogilby's distance measurements. Rodney W. Shirley, *Printed Maps of the British Isles 1650–1750* (Tring, Map Collector Publications; London: The British Library, 1988), 88–89, ascribes the map to Herman Moll. British Library, 577.e.2.

117. For the possible source map, see Shirley, *Printed Maps* (note 115), 16 and 88. The book measures 12 × 9 cm or 4.7 × 3.6 in.

118. *The Kentish Traveller's Companion* was printed in Rochester by T. Fisher and in Canterbury by Kirkby and is usually, but possibly erroneously, attributed to Mostyn John Armstrong.

119. Ohler, *The Medieval Traveller* (note 2), 76–77. According to the holdings of the British Library, the earliest printed phrase books date mainly from the 1530s and were published predominantly in the Netherlands.

120. Noël de Berlaimont's, *Colloquium et dictionariolum septem linguarum, Belicae, Anglicae, Teutonicae, Latinae, Hispanicae Gallicae* (Antwerp, 1586). I am grateful to

Christopher Wells, St. Edmund's College, University of Oxford, for pointing me to these phrase books.

121. The conversation underlines the importance of knowing which towns were walled and which were not. The distinction was made almost as a matter of course on topographical maps printed on the continent in the sixteenth century.

122. See John E. Murdoch, *Album of Science. Antiquity and the Middle Ages* (New York: Charles Scribner's Sons, 1984) for examples of scientific diagrams.

123. Richard Vaughan, *Matthew Paris* (Cambridge: University of Cambridge Press, 1958), 237. The multiple copies arose because Paris inserted the itinerary, and the maps of Palestine and Britain, into the preliminaries of each volume of his monastery's chronicle: Corpus Christi College, Cambridge MS 26, folios i–iii; Corpus Christi College MS 16 (fragment only), folio ii; British Library, Royal MS 14.C.vii, folios 2–4 (the latter volume contains the *Historia Anglorum*, an abstract of the chronicle proper), and British Library, Cotton MS. Nero D. I, folios 183v–184 (a miscellany of related material known as the *Liber Additamentorum*). It is likely that the Corpus Christi version, although stylistically different, was copied from Royal MS. 14.C.vii.

124. Vaughan, *Matthew Paris* (note 123), 247–48, has summarized the very different ways the of the four versions "end" in Apulia.

125. Lewis, *The Art of Matthew Paris* (note 43), 325.

126. Lewis, *The Art of Matthew Paris* (note 43), 324–25.

127. Konrad Miller, *Mappa Mundi. Die ältesten Weltkarten. Vol. III Die Kleineren Weltkarten* (Stuttgart: J. Roth, 1895), 85–90 (text); Vaughan, *Matthew Paris* (note 85), p. 239; Lewis, *The Art of Matthew Paris* (note 43), 324.

128. Lewis, *The Art of Matthew Paris* (note 43), 324–25.

129. Fabri, *The Wanderings* (note 62), vol. 1, part 1, 3; part 2, 49–50.

130. The phrase originated with Bernard's of St Clairvaux and is cited by Daniel K. Connolly, "Imagined Pilgrimage in the Itinerary Maps of Matthew Paris," *Art Bulletin*, 81 (1999), 598–622, esp. 598 and n.9. The importance in the Middle Ages and later surrogate pilgrimage has been underlined in a number of essays published since Connolly's, and since the present essay was written and revised. But see examples cited by Catherine Delano-Smith and Alessandro Scafi, "Sacred Geography," in Zsolt Török, ed., *Szent Helyek a Térképtörténeti [Sacred Places on Maps. A Cartographic Exhibition from the Collections of Pannonhalma (Hungary) and Schottenstift (Vienna)]* (Pannonhalma: Hungary, 2005), 122–54, and 50–52 (figs).

131. Connolly, "Imagined pilgrimage" (note 130), 598.

132. The quotation, from F. M. L. Thompson, *Chartered Surveyors: The Growth of a Profession* (London: Routledge & Kegan Paul, 1968), 24, is cited in J. B. Harley, *John Ogilby. Britannia, London 1675* (Amsterdam: Theatrum Orbis Terrarum, 1970), vi. On Ogliby see also Katherine S. Van Eerde, *John Ogilby and the Taste of His Times* (Folkestone: Dawson, 1976); and Donald Hodson, "Early Printed Road Books" (note 108), vol. 2, 401–91.

133. The use of inverted hill signs, for instance, to indicate descending gradients, is thought to have been Robert Hooke's idea: Harley, *John Ogilby* (note 132), xvi–xvii.

134. Harley, *John Ogilby* (note 132), ix. The designation "English" meant as opposed to "Dutch" (Donald Hodson: personal communication). Various schemes were put forward in England in competition with the Dutch: see "Unfulfilled Projects for County Atlases in the Later Seventeenth Century: A Note" in R. A. Skelton, *County Atlases of the British Isles 1579–1850: A Bibliography* (London: Carta Press, 1970): 184–90.

135. Donald Hodson, "The Making of John Ogilby's *Britannia*," unpublished paper. I am indebted to Dr. Hodson for letting me see this.

136. Hodson, "Early Printed Road Books" (note 108), 424.

137. Bodleian Library, Oxford, Aubrey 4 fol. 220, dated August 24: Harley, *John Ogilby* (note 132), ix.

138. Bodleian Library, Oxford: Rawlinson MS C. 514, cited by Harley *John Ogilby* (note 132), xv.

139. I am grateful to John Goldfinch, of the British Library, for weighing the copy that Ogilby presented to Charles II: British Library, 192.f.1.

140. Delano-Smith and Kain, *English Maps* (note 1), 170.

141. The questionnaire is given in Harley, *John Ogilby* (note 132), xv–xix, and Hodson, "Early Printed Road Books" (note 108), 410.

142. A furlong measures 220 yards or approximately 240 meters.

143. Garrett A. Sullivan, "The Atlas as Literary Genre: Reading the Inutility of John Ogilby's *Britannia*," presented at the Thirteenth Kenneth Nebenzahl, Jr., Lectures in the History of Cartography, Newberry Library, 1999. I am grateful to Dr Sullivan for many stimulating discussions about Ogilby and the status of his strip maps. An early version of his chapter was presented as a lecture "Travelling by Road or Armchair? Reading the Inutility of John Ogilby's *Britannia*," at the Warburg Institute, University of London, November 1998, in the Maps and Society lecture series.

144. Harley, *John Ogilby* (note 132), v.

145. She continues: "A number of such loose strips came to rest in various collections. Others no doubt were lost or destroyed after their immediate use was fulfilled": Van Eerde, *John Ogilby* (note 132), 137. No references identify the "various collections."

146. Donald Hodson reports seeing "butchered strips," but only in collections assembled by antiquarians, typically as part of a county-based collection with maps and sections cut from other works such as, for example, William Camden. Hodson cites British Library, Lansdowne MS 887 fol. 49–51, for the dissected pages taken from Ogilby in a collection relating to Bedfordshire and remarks that they show no signs of having used on the road nor are they likely to have been so used, since they are bound in what is effectively a scrapbook, devoted to the county of Bedfordshire (personal communication).

147. Victoria di Palma has looked at the way Ogilby's book helped foster contemporary understanding of national landscape (a theme also explored by Garrett Sullivan [note 143]) and how image and text together formed a new kind of viewing public: Victoria di Palma, "Reading and Riding: John Ogilby's Ribbon Maps and the Construction of a Landscape in Motion," paper presented to *Text and Image. England 1500–1750*, Fifth Reading Literature and History Conference, University of Reading, Berkshire, U.K., July 10–12, 2002 (publication in Conference Proceedings pending).

148. Sullivan, "The Atlas as Literary Genre" (note 143).

149. Ibid.

150. Thomas Gardner, *Pocket-Guide to the English Traveller: Being a Compleat Survey and Admeasurement of all the Principall Roads and Most Considerable Cross-Roads in England and Wales. One Hundred Copper-Plates* (London, 1719). Gardner was referring to *Itinerarium Angliae: Or, A Book of Roads* (1675), preface, cited by Sullivan, "The Atlas as Literary Genre" (note 143). Gardner's own publication was the first pocket edition of *Britannia*, now "reduc'd to a portable Volume, to render it of general Advantage to an *English* Traveller."

151. Sullivan, "The Atlas as Literary Genre" (note 143).

152. Harley, *John Ogilby* (note 132), xviii.

153. The title of Gardner's edition makes no mention of his debt to Ogilby. In contrast, Senex revealed his debt to Ogilby on the title page of his edition (1719: *An Actual Survey of all the Principal Roads of England and Wales; Described by One Hundred Maps from Copper Plates. On which are delineated All the Cities, Towns, Villages, Churches, Houses, and Places of Note throughout each Road. As Also Directions to the Curious Traveller what is worth observing throughout his Journey . . . First perform'd and publish'd by John Ogilby, Esq; and now improved, very much corrected and made portable by John Senex.* John Owen and Emanuel Bowen also acknowledged their source: *Britannia Depicta or Ogilby Improv'e; Being a Correct Copy of Mr Ogilby's Actuel Survey of all ye Direct and Principal Cross Roads in England and Wales. . . . The Whole for its Compendious Variety & Exactness, preferable to all other Books of Roads hitherto Published or Proposed; and calculated not only for the direction of the Traveller [as they are] but the general use of the Gentlemen and Tradesmen* (1720).

154. Keith Thomas, "Numeracy in Early Modern England," *Transactions of the Royal Historical Society*, 5th series, 37 (1987): 103–32. Arithmetic was only slowly introduced to the school curriculum in the sixteenth century and many people would have been daunted by what to us today seem quite simple sums.

155. It is generally accepted that it was William Smith who gave Norden the idea of adding the keys and graticules missing from Saxton's maps but commonplace on the continent since the early in the sixteenth century. For a brief summary of Smith's cartographical work and his stay in Nuremberg, see Delano-Smith and Kain, *English Maps* (note 1), 74–75, and 186–88.

156. Hodson, "Early Printed Road Books" (note 108), 369–71.

157. Peter H. Meurer, "Zur Frühgeschichte der Entfernungsdreicke," *Cartographica Helvetica* 24 (2001): 9–19. It would not have been impossible for William Smith to have heard about Nefe's work, even conceivably the Silesian map while in Nuremberg.

158. The pocket-book format, together with Norden's evident desire to have tables of uniform size, meant that the matrices for the smaller and/or less densely settled counties (e.g., Oxfordshire, Shropshire, Northamptonshire) include "some confining towns from over the county border." By the same token, the two largest units (Yorkshire, Wales) were printed on larger pages, allowing a large number of places to be listed but meaning that the pages have to be folded to fit between the covers of the book.

159. On Jenner's death, John Garrett produced the *English Traveller's Guide* (1676–80) from the same plates. John Ogilby's *Tables* contains no maps.

160. *A Direction for the English Traviller* (London: Matthew Simmons, 1635), To the Gentle Reader.

161. The Peutinger map comes down to us as a twelfth- or early thirteenth-century transcription of an ancient map on eleven vellum sheets (the twelfth, that containing the British Isles, is missing) glued end to end. See the reduced facsimile edition, Konrad Miller, ed., *Die Peutingersche tafel* (Stuttgart: F.A. Brockhaus, 1962), and O. A. W. Dilke, "Itineraries and Geographical Maps in the Early and Late Roman Empires," in J. B. Harley and David Woodward, eds., *The History of Cartography*, vol. 1, *Cartography in Prehistoric, Ancient, and Medieval Europe and the Mediterranean* (Chicago: University of Chicago Press, 1987), 234–57, esp. 238–42. For a new light on the map, see Salway, "Nature and Genesis of the Peutinger Map" (see note 86) and

Emily Albu, "Imperial Geography and the Medieval Peutinger Map," *Imago Mundi* 57:2 (2005): 136–48. Albu agrees with Salway that the map would have been "a map for display," although she postulates a medieval date for its creation.

162. Dilke, "Itineraries" (note 161), 238.

163. See Adams and Laurence, *Travel and Geography* (note 86).

164. E. J. S. Parsons, *The Map of Great Britain circa A.D. 1360 Known as the Gough Map. An Introduction to the Facsimile* (Oxford: Oxford University Press for the Bodleian Library and the Royal Geographical Society, 1958), 7, notes that, besides latitudes, "Longitudes for London, Hereford, Oxford, Colchester and Berewick had also been worked out by this time."

165. The lines are often referred to as representing roads rather than routes: e.g., Parsons, *The Map of Great Britain* (note 164), 10–11; Stenton, "The Roads of the Gough Map" (note 164); and Brian Paul Hindle, "The Towns and Roads of the Gough Map" (c.1360), *The Manchester Geographer*, 1 (1980): 35–49.

166. See the diagram in Beale, *History of the Post* (note 110), fig. 3.1, which is reproduced in Delano-Smith and Kain *English Maps* (note 1), 47, and the written description by Frank Stenton, "The Roads of the Gough Map," in Parsons, *The Map of Great Britain* (note 164), 16–20 and 36–37.

167. An additional feature of Adams's map is the gazetteer, which is printed along both sides. There are 799 places, not all shown on the map itself, listed alphabetically together with the county in which each place lay and its latitude and longitude. See Shirley, *Printed Maps* (note 116), 18–22; William Ravenhill "John Adams, His Map of England, Its Projection, and His *Index Villaris* of 1680," *Geographical Journal* 144 (1978): 427–37; and Blake Tyson, "John Adams' Cartographic Correspondence to Sir Daniel Flemming of Rydal Hall, Cumbria, 1676–1687," *Geographical Journal* 151 (1985): 21–39.

168. For maps with distance lines between selected places, forming either a route or a small network, see, for example: Luc Antonio degli Umberti's map of Lombardy (c. 1515); Jacob Zeigler's atlas of the Holy Land (1532); the anonymous map of the Holy Land in the Coverdale Bible (1535); Wolfgang Wissemburg's map of the Holy Land (1538); and Claudio Duchetti's map of Tuscany (1602).

169. The story is told in Shirley, *Printed Maps* (note 116) and in Ravenhill in "John Adams" (note 167).

170. I owe this point to Donald Hodson.

171. Maps derived from Adams's include those by William Berry (1671 etc.), Robert Morden (1673 etc.), Robert Walton (1679), John Overton (1685 etc.), Herman Moll (1673 etc.), Thomas Bowles (c. 1710 etc.), and George Willdey (1713 etc.): see the entry "Adams-type distance maps" in the subject index in Shirley, *Printed Maps* (note 116).

172. Mireille Pastoureau, *Les Atlas Françaises XVIe–XVIIe Siècles. Répertoire bibliographic et étude* (Paris: Bibliothèque Nationale, Département des Cartes et Plans, 1984), 477.

173. For example, the 1693 edition, British Library, Maps 14320 (3).

174. On the history of British post roads, see Howard Robinson, *The British Post Office: A History* (Princeton, NJ: Princeton University Press, 1948), and Beale, *A History of the Post* (note 165), chapter 3.

175. Shirley, *Printed Maps* (note 116), 40. It seems that Hick had previously drafted a four-sheet map that was either never printed or has not survived. Carr's map measures 33.5 × 39 cm.

176. Ogilby's map measures 37.5 × 50.0 cm and Willdey's 59.5 × 59.5 cm: Shirley *Printed Maps* (note 116), 104–5 and 151–52 respectively.

177. Signot's map was printed without major alteration in 1515. For a reproduction of the exemplar in the Bibliothèque nationale de France, Paris (Cartes et plans, Rés. Ge. D. 7687), see David Buisseret, "Monarchs, Ministers, and Maps in France before the Accession of Louis XIV," in Buisseret, ed., *Monarchs, Ministers and Maps* (note 33), 99–123, fig. 4.2.

178. Brigitte Englisch, "Erhard Etzlaub's Projection and Methods of Mapping," *Imago Mundi* 48 (1996): 103–23.

179. As paraphrased by Englisch, "Erhard Etzlaub's Projection" (note 178), 113 and, for the German text in full, 54. Etzlaub's map of the environs of Nuremberg (1492) was also provided with a linear scale for measuring distances between places, as had Andreas Walsperger's map of the world of 1448.

180. In 1511, Martin Waldseemüller called his map of Europe *Cartam itinerarium* and his instructions on the measurement of distances between places and along routes shown on the map are essentially the same as Etzlaub's. They were printed not on the map itself but in an accompanying booklet *Instructio Manuductionem Prestans in Cartam Itinerariam Martinii Hilacmili*. For a translation of the relevant passage, see Catherine Delano-Smith, "Cartographic Signs on European Maps and Their Explanation before 1700," *Imago Mundi*, 37 (1985): 9–29, esp. 24.

181. Jörgen Erlinger, *Das heilig Romisch mit allen landstrassen*. Reissues also in 1524 and later. Erlinger also provided instructions for the use of the map.

182. As in so many cases of a route or route-planning map, Stegena's rendering of *itinerarium* as "road map" is thus misleading: Lajos Stegena, ed., *Lazarus Secretarius: The First Hungarian Mapmaker and His Work* (Budapest: Akadémiai Kiadó, 1982), 24–25.

183. Skelton, *County Atlases* (note 134), 89.

184. This particular soft leather-bound bundle of maps had been in Eric Gardner's collection but were sold at Sotheby's in the early 1970s. Their present location is unknown. I am grateful to Laurence Worms and Tony Campbell for bringing the existence of such saddle maps to my attention.

185. British Library, Maps C.7.c.13. According to Skelton, this example is a revised version of State A: R. A. Skelton, *Saxton's Survey of England and Wales with a Facsimile of Saxton's Wall-Map of 1583, Imago Mundi* Supplement no. 6, with a preface by J. B. Harley (Amsterdam: N. Israel, 1974), 21.

186. Skelton, *Saxton's Survey* (note 185), 15.

187. *Book of The Names of All Parishes, Market Towns, Villages, Hamlets, and smallest Places, in England and Wales* (London: printed by Matthew Simmons for Thomas Jenner, 1662), preface.

188. The *Advertisement* continues: "The following Collection forms a PORTABLE ATLAS of NORTH AMERICA calculated in its Bulk and Price to suit the Pockets of Officers of all Ranks." Although "pockets" here refers to the fact that officers had to purchase their own maps and equipment, the atlas has become known as the "Holster Atlas" for its small size and usefulness to mounted officers. The individual, folded, maps cover North America, the West Indies, the Northern Colonies, the Middle Colonies, the Southern Colonies and Lake Champlain. I examined the copy (Atlas H-4) at the William Clements" Library, University of Michigan, to whose staff I am most grateful for all their kind help. The volume measures 24 × 15 cm and is 3.5 cm thick (9.5 × 6 × 1.5 in).

189. John Norden, map of Middlesex (1593) and Hertfordshire (1598); Philip Symonson's map of Kent (1596). There are no roads at all on Norden's printed maps of Sussex and Hampshire or the manuscript maps of Cornwall and Northamptonshire.

190. David Smith, *Maps and Plans for the Local Historian and Collector* (London: Batsford, 1988), 120.

191. Anon, *The Construction of Maps and Globes* (London: for T. Horne, 1717).

192. Skelton, *County Atlases* (note 134), 186; Van Eerde, *John Ogilby* (note 132), 136. Gregory King was aided by Robert Felgate, a local Essex man. Their map was engraved by Francis Lamb and published by William Morgan, Ogilby's step-grandson.

193. For H. G. Fordham the detailing of the road network on county maps signaled the beginning of a phase of revitalization in the mapping of England, to which he applied the term "English School": Fordham, *Carto-Bibliography* (note 79), 15.

194. See J. B. Harley and Yolande O'Donoghue, *The Old Series Ordnance Survey Maps of England and Wales. Scale 1 inch to 1 mile. A Reproduction of the 110 Sheets of the Survey in Early States in 10 Volumes.* Volume 1 *Kent, Essex, E. Sussex and S. Suffolk* (Lympne Caste, Kent: Harry Margary, 1975), figure 10 (page xvii) for a list of all signs used on the maps, inferred from the maps in the absence of a contemporary key.

195. Turnpikes, turnpikes unfenced on one side, gated road, major roads (non-turnpike), minor roads fenced on both sides or on one side only, unfenced; roads under construction; tracks; and footpaths. In addition, road cuttings and canal tow paths are shown. Moreover, the roads connect with river crossings, now detailed as bridges, footbridges, ferries (points and landings), fords and weirs. Information taken from Rodney Fry's analysis of map signs in J. B. Harley and R. R. Oliver, *The Old Series Ordnance Survey Maps of England and Wales. Scale 1 inch to 1 mile. A Reproduction of the 110 Sheets of the Survey in Early States in 10 Volumes.* vol. 7, *North-central England* (Lympne Caste, Kent: Harry Margary, 1989), xvi.

196. Motorways, trunk roads (single carriageway, dual carriageway), main roads (single carriageway, dual carriageway), secondary roads, narrow trunk or main road with passing places, roads with fourteen feet of metalling (not included in the previous category), two categories of roads with under fourteen feet of metalling, (each distinguished as tarred or untarred), minor roads in towns, unmetalled drives or tracks in the country (distinguished as fenced or unfenced), two categories of roads under construction, and paths. In addition, two degrees of steep gradients are marked, as are toll gates. See key to plate 26 in J. B. Harley, *Ordnance Survey Maps. A Descriptive Manual* (Southampton: Ordnance Survey, 1975).

Chapter Three

Part of the research on which this lecture was based was subsequently used for an invited contribution to a conference at the National Maritime Museum, Greenwich, in July 2000. The paper delivered at that conference has meantime been published as Andrew S. Cook, "Establishing the Sea Routes to India and China: Stages in the Development of Hydrographical Knowledge," in H. V. Bowen, Margarette Lincoln and Nigel Rigby, eds., *The Worlds of the East India Company* (Woodbridge: Boydell Press, 2002), pp. 119–36.

1. The standard works are still E. G. R. Taylor, *The Haven-Finding Art: A History of Navigation from Odysseus to Captain Cook* (London: Hollis and Carter, 1956),

and, specifically for the period around 1600, D. W. Waters, *The Art of Navigation in England in Elizabethan and Early Stuart Times* (London: Hollis and Carter, 1958). Taylor has been supplemented by Charles H. Cotter, *A History of Nautical Astronomy* (London: Hollis and Carter, 1968), and W. E. May, *A History of Marine Navigation* (Henley-on-Thames: G. T. Foulis and Co., 1973). J. B. Hewson, *A History of the Practice of Navigation* (Glasgow: Brown, Son, and Ferguson, 1951, revised 1963) continues to have value as a practical mariner's history. Many others have addressed the subject more idiomatically, for example Per Collinder, *A History of Marine Navigation* (London: B. T. Batsford, Ltd., 1954), but to go back much earlier one becomes reliant on the last of the navigation manuals proper, such as S. T. S. Lecky, *"Wrinkles" in Practical Navigation* (London: George Philip and Son, 1881 and subsequent editions).

2. For many years the standard work was Derek Howse and Michael Sanderson, *The Sea Chart* (Newton Abbot: David and Charles, 1973). Advances in color-printing technology have made picture books of sea charts a market proposition, for example Peter Whitfield, *The Charting of the Oceans: Ten Centuries of Maritime Maps* (London: The British Library, 1996). John Blake, *The Sea Chart: The Illustrated History of Nautical Maps and Navigational Charts* (London: Conway Maritime Press, 2004) has developed this genre, but with the laudable concomitant of reproducing many visually unspectacular original surveys from the archives of the United Kingdom Hydrographic Office. An earlier monochrome publication largely of such manuscript surveys was Mary Blewitt, *The Surveys of the Seas: A Brief History of British Hydrography* (London: MacGibbon and Kee, 1957).

3. Assessments of the output of the Thames School of chartmakers in London in the early seventeenth century have been made by Tony Campbell "The Drapers' Company and its school of seventeenth-century chart-makers," in Helen Wallis and Sarah Tyacke, eds., *My Head is a Map: Essays & Memoirs in Honour of R. V. Tooley* (London: Carta Press and Francis Edwards, Ltd., 1973), pp. 81–106, and by Thomas R. Smith, "Manuscript and Printed Sea Charts in Seventeenth-Century London: The Case of the Thames School," in Norman J. W. Thrower, ed., *The Compleat Plattmaker: Essays on Chart, Map, and Globe Making in England in the Seventeenth and Eighteenth Centuries* (Berkeley: University of California Press, 1978), pp. 45–100. Awaited is the chapter on late sixteenth- and early seventeenth-century chartmaking by Sarah Tyacke in vol. 3 of *The History of Cartography* (Chicago: University of Chicago Press, forthcoming).

4. Helen M. Wallis and Arthur H. Robinson, *Cartographical Innovations: An International Handbook of Mapping Terms to 1900* ([Tring?]: Map Collection Publications, 1987), s.n. "Chart," pp. 2–11.

5. The Dutch practice is summarized by Kees Zandvliet, *Mapping for Money: Maps, Plans and Topographic Paintings and Their Role in Dutch Overseas Expansion during the 16th and 17th Centuries* (Amsterdam: Batavian Lion International, 1998), particularly pp. 86–117.

6. Andrew S. Cook, "An Exchange of Letters between Two Hydrographers: Alexander Dalrymple and Jean-Baptiste D'Après de Mannevillette," in Philippe Haudrère, ed., *Les Flottes des Compagnies des Indes, 1600–1857* (Vincennes: Service Historique de la Marine, 1996), pp. 173–82. D'Après had published *Le Neptune Oriental* first in 1745, with a revised edition in 1775 and a posthumous *Supplément* in 1781. D'Après de Mannevillette is currently the subject of more detailed study by Manonmani Filliozat: see

for example her "*Le Neptune Oriental*: Une somme de la cartographie de la Compagnie des Indes," *Cahiers de la Compagnie des Indes*, 3 (1998), pp. 21–30.

7. The British Library, India Office Records [hereafter IOR], L/MAR/A-B. See Anthony Farrington, *Catalogue of the East India Company's Ships' Journals and Logs, 1600–1834* (London: The British Library, 1999). On the equipping of East India Company ships, see also Jean Sutton, *Lords of the Easts: The East India Company and its Ships (1600–1874)* (London: Conway Maritime Press, 1981; revised edition, 2000).

8. See the introduction to Coolie Verner and R. A. Skelton, eds., *John Thornton, "The English Pilot, The Third Book," London, 1703*, facsimile edition (Amsterdam: Theatrum Orbis Terrarum, 1973). H. Cornwall, *Observations upon Several Voyages to India* (London, 1720).

9. William Herbert, *A New Directory for the East Indies* (London, 1758; 2nd edition, 1759; 3rd edition, 1767; 4th edition, 1776). J. B. N.-D. D'Après de Mannevillette, *Le Neptune Oriental ou routier general des cotes des Indes Orientales et de la Chine*... (Paris, 1745).

10. C. F. Noble, *The French, and English, Marine Regulations Compared* ([1755]; reprinted [London, 1793]).

11. Alexander Dalrymple to Vansittart, undated but marked as received April 17, 1762 (Robert Orme Collection [IOR: MSS Eur Orme O.V.67(24)], pp. 107–119, especially p. 117.

12. Alexander Dalrymple to Robert Orme, September 8, 1775 (Orme Collection [IOR: MSS Eur Orme O.V.171(1)]).

13. John Gould, "James Rennell's View of the Atlantic Circulation: A Comparison with Our Present Knowledge," *Ocean Challenge: The Magazine of the Challenger Society for Marine Science*, 4, nos. 1/2 (1993), pp. 26–32, figure 4 (p. 29): "Schematic representation of the Atlantic current system."

14. From the 1723 edition onward, the equatorial part of the Atlantic Ocean chart in *The English Pilot, The Third Book*, was marked with a diagram of lines suggesting east and west longitude limits for a favorable passage.

15. Alexander Dalrymple, "Journal of a Voyage to the East Indies... in the Year 1775," *Philosophical Transactions of the Royal Society*, 68, part 1 (1778), pp. 389–418.

16. For Dalrymple's early plans of ports, see Andrew S. Cook, "Alexander Dalrymple's *A Collection of Plans of Ports in the East Indies* (1774–1775): A Preliminary Examination," *Imago Mundi*, 33 (1981), pp. 46–64. For Dalrymple generally, see the introduction to Andrew S. Cook, "Alexander Dalrymple (1737–1808), Hydrographer to the East India Company and to the Admiralty, as Publisher: A Catalogue of Books and Charts," Ph.D. thesis, St. Andrews, 1992.

17. Andrew S. Cook, "Alexander Dalrymple and John Arnold: Chronometers and the Representation of Longitude on East India Company Charts," *Vistas in Astronomy*, 28 (1985): 189–95.

18. Alexander Dalrymple, untitled pamphlet "A Comparison of the several tracks..." ([London,] 1778), example at IOR: MSS Eur Orme O.V.88(8), pp. 101–8.

19. East India Company, Court Minutes, January 21, 1778 (IOR: B/93), p. 513.

20. Alexander Dalrymple, untitled pamphlet "The East India Company having thought proper to employ me..." ([London,] 1779), examples at Washington, Library of Congress: G1059.D23 Text 4(5–6), Paris, Archives Nationales: Marine 3JJ 1(29), and Stockholm, Universitetsbibliotek: Rf.142.

21. London: Admiralty Library, Vf 1/13.

22. Appended to his 1778 and 1779 pamphlets.

23. *General Introduction to the Charts and Memoirs Published by Alexander Dalrymple, Esq.* (London, 1772), p. xi.

24. Dalrymple printed the text in the account of his appointment in *Collection of Plans of Ports in the East Indies* (1775; second edition 1782), introduction, pp. 26–34, particularly pp. 27–28. His memorial had been considered by the Court of Directors on February 3, 1779 (East India Company, Court Minutes [IOR: B/94], p. 502).

25. "Angles for determining the reciprocal Positions of the Lands around False Bay at the Cape of Good Hope, and especially intended to assist in compleating a Survey of Simon's Bay. Taken in August and September 1775 by AD...," in Alexander Dalrymple, ed., *Collection of Views of Land and Plans of Ports in the East Indies* (London, 1781), pp. 1–24, particularly p. 18: "In 1775 when I was at Simon's Bay I intended to have gone upon the Hills, towards the Cape Good Hope, to have endeavoured to have got sight of the Anvil from thence, but unluckily I was prevented by the weather."

26. East India Company, Court Minutes, 1781–1808, s.n. Dalrymple (IOR: B/97–146); East India Company, Cash Journals, 1779–1811, passim (IOR: L/AG/1/5/21–27).

27. Alexander Dalrymple, untitled pamphlet "Notwithstanding the many years that the Europeans have navigated to India..." ([London] 1779), examples at the British Library: G.2198(13.), Paris, Archives Nationales: Marine 3JJ 341(19), and Stockholm, Universitetsbibliotek: Rf.142.

28. Charles Wilkins, East India Company Librarian, reported complying with the executors' request (East India Company, Court Minutes, June 28, 1809 [IOR: B/149, p. 449]).

29. Dalrymple, untitled pamphlet "Notwithstanding...," p. 3.

30. Alexander Dalrymple to East India Company Court of Directors, April 8, 1779 (East India Company, Miscellaneous Letters Received 1779 [IOR: E/1/64], p. 77).

31. Alexander Dalrymple, *Collection of Charts, Views of Land and Plans of Ports in the East Indies* (London, 1783), introduction, pp. 4–6. For the appellation "Eastern Sea charts by special order," see East India Company, General Commerce Journal, 1779–1785 [IOR: L/AG/1/6/18].

32. Alexander Dalrymple, *Memoir Concerning the Passages to and from China* ([London,] 1782). The first "edition" was printed in restricted numbers for issue by the East India Company Secret Committee in 1782 to China ships: it is now known only from the copy sent to the Governor-General in Bengal in January 1783, now in the National Archives of India, and published in C. H. Philips and B. B. Misra, eds., *Fort William-India House Correspondence, vol. XV: Foreign and Secret, 1782–1786* (Delhi, 1963), pp. 20, 24–40. The second (1785) and third (1788) editions enjoyed a more open peacetime circulation.

33. Alexander Dalrymple, *General Collection of Nautical Publications* (London, 1783), pp. 9–10.

34. Alexander Dalrymple, *Memoir of a Chart of the Straits of Sunda and Banka* (London, 1786).

35. Alexander Dalrymple, *Proposition for a Survey of the Coast of Choromandel* (London, 1784), example at the British Library: G.2197(22). Later reissued in the preliminary pages of Alexander Dalrymple, ed., *An Hydrographical Journal of a Cursory Survey of the Coasts and Islands of the Bay of Bengal by Capt. John Ritchie 1770 and 1771* (London, 1784), itself reissued in Alexander Dalrymple, *Collection of Nautical Papers concerning the Bay of Bengal* (London, 1785).

36. [Alexander Dalrymple,] *Instructions concerning the Chronometers, or Time-Keepers, sent to Bombay, 1786* ([London, 1786]), only known example in a private collection.

37. Alexander Dalrymple, ed., *An Account of the Navigation between India and the Gulph of Persia, at all Seasons, with Nautical Instructions for that Gulph, by Lieutenant John McCluer* (London, 1786).

38. Cook, "Alexander Dalrymple and John Arnold."

39. Dalrymple, *General Collection of Nautical Publications*, p. 13.

40. Alexander Dalrymple to D'Après de Mannevillette, August 10, 1779, Paris: Archives Nationales, Marine 3JJ 341(19).

41. Dalrymple, *General Collection of Nautical Publications*, pp. 14–15.

42. Cook, "An Exchange of Letters." [Alexander Dalrymple], untitled pamphlet "Some Notes of the Islands to the Northward of Madagascar, extracted from a letter of M. D'Après de Mannevillette . . . " ([London] 1772), example in Canterbury Cathedral Library: H/U-13-19(1).

43. Untitled chart, showing part of Madagascar and islands to the Northward (Taunton: United Kingdom Hydrographic Office: A13 in folio A). This is an impression made for archive purposes when the plate, which the Hydrographic Office acquired with others after Dalrymple's death, was cancelled and destroyed as part of normal working practices.

44. Dalrymple, *General Collection of Nautical Publications*, pp. 3–4.

45. "Chart of the Mozambique Channel and Island Madagascar by N. Bellin, 1767," published by Dalrymple, May 7, 1791; "Chart of the Mozambique Channel and of Madagascar by John Thornton 1703," published May 7, 1791; "Chart of the Coasts of Suffalo and Moçambique with the Island Madagascar by John Van Keulen," published May 20, 1791; "Chart of the Mozambique Channel with the Island Madagascar and the opposite Coast of Africa by M. D'Après de Manevilletter, 1753," published May 27, 1791; "Chart of the Moçambique Channel with Madagascar and the opposite Coast of Africa from the New Edition of the Neptune Oriental by M. D'Après de Mannevillette 1775," published May 29, 1791. "Charts of the Malabar Coast comparing to various published and MS. Charts from Mangalore to Bombay," published by Dalrymple, January 4, 1789, with the parallel sources given as "M. D'Après 1775," "M. D'Après 1745," "Van Keulen," "a Dutch MS.," "an old English MS. by Augustus Fitzhugh," and "John Thornton 1703."

46. "Chart of the Indian Ocean with the Coasts, Islands, Rocks and Shoals from Madagascar to India, Sumatra and Java composed from various Materials explained in a Memoir by A. Dalrymple 1787," with "publication" date January 24, 1793, examples in Washington, Library of Congress: G1059.D23 Maps 3a/49r and G1059.D24 Maps 5/7. Alexander Dalrymple, *Memoir of a Chart of the Indian Ocean . . .* (London, 1787).

47. For the inferences from an extensive examination of impressions of Dalrymple's chart papers and sizes, see Cook, "Alexander Dalrymple . . . as Publisher," Ph.D. thesis, St. Andrews, 1992, chapter 7.

48. Alexander Dalrymple, Practical Navigation, unpublished treatise c. 1790 (two incomplete proof copies known, in Paris: Bibliotheque Nationale de France, and Edinburgh: National Library of Scotland), pp. 49–50.

49. *List of Charts, Plans of Ports, &c. published by A. Dalrymple, before 1st of June 1789* (London, 1789).

50. An example of the 1783 issue is in Paris: Bibliotheque Nationale de France, Department des Cartes et Plans; an example of the 1788 issue is in London: British Library, India Office Records (IOR: X/3626/1–2).

51. Alexander Dalrymple to James Horsburgh, Cheltenham, October 3, 1805 (Letter-book of James Horsburgh [IOR: MSS Eur F305(II)], p. 147).

52. For the hydrographic surveying activities of Bombay Marine officers and vessels, see C. R. Low, *History of the Indian Navy*, 2 vols (London: Sampson Low, 1873), passim.

53. Alexander Dalrymple, ed., *Memoir of a Chart of the Passage to the Eastward of Banka, with the Relative Positions of Batavia, and the several Places from the Strait of Sunda, to Canton, by Capt. Lestock Wilson, 1789* (London, 1806).

54. Order-in-Council, August 12, 1795 (text published in Sir Archibald Day, *The Admiralty Hydrographic Service, 1795–1919* [London: Her Majesty's Stationary Office, 1967], pp. 334–35). The summary account of Dalrymple's Hydrographic Office work that follows is based on chapter 5 ("A Proper Person for that Office": The Additional Responsibility for the Hydrographical Office of the Admiralty") of my Ph.D. thesis, and subsequently amplified in Andrew S. Cook, "Alexander Dalrymple and the Hydrographic Office," in Alan Frost and Jane Samson, eds., *Pacific Empires: Essays in Honour of Glyndwr Williams* (Melbourne: Melbourne University Press, 1999), pp. 53–68.

55. Alexander Dalrymple to William Wellesley Pole, Secretary to Board of Admiralty, December 23, 1807 (London: National Archives, Admiralty, Hydrographer's Correspondence: ADM.1/3522).

56. J. C. Sainty, *Office-holders in Modern Britain, IV: Admiralty Officials 1660–1870* (London, 1975).

57. Alexander Dalrymple to Evan Nepean, February 27, 1795 (London: National Archives, ADM.1/3522).

58. Order-in-Council, August 12, 1795 (Day, *The Admiralty Hydrographic Service*, pp. 334–35).

59. Admiralty office note of February 22, 1797, endorsed on Dalrymple's account of "Disbursements in the Hydrographical Office," February 7, 1797 (London: National Archives, ADM.1/3522): "Mr. Arrowsmith's Salary at the rate of £100 from 7th Sept. 1795 to 10 Nov. 1796."

60. Alexander Dalrymple to East India Company, June 10, 1795 (East India Company, Miscellaneous Letters Received, 1795 [IOR: E/1/92], p. 101).

61. Dalrymple's account of "Disbursements in the Hydrographical Office," February 20, 1797, and his letter of May 18, 1797 to Nepean (London: National Archives, ADM.1/3522).

62. Dalrymple gave a receipt for the charts and journals on November 9, 1795, on the list "Etat des Calques contenues dans ce Paquet," dating it "Hydrographical Office, Admiralty" (Paris, Archives Nationales: Marine BB[4] 993). For the copy of this list and receipt kept in the Hydrographic Office, and bearing William Marsden's subsequent receipts of August 17, 1796, see the paper of the same title in London, National Archives, ADM.1/3523. Dalrymple supplied a certificate of Rossel's employment on August 25, 1799 (London: National Archives, ADM.1/3522). For the circumstances of the "deposit" of D'Entrecasteaux's charts in 1795, see Hélène Richard, *Le Voyage de D'Entrecasteaux à la Recherche de Lapérouse* (Paris, 1976), pp. 210 and 215–16, and Gavin De Beer, *The Sciences Were Never at War* (London, 1960), p. 50.

63. Dalrymple to Nepean, March 24, 1796, and March 8, 1799, and paper of November 12, 1799 (London: National Archives, ADM.1/3522).

64. Dalrymple to Marsden, June 17, 1798, and list of charts, plans, and views, June 19, 1798 (London: National Archives, ADM.1/3522).

65. Dalrymple to Pole, December 23, 1807 (London: National Archives, ADM.1/3522).

66. Dalrymple's memorandum of April 26, 1808, enclosed in his letter of April 29, 1808, to Viscount Melville (Edinburgh, Scottish Record Office: GD.51/2/399/1).

67. Dalrymple to Nepean, March 22, 1800 (London: National Archives, ADM.1/3522); Dalrymple to Matthew Boulton, March 31, 1800 (Birmingham Reference Library, Archives Department: Matthew Boulton Papers, Letter D22).

68. Dalrymple to Nepean, January 15, 1802 (London: National Archives, ADM.1/3522). The only known example of the printed chart is in the Mitchell Library (State Library of New South Wales, Sydney).

69. Dalrymple to Marsden, June 13 and June 19, 1804, and supplementary "Statement" June 21, 1804 (London: National Archives, ADM.1/3522). Dalrymple settled with the Board of Admiralty for a payment of 1000 guineas instead of the calculated retail price of £3006, presumably because the Hydrographic Office copperplate printer carried out the work of printing.

70. Dalrymple to Pole, October 10, 1807 (London: National Archives, ADM.1/3522), pp. 13–14.

71. Dalrymple to Nepean, February 6, 1804 (London: National Archives, ADM.1/3522), enclosing a proof of the "Form of Remark-Book" that had been previously discussed between Nepean and Dalrymple, though no correspondence survives.

72. Bligh was requesting payment of an invoice on May 10, 1804 (London: National Archives, ADM.1/3522).

73. "List of English Charts," enclosure in letter from Dalrymple to Pole, October 10, 1807 (London: National Archives, ADM.1/3522). This sixty-one-page catalogue is a neglected synopsis of the state of commercial chart publication in London in the early nineteenth century.

74. Dalrymple to Pole, November 24, 1807 (London: National Archives, ADM.1/3522).

75. Chart Committee to Pole, May 26, 1808, extract (London: National Archives, ADM.1/3523).

76. Pole to Dalrymple, May 28, 1808, copied in letter from Dalrymple to Melville, May 30, 1808 (SRO: GD.51/2/399/2; draft in London: National Archives, ADM.1/3523).

77. W. Ramsay, secretary, East India Company, to Horsburgh, November 2, 1810; Horsburgh to Committee of Shipping, East India Company, November 8, 1810, IOR: MSS Eur F305(I), pp. 196–97.

78. On June 10, 1808, Pole gave Hurd executive instructions on a proposal of the Chart Committee of April 22 to purchase charts.

79. Published in 1811, and known as Hurd's *Channel Atlas*.

80. Will of Alexander Dalrymple, December 24, 1798, with cancellations and codicils of October 29, 1805 (London: National Archives, PROB.10/3854), paras. 28–29.

81. For developments in the Hydrographic Office in the nineteenth century, see Day, *Admiralty Hydrographic Service*.

82. In four volumes of letterpress memoirs and four volumes of charts, plans, and views of land, now in Washington, D.C., Library of Congress, Geography and Map Division (G1057.D23–24), for which contents lists were drawn up by Beaufort on the unused parts of letter sheets received by him in *Woolwich* in November and December 1805.

Chapter Four

1. Andrew M. Modelski, *Railroad Maps of North America: The First Hundred Years* (Washington, D.C.: Library of Congress, 1984), p. vii.

2. "Lithographed from a Plan drawn by E. S. Chesbrough" and printed by Pendleton's Lithography, Boston.

3. The profiles were also the work of McNeill and his assistants and were printed by Pendleton's Lithography, Boston. A slightly later, midwestern example is the 1850 "*Map of the Bellefontaine and Indiana Railroad and Connecting Lines Accompanying the Report on the Preliminary Surveys, W. Milnor Roberts, Chief Engineer*," bound into a forty-page *Report on the Preliminary Surveys of the Bellefontaine & Indiana Rail Road Company*, authored by W. Milnor Roberts, Chief Engineer.

4. While "air line" was the common name for such railroads, a few used other terminology. For example, a railroad proposed in 1853 to run from the Ohio River at Evansville, Indiana to the Ohio-Indiana state line took the name Evansville, Indianapolis & Cleveland Straight Line Railroad Company.

5. The report was printed by Casper C. Childs of New York. The maps are untitled and do not list the lithographer. However, the report was written by Richard P. Morgan, an engineer who conducted surveys for several midwestern railroads and who may have had a hand in drawing at least the proposed route of the Rock Island & LaSalle.

6. A subtler motive may also have been at work. In 1853–54, the G&CU had beaten back the threat of a potential rival, the Chicago, St. Charles & Mississippi Air Line, a new company that threatened to steal away half of the territory the G&CU considered to be its service area. Having driven the upstart CStC&M into bankruptcy, the older line may have been reveling in its success and, for the benefit of any other potential competitors, marking large portions of northern Illinois, southern Wisconsin, eastern Iowa, and eastern Minnesota as its own.

7. The maps are undated but were surely printed in 1875. They are the work of A. Meisel, Lithographers, of Boston. Each measures 12 ¼ × 14 ⅔ inches.

8. The Southern Pacific's seventeenth annual report for 1901 includes a foldout map, "Southern Pacific Railway and Steamship Lines, 1901." Executed by Rand, McNally & Co. of Chicago, it measures 35 × 23 inches, depicts the railroad's lines in red and other railroads in black, and indicates mountain ranges in light green. The map in the Southern Pacific's twenty-fourth annual report June 30, 1908, is even larger, measuring 42 × 30 inches. "Southern Pacific and Union Pacific Systems, 1908" was also the product of Rand, McNally & Co., Chicago. However, its typography appears more mechanical, having lost the artistic air of the maps of a few years earlier.

Another example is the large fold-out map, created by McGill-Warner Co. of St. Paul, contained within the CB&Q annual report for 1922. The routes of five railroads—the CB&Q; the Great Northern; the Northern Pacific; the Spokane, Portland & Seattle; and the Colorado & Southern—are shown in color.

9. Italics are in the text.

10. The "Map of Northern Pacific Lands in Minnesota" was printed by the National Railway Publication Company, Philadelphia.

11. *Harvest Excursions via the Chicago, Milwaukee, and St. Paul Railway* (Chicago: Chicago, Milwaukee, and St. Paul Railway, 1895). The map measures 7 ½ × 3 ½ inches, was the product of Rand, McNally & Co., and is in the Newberry Library collection.

12. Before the railroad, the U.S. Government was unable to sell nearby land for as little as 12.5 cents an acre. See E. H. Talbott, *Railway Land Grants in the United States: Their History, Economy and Influence Upon the Development and Prosperity of the Country* (Chicago: The Railway Age Publishing Company, 1880).

13. Hopkins Rowel, *The Great Resources and Superior Advantages of the City of Joliet, Illinois* (Joliet: Republican Steam Press, 1871). The map in this thirty-two-page booklet was the work of Darling S.C. of Lockport, Illinois.

14. This fold-out map was drawn by H. Von Minden of St. Paul and printed by Benham & Rice, St. Paul.

15. *History of Allen County* (Chicago: Kingman Brothers, 1880), p. 89.

16. This image of a farm with a train in the foreground is drawn from the *Illustrated Historical Atlas of Elkhart County, Indiana* (Chicago: Higgins, Belden & Co., 1874), p. 58.

17. Both maps are attributed to Rand, McNally & Co., Map Engravers and Printers, Chicago. Unfolded, this publication measures 28 ½ × 16 ½ inches.

18. George H. Douglas, *All Aboard! The Railroad in American Life* (New York: Paragon House, 1992), p. 155.

19. Henry Schenk Tanner, *A New Universal Atlas* (Philadelphia: Carey & Hart, 1845).

20. J. H. Colton and A. J. Johnson, *Johnson's New Illustrated Family Atlas* (New York: Johnson & Browning, 1862).

21. This map, contained in the 1919 Time Table for the Buffalo Division, was drawn in November 1909 by V. F. L.

22. This information was provided by Tom Hoback, president of the Indiana Rail Road Company.

23. The scale of this 8 ½ × 14 inches map is 1 inch to 300 feet. Produced by the Office of the Engineer of Maintenance-of-Way for the Pennsylvania Railroad, Richmond Division of the Lines West of Pittsburgh, the map is dated "10-3-[18]92."

24. From "A Brief Statement as to the Purposes and Uses of the Atlas of Traffic Maps" in *Atlas of Railway Traffic Maps* (Chicago: LaSalle Extension University, 1913). Shelton describes himself as "sometime of the Tariff Bureau of the C. & N. W. Railway" [Chicago & North Western].

25. Harold Hathaway Dunham, *Government Handout: A Study in the Administration of the Public Lands, 1875–1891*, (New York: Da Capo Press, 1970), p. 113.

26. This movement toward regulation eventually resulted in the establishment of the federal Interstate Commerce Commission.

27. This 30 × 24 inch map was prepared under the direction of Van R. Richmond, state engineer and surveyor, and his deputy, S. H. Sweet. Argus Company of Albany was the lithographer.

28. Modelski, *Railroad Maps of North America: The First Hundred Years*, p. 88.

29. John S. Wright, *Chicago: Past, Present, Future: Relations to the Great Interior, and to the Continent*, 2nd edition (Chicago: Chicago Board of Trade, 1870), pp. 364–65.

Wright's tables note that 10,000 of the 22,000 miles of new track built in the 1850s were built in just seven midwestern states.

30. George Rogers Taylor and Irene D. Neu, *The American Railroad Network: 1861–1890* (Cambridge: Harvard University Press, 1956), pp. 4–5.

31. This fold-out map was engraved by Alonzo Lewis.

32. This fold-out map is bound in a booklet entitled *Report on the Preliminary Surveys for the Bellefontaine & Indiana Rail Road Company* (Pittsburgh: Johnston and Stockton for the Bellefontaine & Indiana Rail Road Co., 1850).

33. Henry William Ellsworth, *Valley of the Upper Wabash, Indiana, with Hints on Its Agricultural Advantages* (New York: Pratt, Robinson & Co., 1838), pp. 22–27.

34. *The Railroad Jubilee: An Account of the Celebration Commemorative of the Opening of Railroad Communication Between Boston and Canada* (Boston: J. H. Eastburn, 1852), part of the text of the map.

35. John F. Stover, *History of the Baltimore and Ohio Railroad* (West Lafayette, IN: Purdue University Press, 1987), p. 52.

36. "An Act to amend the charter of the Green Bay, Milwaukee, and Chicago Rail Road Company, and to authorize an extension of said road," *Laws of Wisconsin: Private & Local*, March 23, 1852.

37. See especially Andrew M. Modelski, *Railroad Maps of the United States* (Washington, D.C.: Library of Congress, 1975), pp. 2–4.

38. "Railway Advertising," *American Railway Times*, 4, no. 15, April 8, 1852, p. 2.

39. For a summary of their publication histories, see Walter W. Ristow, *American Maps and Mapmakers: Commercial Cartography in the Nineteenth Century* (Detroit: Wayne State University Press, 1985).

40. J. Calvin Smith, "Guide Through Ohio, Michigan, Indiana, Illinois, Missouri, Wisconsin, and Iowa." In *The Western Tourist and Emigrant's Guide* (New York: J. H. Colton, 1840).

41. In this regard, English railways and independent publishers were well ahead of the Americans. By the late 1830s, Francis Coghlan was publishing his *Iron Road Book and Railway Companion from London to Birmingham, Manchester, and Liverpool* (London: A. H. Bailey & Co., 1838). This 180-page pocket-sized book included a series of single page maps of the route, along with accounts of the towns the line passed. Tebbutt's 1841 *Guide to the North Midland, Midland Counties, and London & Birmingham Railways* includes a fold-out map and a fold-out timetable. The Continent apparently also saw rail travel maps in advance of America. One example is "Carte-Itinéraire du Chemin de Fer de Paris à Rouen," a fold-out map bound into the 1845 booklet *Itinéraire du Chemin De Fer De Paris à Rouen, Description Historique Et Pittoresque*, Ernest Bourdin, editor. The booklet includes thirty vignettes of villages and towns along the railway's line.

42. The idea of the Illinois Central, of course, dates back to the State of Illinois' grandiose plans of 1836. In that early version, the line was to connect the northwestern corner of Illinois, at the Mississippi River, with the Mississippi and the Ohio in southern part of the state. In so doing, the railroad was to address problems of shipping via the Mississippi: the slow pace of shipping on the winding river, the dangers to navigation of snags and shoals, and the problems winter posed for river navigation. In its 1850 reappearance the Illinois Central now involved a line from Chicago and a line from northwest Illinois meeting in the center of the state and

running to the southernmost part of the state. It was now to serve the greater portion of the state.

43. See Frank Walker Stevens's *The Beginnings of the New York Central Railroad: A History* (New York: G. P. Putnam's Sons, 1926).

44. *Weekly Chicago Democrat*, January 1, 1853, p. 1.

45. "Interesting Railway Statistics," *Weekly Chicago Democrat*, July 30, 1859, p. 2.

46. "The Air Line Railway," *American Railway Times*, May 20, 1852, p. 1.

47. John Ashcroft, *Ashcroft's Railway Directory: 1862* (New York: John Ashcroft, Publisher, 1862).

48. "Canal and Railway Transportation," *Weekly Chicago Democrat*, January 1, 1859, p. 3. See also the *Weekly Chicago Democrat*, April 2, 1859, p. 1.

49. See, for example, John F. Stover, *Iron Road to the West: American Railroads in the 1850s* (New York: Columbia University Press, 1978), pp. 190–92.

50. See especially John H. White, Jr., *The American Railroad Passenger Car* (Baltimore: Johns Hopkins University Press, 1978). Another good source is Edwin A. Pratt, *American Railways* (New York: Macmillan Co., 1903), pp. 86–103.

51. "Finished to Janesville," *Weekly Chicago Democrat*, January 1, 1853, p. 1.

52. *Fifth Annual Review of the Prospects, Condition, Traffic, Etc., of the Railroads Centering on Chicago . . . for the Year 1856*, pp. 62–63, as quoted by Wyatt Winton Belcher, *The Economic Rivalry Between St. Louis and Chicago, 1850–1880* (New York: Columbia University Press, 1947), p. 70. This 3.5 million figure is somewhat inflated by the fact that passengers arriving by one line and departing by another appear to be counted twice. Nonetheless, the growth in passenger traffic was dramatic.

53. Brad S. Lomazzi, *Railroad Timetables, Travel Brochures & Posters: A History and Guide for Collectors* (Spencertown, NY: Golden Hill Press, 1995), p. 73.

54. *The Locomotive Courant: A Monthly Record of Material Progress*, vol. 2, no. 4, April, 1857.

55. *American Railroad Network*, p. 10.

56. *Ohio Railroad Guide* (Columbus, OH: Ohio State Journal Co., 1854).

57. *American Railway Times*, vol. 10, no. 27, July 3, 1858, p. 4 (emphases mine).

58. David Woodward discusses how wax engraving permitted publishers to easily make changes or additions to the face of maps. *The All-American Map: Wax Engraving and Its Influence on Cartography* (Chicago: University of Chicago Press, 1977), see especially pp. 30–36 for the effect of wax-engraving on railroad maps.

59. One each of the 1868, 1869, and 1870 editions have been reprinted. The publisher of the reprint of the June 1870 issue, the first published by the National Railway Publication Company, is Edwards Brothers, Inc., Ann Arbor, Michigan.

60. The 1870 map, measuring 24 × 36 inches, was engraved by Fisk, Russell & Ames, New York.

61. By the time this 1882 *Guide* was published, still by the National Railway Publication Company, W. F. Allen had replaced Vernon as editor. It is worth noting that the title page of this October 1882 issue lists A. McNally of Chicago as the vice president of the publishing company.

62. The book was edited by N. J. Watkins and published by J. D. Ehlers & Co.'s, Engraving and Printing House. The mutual trips were under "the leadership of Maj. N. H. Hotchkiss, Traveling Agent of Chesapeake & Ohio and Richmond & York River Railroads."

63. The map was a product of the Engineer Office of Jed. Hotchkiss, Staunton, Va. and was drawn by D. C. Humphreys. The railroad line had just reached Huntington, West Virginia, and the distance from Huntington to Cincinnati was traversed by steam packet.

64. *Chicago: Relations to the Great Interior & Continent*, p. 53.

65. Ibid., p. 355.

66. See especially Woodward, *The All-American Map*.

67. *Railroad Timetables, Travel Brochures & Posters*, p. 13.

68. Though others no doubt were being produced at the time or even a little earlier.

69. Timetable is in the collection of the Indiana Historical Society, Indianapolis.

70. Modelski, *Railroad Maps of North America: The First Hundred Years*, pp. 88–89. See also the 1876 Baltimore & Ohio map Modelski includes on pp. 104–5. The Erie Railway produced such timetables with maps at least as early as 1863. Its March 1865 timetable includes a 9 ⅔ × 8 ½ map entitled "Great Direct Broad Gauge Route," the map being attributed to *Appleton's Railway Guide*.

71. Ibid., p. 88.

72. The map is the product of Poole Bros. of Chicago.

73. See Julius Grodinsky, *Transcontinental Railway Strategy, 1869–1893: A Study of Businessmen* (Philadelphia: University of Pennsylvania Press, 1962); Wyatt Winton Belcher, *The Economic Rivalry Between St. Louis and Chicago, 1850–1880* (New York: Columbia University Press, 1947), p. 73ff.; and H. Craig Miner, *The St. Louis-San Francisco Transcontinental Railroad: The Thirty-fifth Parallel Project, 1853–1890* (Lawrence, KS: University Press of Kansas, 1972).

74. Lewis H. Haney, *A Congressional History of Railways in the United States*, vol. 2 (Madison, WI: Democrat Printing Co., 1910), pp. 49–64.

75. *Reports of Explorations and Surveys, to Ascertain the Most Practicable and Economical Route for a Railroad from the Mississippi River to the Pacific Ocean* (Washington, 1856–60).

76. See Paul E. Cohen, *Mapping the West: America's Westward Movement 1524–1890* (New York: Rizzoli, 2002), pp. 172–75.

77. Pratt, *American Railways*, pp. 87–88.

78. Ibid., p. 89.

79. See White, *American Railroad Passenger Car*, pp. 130–45.

80. Alfred Runte, *Trains of Discovery: Western Railroads and the National Parks* (Niwot, Colorado: Roberts Rinehart Publishers, 1994), p. 9. Other studies of the relationship between railroads and the development of the national parks include Anne Farrar Hyde, *An American Vision: Far Western Landscape and National Culture, 1820–1920* (New York: New York University Press, 1990), and Marguerite S. Shaffer, *See America First: Tourism and National Identity, 1880–1940* (Washington, D.C.: Smithsonian Institution Press, 2001).

81. *Railroad Maps of the United States*, p. 2.

82. This timetable is in the collection of the Indiana Historical Society. The map is the work of Rand, McNally & Company of Chicago.

83. Produced by Poole Bros., Chicago, in February 1897, the map measures 5 × 2 ¼ inches.

84. This 8 ½ × 31 ½ inch brochure, executed by H. H. Green, is undated, but it likely comes from the same decade of the 1890s as our previous map. My thanks to Jeff Darbee for calling my attention to this brochure and making a color copy for my use.

85. MLS&W, *Gegobic* (Chicago: Poole Brothers, 1886).

86. Ernest Ingersoll, *Down East Latch Strings* (Boston: Passenger Department Boston & Maine Railroad, 1887), p. vii.

87. Woodward, *All-American Map*, p. 33.

88. "Trail of the Olympian: Two Thousand Miles of Scenic Splendor, Chicago to Puget Sound" (Chicago: Poole Bros. for Chicago, Milwaukee, and St. Paul Railway, 1924), pp. 13 and 39.

89. H. Roger Grant, "Electric Traction," in *Encyclopedia of American Business History and Biography: Railroads in the Age of Regulation, 1900–1980*, ed. Keith L. Bryant, Jr. (New York: Facts On File Publications, 1988), p. 129.

90. *All Aboard! The Railroad in American Life*, p. 245.

91. John R. Stilgoe, *Metropolitan Corridor: Railroads and the American Scene* (New Haven: Yale University Press, 1983), pp. 293–97.

92. The map is credited to Matthews-Northrup Works, Buffalo, NY. Neither the brochure nor the maps are dated. The latest date mentioned in the text of the brochure is 1902, and the photographs of street scenes suggest the date of publication was during the first decade of the 1900s.

93. "Tri-State Trolley Map showing Boston & Northern and Old Colony Street Railway Companies' Systems and Connecting Lines" (Boston: C. J. Peters & Son Co., engravers, 1907).

94. *Trolley Trips*, p. 2.

95. *All Aboard! The Railroad in American Life*, p. 155.

96. An example is Clason's *Kansas Green Guide*. While undated, it appears to be from the 1920s. The fold-out sheet measures 28 × 21 inches.

Chapter Five

1. John Stilgoe, *The Metropolitan Corridor: Railroads and the American Scene* (New Haven: Yale University Press, 1983), 249.

2. On the contrast between the experience railroad travel and early motoring, see John A. Jakle, *The Tourist: Travel in Twentieth-Century North America* (Lincoln: University of Nebraska Press, 1985), 84–95, 101–4.

3. Douglas Brinkley, *The Magic Bus: An American Odyssey* (New York: Anchor/Doubleday, 1993), 14–15.

4. Henry B. Joy, "The Traveller and the Automobile," *Outlook* 115, 17 (April 25, 1917): 740.

5. David Halberstam, "An American Romance," *Popular Mechanics*, May 1989; quoted in K. T. Berger, *Where the Road and Sky Collide: America through the Eyes of Its Drivers* (New York: Henry Holt, 1993), 78.

6. John A. Jakle, "Landscapes Redesigned for the Automobile," in *The Making of the American Landscape*, ed. Michael P. Conzen (Boston: Unwin Hyman, 1990), 293.

7. Andrei Codrescu, *Road Scholar: Coast to Coast Late in the Century* (New York: Hyperion, 1993), 3.

8. See, for example, Henry Schenk Tanner, *A New American Atlas* (Philadelphia: H.S. Tanner, 1823).

9. See, for example, *Cram's Bankers' and Brokers' Railroad Atlas* (Chicago: George F. Cram, 1890?) and *Rand McNally & Co.'s Business Atlas* (Chicago: Rand McNally, 1880).

10. Douglas A. Yorke, Jr., John Margolies, and Eric Baker, *Hitting the Road: The Art of the American Road Map* (San Francisco: Chronicle Books, 1996), 6.

11. *Auto Trails of Minnesota* (Chicago: Rand McNally for the Minnesota Hardware Association, 1923?). Copy in the Newberry Library.

12. The author refers to his mother-in-law, who retained all of her family's *Triptiks* as souvenirs until they passed into his personal collection. For more on the phenomenon of map annotation, see James R. Akerman, "Private Journeys on Public Maps: A Look at Inscribed Road Maps," *Cartographic Perspectives* 35 (Winter 2000): 27–47.

13. For the broad outlines of the history of promotional road mapping, see "American Promotional Road Mapping in the Twentieth Century," *Cartography and Geographic Information Science* 29, 3 (July 2002): 175–91.

14. General Drafting's major contribution to the industry was the maps it prepared for Standard Oil of New Jersey (Esso) and its affiliates, more recently known as Exxon. This particular issue was sold by the American News Company for 25 cents.

15. *Pictorial Map of the United States with Trip-Planning Guide* (Convent Station, NJ: General Drafting for American News Co., 1952).

16. See Marguerite S. Shaffer, *See America First: Tourism and National Identity, 1880–1940* (Washington: Smithsonian Institution Press, 2001).

17. See Anne Farrar Hyde, *An American Vision: Far Western Landscape and National Culture, 1820–1920* (New York: New York University Press, 1990), 107–46.

18. I borrow the term from John F. Sears, *Sacred Places: American Tourist Attractions in the Nineteenth Century* (New York: Oxford University Press, 1989).

19. "U.S. 60-66-70-80 Southern Routes West-East Standard Oil Interstate Route Map" (Chicago and San Jose: H. M. Gousha for Standard Oil Company of California, 1941).

20. Maps exhibiting this common cartographic strategy include "1931 Official Road Map, Indiana, with Compliments of Richman Brothers Co." (Chicago: Rand McNally for Richman Brothers Co., 1931); "Motor Court and Cabin Guide Map of New England, New York, Quebec, 1953" (Chester, VT: The National Survey for Northeastern Cabin Owners' Association, 1953); and Road Map of Southeastern United States (Chicago: Tempo Designs for Morrison's Cafeterias and Admiral Benbow Inns, 1974).

21. The author's personal copy is dated 1951.

22. Other examples of this genre include: *New! US 40, the Main Line of America!* (Denver: National U.S. 40 Association, c. 1955); *US 20, Your Key Route to America* (Chadron, NE: U.S. Highway 20 Association, 1957); *Travel US Route 54, the Safe Family Highway* (Greensburg, KA: U.S. 54 National Headquarters, c. 1955); *See and Enjoy American Coast to Coast, Travel US 60* (Hereford, TX: U.S. Highway 60 Association, c. 1965); and *U.S. 6, Roosevelt Highway, the Scenic Way from the Atlantic to the Pacific* (Des Moines: U.S. 6, Roosevelt Highway, n.d.).

23. New Holland, PA: Edward C. Procter for the Amish Farm and House, Lancaster, PA.

24. "Ozarks—White River Lakes Country . . . Home of Marvel Cave," in "White Lakes Area Map Compliment of Marvel Cave Park" (Aurora: MWM Color Press for Marvel Cave Park, c. 1965)

25. See Sears, *Sacred Places.*

26. See for example, Carl J. Hals and A. Rydström, "Map of the Yellowstone National Park Compiled from Different Official Surveys and Our Personal Survey, 1882,"

in *Alice's Adventures in the New Wonderland, the Yellowstone National Park* (Chicago: Poole Bros. for the Northern Pacific Railroad [1884]).

27. "Map of Grand Canyon of Arizona" (Chicago: Poole Bros., n.d.), in *The Grand Canyon of Arizona* ([Chicago]: Passenger Department, Atchison, Topeka & Santa Fe Railroad, 1902).

28. I note that this cherished flexible access to major portions of national parks embraced and fostered by the first director of the National Park System, Stephen Mather, has been abandoned of necessity more recently to alleviate pollution and other environmental problems. In Grand Canyon National Park, for example, private automobile access to some of the rim roads has been severely restricted in favor of shuttle buses and, perhaps in the near future, a light rail system.

29. Robert Shankland, *Steve Mather of the National Parks* (New York: Alfred A. Knopf, 1954), 160. On highway construction and the motorization of national parks, see also Alfred Runte, *National Parks: The American Experience* (Lincoln: University of Nebraska Press, 1979), 155–79; James J. Flink, *The Automobile Age* (Cambridge, MA: MIT Press, 1988), 171–82; and Ethan Carr, *Wilderness by Design: Landscape Architecture and the National Park Service* (Lincoln: University of Nebraska Press, 1998).

30. *Hi-way map to Zion, Bryce Canyon, Grand Canyon National Parks* (Chicago: Rand McNally for Utah Parks Company, 1936).

31. Runte, *National Parks,* 171. I treat the railroads' cartographic response to incursions by the automobile more fully in "Maps as Promotional Tools for American Railroads," presented at the *Maps and Popular Cartography,* the Second Biennial Virginia Garrett Lectures on the History of Cartography, October 2000.

32. John Erastus Lester, *The Atlantic to the Pacific: What to See and How to See It* (London, 1873), 224–25, paraphrased in Hyde, *An American Vision,* 108.

33. Hyde, *An American Vision,* 108.

34. On the gradual spread of leisure travel among the middle and working classes in the United States during this period, see Cindy S. Aron, *Working at Play: A History of Vacations in the United States* (New York: Oxford University Press, 1999).

35. *The Pacific Tourist . . . A Complete Traveler's Guide of the Union and Central Pacific Railroads* (New York: Henry T. Williams, 1876), 8–9.

36. "Williams' New Trans-Continental Map of The Pacific R.R." (New York: Henry T. Williams, 1876), in *The Pacific Tourist.*

37. See T. R. Nicholson, *Wheels on the Road: Maps of Britain for the Cyclist and Motorist, 1870–1940* (Norwich, England: Geo Books, 1983), 1–44.

38. See Philip P. Mason, "The League of American Wheelmen and the Good-Roads Movement," Ph.D. dissertation, University of Michigan, 1957.

39. No bibliography of LAW road books and maps exists; this information was gleaned from entries in *The National Union Catalogue, Pre-1956 Imprints* (London: Mansell, 1968–81).

40. W. W. Randall and Carl Hering, *Road Book of Pennsylvania, Western Section,* 1898 (Philadelphia: Pennsylvania Division, LAW, 1898), 3.

41. L. W. Conklin and C. L. Steen, *The Official Cyclists' Road Book of Illinois* (New York: J. B. Beers & Co. for the Illinois Division, League of American Wheelmen, 1892).

42. Published in two 2 vols. by New York: Scribner's, 1887–88.

43. On the bicycle craze of the late nineteenth century, see Gary Allan Tobin, "The Bicycle Boom of the 1890s: The Development of Private Transportation

and the Birth of the Modern Tourist," *Journal of Popular Culture* 7 (1974): 838–49.

44. Flink, *The Automobile Age*, 5–6.

45. See Walter W. Ristow, "A Half-Century of Oil-Company Road Maps," *Surveying and Mapping* 24, 4 (1964): 620, 623.

46. According the back cover of Walker's *Latest Map of Southern New Hampshire*, by 1917, as Walker Lith. & Pub. Co., it was selling thirty-five "automobile maps" for $2 each.

47. Worcester, MA: F. S. Blanchard for the Home Educator Company, 1905.

48. *Scarborough's Complete Road Atlas of Massachusetts and Rhode Island* (Boston: Scarborough Co., 1905).

49. *Highway Statistics: Summary to 1975* (Washington, D.C.: U.S. Department of Transportation, 1977), table MV 213.

50. See Dayton Duncan and Ken Burns, *Horatio's Drive* (New York: Knopf, 2003).

51. *Twin City—Glacier Park Auto Tour, July 11th to 19th, 1913, National AAA Reliability Run* (American Automobile Association, 1913).

52. Emily Post, *By Motor to the Golden Gate* (New York and London: D. Appleton, 1916).

53. Carey S. Bliss, *Autos Across America: A Bibliography of Transcontinental Automobile Travel, 1903–1940* (Austin & New Haven: Jenkins & Reese, 1982).

54. Hrolf Wisby, "Camping Out in an Automobile," *Outing* 45 (March, 1905), 739–40. I am grateful to Robert Buerglener for locating this article. I am indebted to Mr. Buerglener for our discussions of early pathfinders.

55. H. B. Haines, "How to Tour in a Motor Car," *Scientific American* 97 (Nov. 9, 1907): 330–31.

56. See Virginia Rishel, *Wheels to Adventure: Bill Rishel's Western Routes* (Salt Lake City, UT: Howe Brothers, 1983).

57. For example, the "System Log . . . Las Vegas to Raton, N.M." issued by the Transcontinental Garage Service Inc. and the National Garage in Raton, N.M., describes 112 miles of road on four typeset cards. This log has been reproduced in Doris Wiant Harvey, *Wiant-Bogart Cross-Country Odyssey 1915* (Sweet Valley, PA: privately published, 1982), 7–8.

58. *Automobile Official AAA 1910 Blue Book, volume 2, New England* (New York and Chicago: Automobile Blue Book Publishing Co., 1910), route 346, pp. 454–55.

59. *The Live-Map* (New York: United Manufacturers, 1909). See also the discussion by Robert French in chapter 6.

60. *Photo-Auto Maps . . . New York to Chicago, Chicago to New York* (Chicago and New York: Rand McNally, 1909). The concept for these route guides was developed by H. Sargent Michaels in 1905 and copyrighted by G. S. Chapin in 1907; Rand McNally took over publication of the series in 1909. See Ristow, "A Half-Century of Oil-Company Road Maps," 619–20.

61. Beth O'Shea, *A Long Way from Boston* (McGraw-Hill, 1946); quoted in Drake Hokanson, *The Lincoln Highway: Main Street Across America* (Iowa City: University of Iowa Press, 1988), 91.

62. Estella M. Copeland, *Overland by Auto in 1913* (Indianapolis: Indiana Historical Society, 1981), 18–19.

63. Post, *By Motor to the Golden Gate*, 3–7.

64. The best summary of the good roads movement from the 1880s to the early 1920s is Peter J. Hugill, "Good Roads and the Automobile in the United States, 1880–1929," *Geographical Review* 72 (1982): 327–49; see also *America's Highways, 1776–1976: A History of the Federal-Aid Program* (Washington, D.C.: U.S. Dept. of Transportation, 1979).

65. Harvey, *Wiant-Bogart Cross-Country Odyssey,* 6.

66. The 1921 edition of the *Rand McNally Official Auto Trails Map of the United States* lists 92 auto trails, but perhaps only a third to a half of these were interregional in character.

67. *Colorado to Gulf Highway: The Great Automobile Route from the Gulf Coast to the Rocky Mountains* ([Amarillo, TX]: The Colorado-to-the-Gulf Highway Association, 1914?).

68. Yellowstone Trail Association, *The Yellowstone Trail* (Ipswich, SD: Joseph W. Parmley, 1914).

69. *Appalachian Scenic Highway, Roscoe A. Marvel, President, Kenilworth Inn, Asheville, N.C.* (Villa Tassom, FL: F. V. Orr for the Appalachian Scenic Highway, 1925).

70. The maps show an Appalachian Highway in North and South Carolina, but this seems to be an altogether different route.

71. *Colorado to Gulf Highway,* 2.

72. *State of Maryland Showing Seven Hundred Miles of National Highways Proposed by the National Highways Association* (Washington, D.C.: National Highways Association, 1914).

73. See James R. Akerman, "Maps of the National Highways Association from a Recent Gift," *Mapline* 72–73 (Winter/Spring 1994): 8–9. On the NHA's position in the good roads movement, see Richard F. Weingroff, "Good Roads Everywhere: Charles Henry Davis and the National Highways Association," online publication of the U.S. Department of Transportation, Federal Highway Administration, modified to April 28, 2003 (www.fhwa.dot.gov/infrastructure/davis.htm).

74. *America's Highways, 1776–1976: A History of the Federal-Aid Program* (Washington, D.C.: U.S. Dept. of Transportation, 1979), 90–107.

75. Joy, "The Traveller and the Automobile," 740.

76. Warren Belasco, *Americans on the Road, 1895–1910: From Autocamp to Motel, 1910–1945* (Cambridge, MA: MIT Pres, 1979), 72–74.

77. This evolution is thoroughly documented by Belasco, *Americans on the Road;* and John A. Jakle, Keith A. Sculle, and Jefferson S. Rogers, *The Motel in America* (Baltimore: Johns Hopkins, 1996).

78. Flink, *The Automobile Age,* 360.

79. *Location of U.S. Forces in Training and Our Important Motoring Highways* (Chicago: Motor Age, 1918).

80. "Rand McNally Official Auto Trails Map of the United States" (Chicago: Rand McNally, 1921).

81. Ristow, "A Half-Century of Oil-Company Road Maps," 619–23.

82. See Roderick Clayton McKenzie, "The Development of Automobile Guides in the United States," M.A. thesis, University of California–Los Angeles, 1963, 19, 29, 32; and Ristow, "A Half-Century of Oil-Company Road Maps," 619.

83. This development may be contrasted with the decision of the French tire company, Michelin, to establish its cartographic offices, which continue to sell their highly regarded maps independently from their sales of tires. See "Les Services de Tourism

du Pneu Michelin: Histoire et évolution des publications cartographiques Michelin," *International Yearbook of Cartography* 3 (1963): 170–80.

84. Ristow, "A Half-Century of Oil-Company Road Maps," 625–28.

85. For the early history of these firms, see Otto G. Lindberg, *My Story* (Convent Station, NJ: General Drafting, 1955); and "The National Survey," *Vermont Life,* spring 1951, pp. 26–35.

86. Aside from its sponsorship of the *Automobile Blue Books,* the AAA contracted its first map in 1905 from the Survey Map Company. See American Automobile Association, "A Journey in Road Maps" (http://www.ouraaa.com/aaainfo/100/nation/maphistory. html).

87. A comprehensive chronological listing of these issues is Mark Greaves, "Official Maps Master List" (Road Map Collectors Association, updated 11-25-02): www.roadmaps.org/omml.

88. James R. Akerman, "Selling Maps, Selling Highways: Rand McNally's 'Blazed Trails' Program," *Imago Mundi* 45 (1993): 77–89; John G. Brink, 1926, "History of Rand McNally Auto Trails, 1908–1926," unpublished typescript, Rand McNally Collection, the Newberry Library.

89. It is probably through the popular road atlas that the name Rand McNally became synonymous with road maps, rather than through its oil company maps, which consumers identified with the clients.

90. For further details on the general outlines of the emergence of promotional road mapping, see James R. Akerman, "American Promotional Road Mapping in the Twentieth Century," *Cartography and Geographic Information Science* 29, 3 (July 2002): 175–91.

91. Though some publishers inserted innocuous phantom features and towns as traps for unwitting copyright violators.

92. *Official Automobile Blue Book 1920,* vol. 5 (Chicago and New York: Automobile Blue Book Publishing Company, 1920), routes 596, 691, 741, 744, 751, and 756; *Rand McNally Official Auto Trails Map, District No. 10* (Chicago: Rand McNally, 1920).

93. This assurance may be found, for example, on page 4 of the booklets that accompanied the 1920 editions of the maps.

94. Richard F. Weingroff, "From Names to Numbers: The Origins of the U.S. Numbered Highway System," online publication of the U.S. Department of Transportation, Federal Highway Administration, modified to April 28, 2003 (www.fhwa.dot.gov/infrastructure/numbers.htm).

95. William H. Thompson, *Transportation in Iowa: A Historical Summary* ([Ames]: Iowa Department of Transportation, 1989), 105.

96. See Weingroff, "From Names to Numbers," for a full account of the political wrangling that produced this system.

97. *America's Highways,* 126.

98. *Good Roads Atlas of the United States* (Chicago and New York: The G.F. Cram Co., [1921]); *Rand McNally Junior Auto Trails Atlas of the United States* (Chicago: Rand McNally, 1924); *Langwith's Folding Road Map Atlas of the United States* (Minneapolis: Langwith Publishing Co., 1928); *Clason's Atlas, Best Roads of the United States* (Denver and Chicago: Clason Map Co., 1925?); *Highway Atlas of the United States and Canada* (Kansas City, MO: Gallup, 1928).

99. Brink, "History of Rand McNally Auto Trails," 48.

100. Unpublished inventory of the H. M. Gousha Collection, Newberry Library.

101. See John A. Jakle and Keith A. Sculle, *The Gas Station in America* (Baltimore: Johns Hopkins University Press, 1994), 108–10. The earliest nationwide commercial series of separately issued maps may be the 1929 series published by Rand McNally for Texaco and the series published in the same year by Gousha for Firestone.

102. Based on a survey of editions at five-year intervals: *Rand McNally Road Atlas of the United States* (Chicago: Rand McNally, 1926); *Rand McNally Road Atlas of the United States* (Chicago: Rand McNally, 1931); *Rand McNally Road Atlas of the United States, Canada, and Mexico* (Chicago: Rand McNally, 1936); *Rand McNally Road Atlas of the United States, Canada, and Mexico* (Chicago: Rand McNally, 1941).

103. "Iowa Primary Road Map" (Ames: Iowa State Highway Commission, 1931); see Daniel Block, *Romantic and Modernist Images on Twentieth Century Iowa Official State Highway Maps*, Newberry Library Slide Set 28 (Chicago: Newberry Library, 2002).

104. Block, *Romantic and Modernist Images.*

105. "Map of the State of New Mexico" (Santa Fe: State Highway Commission, 1927); "Road Map of New Mexico" (Santa Fe: State Highway Department, 1929); "Official Road Map of New Mexico, 'the Sunshine State'" (Santa Fe: State Highway Department, 1936).

106. Santa Fe: New Mexico State Tourist Bureau, State Highway Department, 1940.

107. Daniel Block and I discuss a similar shift the promotional rhetoric of official state highway maps in "The Shifting Agendas of Midwestern Official State Highway Maps," *Michigan Historical Review* 31, 1 (Spring 2005): 123–65.

108. See Akerman, "American Promotional Road Mapping in the Twentieth Century," 181–85.

109. Ristow, "A Half-Century of Oil-Company Road Maps," 623, estimates the 1964 output of "gas maps" at 200 million and the total production for the fifty years previous at 5 billion copies. The addition of the production by other sources yields an estimate approaching a half billion road maps published annually during in the 1950s and 1960s.

110. Jakle and Sculle, *The Gas Station in America,* 52–60.

111. Richard Ohmann, *Selling Culture: Magazines, Markets, and Class at the Turn of the Century* (London: Verso, 1996), 180–206.

112. Roland Marchand, *Advertising the American Dream: Making Way for Modernity* (Berkeley: University of California Press, 1985).

113. Christina Dando, *Going Places? Gender and Map Use on 20th Century Road Maps*, Newberry Library Slide Set No. 33 (Chicago: The Newberry Library, 2002). I wish to acknowledge the frequent discussions Dr. Dando and I have had on the subject of gender issues in map cover art.

114. "Texaco Road Map, Indiana" (Chicago: Rand McNally for the Texas Company, 1930). On the "map within the map" trope, see Yorke, Margolies, and Baker, *Hitting the Road,* 30–33. Illustrations of much of the map art discussed here may be found in this source, along with thematic discussions similar to those pursued here.

115. Hugill, "Good Roads and the Automobile in the United States," 337–40.

116. Roy D. Chapin, "The Motor's Part in Transportation," in Clyde L. King, ed., *The Automobile: Its Province and Its Problems, the Annals of the American Academy of Political and Social Science*, vol. 116 (1924): 2; Jakle, *The Tourist,* 171–72; John F. Stover, *American Railroads,* 2nd ed. (Chicago: University of Chicago Press, 1997), 219.

117. Chapin, "The Motor's Part in Transportation." For a lengthy discussion of the contest between automobile interests and rail interests, see Stephen B. Goddard, *Getting There: The Epic Struggle between Road and Rail in the American Century* (New York: Basic Books, 1994).

118. Conoco's road-map covers for 1933 featured a simplified pictorial map of the United States. In the center of the map is a cartouche bearing the slogan "The Land of Pleasure" framed by two lithe mermaids.

119. See Virginia Scharff, *Taking the Wheel: Woman and the Coming of the Motor Age* (Albuquerque: University of New Mexico Press, 1991), 135–64; Dando, "Going Places"; and Yorke, Margolies, and Baker, *Hitting the Road,* 63–70.

120. For example, Gulf Oil covers for 1934. An atypical map cover showing a family traveling by auto before 1945 is the "Tydol Trails thru New England" (Chicago: Rand McNally for Tydol, c. 1928).

121. As we have seen, images of Indians figured prominently in official road covers for New Mexico in the 1930s and in official South Dakota highway maps in the 1970s and 1980s.

122. See the generic Colonial Beacon covers for 1930 and the generic Sunoco covers for 1939.

123. See for example, the generic Texaco covers for 1953–56.

124. "Official Highway and Touring 1967 Vermont Map" (Montpelier: Vermont Department of Highways and Vermont Development Department, 1967). Official state-map covers promoting the construction of modern interstate highway abound, especially from the late 1950s through the early 1980s. Daniel Block discusses this point in connection with Iowa state highway maps in *Romantic and Modernist Images.* See also Akerman and Block, "The Shifting Agendas of Midwestern Official State Highway Maps."

125. Shaffer, *See America First.*

126. Frederic F. Van de Water, *The Family Flivvers to Frisco* (New York and London: D. Appleton, 1927), 6.

127. Sunoco road map series for 1941 (Chicago: Rand McNally for Sun Oil, 1941).

128. See, for example, "A Good-Natured Map of the United States Setting forth the Services of the Greyhound Lines" (n.p., c. 1940) and "Sinclair Pictorial United States," published with all 1933 Sinclair maps.

129. "Indian Country" (Los Angeles: Automobile Club of Southern California, c. 1950).

130. "Hex Highway, Linking Lehigh and Northern Berks Counties" (Allentown: Allentown-Lehigh County Tourist & Convention Bureau, c. 1980).

131. "Williamsburg, Jamestown, Yorktown Specially Prepared for the 350th Anniversary of the Founding of Jamestown, Virginia, 1607" (Convent Station, NJ: General Drafting for Standard Oil of New Jersey, 1957); "Williamsburg, Jamestown, Celebrating America's 350th Birthday" (Chicago: Rand McNally for American Oil Company, 1957); "Through Tidewater Virginia, Historyland Highways" (Hampton, VA: Division of Toll Facilities, Hampton, Va., 1965); "Tidewater Virginia, US 17, George Washington Memorial Highway into Colonial National Historic Park" (n.p., 1949).

132. "Map of the Principal Events in the Life of George Washington" (Convent Station, NJ: General Drafting for Standard Oil Company of New Jersey, 1932).

133. "The Blazed Trails" (Chicago: Rand McNally, 1922).

134. For example: Hugo Alois Taussig, *Retracing the Pioneers of from West to East in an Automobile* (San Francisco: Privately printed, 1910); and Victor Eubank, *Log of an Auto Prairie Schooner: Motor Pioneers on the "Trail to Sunset," Sunset,* February 1912, pp. 188–95.

135. "The National Highway and Transcontinental Connections at Washington, Penna" (Buffalo, Cleveland, and New York: Matthews Northrup for the Williams Foors Hotels Co., 1925).

136. Ezra Meeker, *Story of the Lost Trail to Oregon* (Seattle [1915]).

137. "Old Oregon Trail Information, Federal Route 30" (n.p.: Inland Empire Hotel Association, 1930?).

138. "Greater Kansas City Historical Map" (Kansas City: Greater Kansas City Chamber of Commerce, 1967). "Historic Jackson County Missouri," in *Independence, Missouri Welcomes You . . . Queen City of the Trails* (Independence, MO: Independence, Missouri Chamber of Commerce, 1961).

139. For example, "Nebraska 1980 Official State Highway Map" (Lincoln: Nebraska Department of Roads, 1980).

140. *The Oregon Trail: The Missouri River to the Pacific Ocean,* sponsored by the Oregon Trail Memorial Association (New York: Hastings House for the Oregon Trail Memorial Association, 1935).

141. Gregory M. Franzwa, *Maps of the Oregon Trail* (Gerald, MO: The Patrice Press, 1982).

142. "Lincoln Heritage Trail" (Champaign, IL: Lincoln Heritage Trail Foundation, 1969). This was one of only several efforts to map Lincoln's migrations for recapitulation by motor tourists. Other examples include "Historic Proofs and Dates in Support of The Lincoln Way, Being the Route Traveled by the Thomas Lincoln Family in Coming from Indiana to Illinois in the Year 1830" (Greenup, IL: Abraham Lincoln Memorial Highway Association, 1929); and "The Lincoln Legacy in Central Illinois" (Springfield, IL: Central Illinois Tourism Council, 1988).

143. For example: John B. Bachelder, *Gettysburg: What to See, and How to See It* (Boston: John B. Bachelder, 1873).

144. "Auto Trails to and from Gettysburg" (Chicago: Rand McNally for Gettysburg Chamber of Commerce, c. 1920).

145. "Civil War Centennial Events 1861–1961" (Chicago: Rand McNally, 1961).

146. *Dave Hunter's Along the Interstate 75 Information & Services Guide* (North Tonawanda, NY: Mile Oak, 1992).

147. Michael Wallis, *Route 66: The Mother Road* (New York: St. Martin's Press, 1990).

148. A fictitious trip on Route 66 forms the basis of a costume and artifacts sold to accompany "Molly," a doll intended to represent a typical girl of the 1940s, one of the popular historical American Girl dolls marketed by Mattel. A scrapbook complete with maps and postcards is one of the artifacts included in the kit.

149. Tom Snyder, *The Route 66 Traveler's Guide and Roadside Companion* (New York: St. Martin's Press, 1990).

150. See, for example, "Blue Ridge Parkway" (Washington, D.C.: National Park Service, 1963). On the meaning of the Blue Ridge Parkway, see Alexander Wilson, *The Culture of Nature: North American Landscape from Disney to the Exxon Valdez* (Cambridge, MA: Blackwell, 1992).

151. Bill Bryson, *The Lost Continent* (New York: Harper & Row, 1989), 77–78.

152. *On the Road, The Dharma Bums, The Subterraneans* (New York: Quality Paperback Book Club, 1993), 12–13.

153. Joy, "The Traveller and the Automobile," 740.

Chapter Six

The author is grateful to the following persons for their suggestions and assistance during the course of his research: Stuart Butler, George Chalou, Gary Morgan, Robert Richardson, Richard Smith, Aloha South, and Mitchell Yockelson of the National Archives and Records Administration; James Flatness, Edward Redmond, and Carlin Rene Sayles at Library of Congress; Clifford Nelson with the U.S. Geological Survey; and Herman Viola, Smithsonian Institution.

1. C. J. Zimmermann, "Developing Aeronautical Maps," *Aircraft Journal* 6 (May 1, 1920), 5.

2. The basic reference work on aviation cartography is Walter W. Ristow's invaluable and comprehensive bibliography, *Aviation Cartography. A Historico-Bibliographic Study of Aeronautical Charts* (Washington, D.C.: Library of Congress Map Division, 1960). Also useful is James Flatness, *Aviation Cartography. A Map Exhibit Commemorating the 50th Anniversary of Charles A. Lindbergh's 1927 New York to Paris Flight* (Geography and Map Division, 1977). This unpublished exhibit catalogue describes fifty-six aviation maps. It is available from the Geography and Map Division, Library of Congress.

The original correspondence, documents, memoranda, and flight reports relating to the development of the air navigational chart in the United States are found principally among the records of the National Archives and Records Administration, which are described in *Guide to Federal Records in the National Archives of the United States* (Washington, D.C.: National Archives and Records Administration, 1995, 3 vols.). The Smithsonian Institution's National Air and Space Museum Records Management Division and the Smithsonian's Postal Museum hold a few pertinent early documents. Regrettably, few contemporary primary sources were found relating to the development of commercial aviation maps.

The two largest collections of aeronautical maps are found in the Geography and Map Division of the Library of Congress and the Cartographic Section, Media Branch of the National Archives and Records Administration in College Park. For a general introduction to these collections and other repositories in the nation's capital, see Ralph E. Ehrenberg, *Scholars' Guide to Washington, D.C. for Cartography and Remote Sensing Imagery* (Washington, D.C.: Smithsonian Institution, 1987).

3. Airline souvenir maps were similar in design and purpose to American oil company road maps. Distributed free to passengers by airlines from about 1928 to the 1980s, they promoted air travel and individual air carriers, advertised oil companies and car rental agencies, instructed readers in the safety features of aircraft, and occupied and educated passengers in map reading, meteorology and geography before the introduction of inflight radio and movies.

4. William Mitchell, *Skyways. A Book on Modern Aeronautics* (Philadelphia: J. B. Lippincott Company, 1930), 153.

5. St. Clair Streett to Maj. Jouett, November 6, 1924, entry 143, box 15, Records of the Army Air Force, Record Group 18, National Archives and Records Administration (hereafter cited as RG 18, NARA).

6. Norris B. Harbold, *The Log of Air Navigation* (San Antonio, TX: The Naylor Company, 1970), 19.

7. Raymond L. Ross, "Mapping United States Airways," *Military Engineer* 20 (Nov.–Dec., 1928), 476.

8. H. W. Sheridan, "Transcontinental Reliability and Endurance Contest. Flight [Report] Entry No. 40, October, 1919," Entry 187, Correspondence Relating to Transcontinental Reliability Test Flights and to Station at Columbus, N. Mex. 1919, RG 18; U.S. Army, Air Service, "Part V. Aerial Navigation," *Air Service Manual* (Air Service Information Circular, no. 84, Sept. 20, 1920) (Washington, D.C.: Government Printing Office, 1920), 53.

9. D. G. Jeffrey, "Navigating the Air," *Air Travel News* 3 (November, 1929), 21.

10. P. V. H. Weems, *Air Navigation* (New York: McGraw-Hill Book Company, 1943), 3–4; Norris B. Harbold, *The Log of Air Navigation* (San Antonio, TX: The Naylor Company, 1970), 199.

11. John F. Welch, ed., *Van Sickle's Modern Airmanship* (Blue Ridge Summit, PA: TAB Books, 1990), 478; Terry L. Lankford, *Understanding Aeronautical Charts*, (New York: TAB Books, 1996), 79–186; Donald J. Clausing, *The Aviator's Guide to Modern Navigation* (Blue Ridge Summit, PA: TAB Books, 1987), 107.

12. Irwin Amberg to chief of Air Service, January 6, 1923, E147, box 2, folder 147, RG 18; William D. Coney, "Flying Across the Continent in Twenty-two Hours," *U.S. Air Service* 5 (April 1921), 13–14.

13. Aircraft specifications are taken from R. E. G. Davies, *Charles Lindbergh: an Airman, His Aircraft, and His Great Flights* (McLean, VA: Paladwr Press, 1997), 9, 13; John W. R. Taylor and Kenneth Munson, *History of Aviation* (New York: Crown Publishers, 1976), 242.

14. Sidney D. Waldon to George O. Smith, May 8, 1917, E29, box 74, file 110.11.2 NACA, Records of the U.S. Geological Survey, Record Group 57, NARA (hereafter cited as RG 57).

15. Harold. E. Hartney to chief, Training and Operations Group, Dec. 2, 1920. E166, 361.A1, RG 18; Neil D. Van Sickle, *Modern Airmanship* (Princeton, NJ: D. Van Nostrand Company, 1957), 1.

16. From the air, according to the officer in charge of the plowing project, this furrow gave "the impression of a pencil having been drawn across the desert." Roderic Maxwell Hill, *The Baghdad Air Mail* (London, 1929), 19–25, 66, 138–139; J. J. Lloyd-Williams, "Re-marking the Air Route from Ramadi to Landing Ground R," *Geographical Journal* LXII (1922): 358–359. Dr. Ian Mumford first brought this interesting aviation phenomenon to the author's attention.

17. William M. Leary, ed., *Pilot's Directions: The Transcontinental Airway and Its History* (Iowa City: University of Iowa Press, 1990), 2–4; George L. Vergara, *Hugh Robinson Pioneer Aviator* (University of Florida Press, 1995), 53.

18. George B. Harrison, "Aerial Map Making," *The Air Scout*, 1 (December 1910): 42.

19. Ristow, *Aviation Cartography*, 2–4; "Aero Maps," *American Aeronaut* (August 1909): 33; "The Colouring of Aeronautical Maps," Aeronautics 4 (May 1911): 52; Charles C. Turner, *Aerial Navigation of To-Day: A Popular Account of the Evolution of Aeronautics*. (London, 1910), 185–89.

20. "Meeting of the International Commission for Aeronautical Maps," *The Aëronautical Journal*, 15 (July 1911): 123–24.

21. In 1911, France had 353 licensed pilots, Great Britain, 57; Germany, 46; Italy, 32; Belgium, 27; and the United States, 26. Ristow, *Aviation Cartography*, 5.

22. Charles Lallemand, "International Air Map and Aeronautical Marks," *Geographical Journal,* 38 (November 1911): 469–83; Bertram G. Cooper, "Aeronautical Maps and Signs, Meeting of the British Association, Portsmouth, September 5th, 1911," *Aëronautical Journal* 15 (October 1911): 142; and *Annual Report of the Board of Regents of The Smithsonian Institution, 1911* (Washington, D.C.: Government Printing Office, 1912), 295–302.

23. Henry Woodhouse, ed., *The Aero Blue Book and Directory of Aeronautic Organizations* (New York: The Century Company, 1919) 169; Herbert A. Johnson, *U.S. Army Aviation through World War I* (Chapel Hill: University of North Carolina Press, 2001), 13–18.

24. Williams Welch, "[Map of] Courses of Balloons in International Race 1907/Showing Positions and Speed for Each Hour/Compiled in the Signal Corps Office, War Department for/The Aero Club of America/From aeronaut's logs and messages received," *Aero Club of America 1908* (New York: Aero Club of America, 1908), fold-out. A blueline copy with same title indicates that the map was "Compiled under the direction of Captain Charles DeF. Chandler." See folder "Airways, 1907," Geography and Map Division, Library of Congress (cited hereafter as G&M).

25. "The Aerial 'Highways'" *New York Times* (April 6, 1911), 10; "Maps for Aviators Soon to be Ready" *New York Times* (June 22, 1911), 13. The author is indebted to James Flatness for bringing these articles to his attention.

26. E. Adrian von Muffling, "A Discussion of Systems of Aeronautical Maps, *Fly Magazine* (October 1911): 20; "Aeronautic Maps Produced," *Aero,* 3 (October 28, 1911): 83.

27. Henry Woodhouse, "Aëronautical Maps and Aërial Transportation," *Geographical Review* 4 (November 1917): 339. A picture of the raised relief model appeared in *Aircraft,* 2 (November 1911), 306; a contemporary photograph is filed as G3802.L6P6 1911. A Vault, G&M.

28. Woodhouse (1884–1970) is one of the more fascinating figures associated with aviation cartography. Educated in France, Switzerland, England, and Belgium, Woodhouse immigrated to the United States in 1905. Convicted of murder shortly after his arrival, he spent four years in prison before being befriended by Robert J. Collier, heir to the Collier's publishing empire, who funded several of Woodhouse's aviation publishing efforts. Following World War I Woodhouse became a land speculator and art and antique dealer in New York City. His legacy includes many George Washington documents donated to the Library of Congress and Huntley Meadows Park, a wildlife preserve in Fairfax County, Virginia. See www.rcls.org/jkuntz/woodhouse.html.

29. Giovanni Roncagli, "To Make an Aeronautical Map of the World," *Flying* 2 (September 1913): 16–17, 24; (October 1913): 8–9, 30.

30. "Aero Club of America Appoints Committee to Make Aeronautical Map of the World," nd, E 1 H, box 1, folder, "Committees and Associated Press Releases—Aero Clubs," Papers of Robert E. Peary, National Archives Gift Collection of Materials Relating to Polar Archives, RG 401, NARA (hereafter cited as RG 401); Henry Woodhouse to Peary, May 19, 1914, Entry 1H, box 55, Letters Received, Papers of Robert E. Peary, RG 401; Woodhouse, "To Make a Standard Aeronautical Map of the World and an Efficient Aeronautical Map of the U.S.," *Flying* 3 (July 1914): 172.

31. Henry Woodhouse, "To Establish a Chain of Landing Stations for Aircraft," *Flying* 3 (January 1915): 357–58.

32. The following two works provide background information on Lawrence Sperry and the Sperry Gyroscope Company, but neither mention his contribution to the development of air navigation maps: William Wyatt Davenport, *Gyro! The Life and Times of Lawrence Sperry* (New York, 1978), and Thomas Parke Hughes, *Elmer Sperry Inventor and Engineer* (Baltimore: Johns Hopkins Press, 1971).

33. No original copies of these maps have been located but they are described in: "Sperry takes steps to supply much Needed Aeronautic maps," *Aerial Age Weekly* (January 1, 1917): 410; Earl Hamilton Smith, "New York—Chicago Aerial Mail Routes," Flying 5 (January 1917): 504–505; Henry Woodhouse, *Textbook of Military Aeronautics* (New York: Century, 1918), pp. 198–199; "Aeronautic Map for New York-Newport News Trip. Prepared by Lawrence B. Sperry," *Aerial Age Weekly* 4 (January 8, 1917): 438; and Omar B. Whitaker, "Aëronautical Charts," *Geographical Review* 4 (July 1917): 3–4.

34. Woodhouse, "Aëronautical Maps and Aërial Transportation," 332–33; "Aerodrome to Aerodrome' Maps," *Flight* 3 (October 28, 1911): 934; Woodhouse, "To Make a Standard Aeronautical Map of the World and an Efficient Aeronautical Map of the U.S.," 170. A reproduction of the Italian model is reproduced in Woodhouse (1911), figure 3. Clift's map is reproduced in *Flight* (June 8, 1912): 509. A set of eight Clift "Aerodrome to Aerodrome" aviation maps is on file in the British Map Library (file "1190 [56]"). Each map is drawn at a scale of 1:500,000 and measures 4 ½ × 13 inches.

35. Whitaker, "Aëronautical Charts," 3.

36. Woodhouse, "To Make a Standard Aeronautical Map of the World and an Efficient Aeronautical Map of the U.S.," 170; "Maps for Aviators Soon to be Ready," *New York Times* (June 23, 1911), 13.

37. For examples of these advertisements, see: "Always Keeps the Chart in Sight," *Flying* 6 (February, 1917), 63; "Aeronautic Maps," *Aerial Age Weekly,* (April 9, 1917), 127; and "Aeronautic Maps," *Flying* (December, 1917), 987.

38. "Aeronautical Library" and "Sperry Exhibit," *Aerial Age Weekly* (February 12, 1917): 81, 591–92. A list of speakers is found in *Preliminary Program. Addresses to Be Delivered During the First Pan-American Aeronautic Exposition. Grand Central Palace. New York, February 8–15, 1917,* (4 pp.), Papers of Robert E. Peary, Entry 1L, RG 401. Sperry's paper was not published and no copy has been located. Sperry was one of two speakers who spoke on the subject of aeronautic maps. The other was E. Lester Jones, superintendent of the U.S. Coast and Geodetic Survey; his presentation was published under the title "Aeronautic Maps" in *Aerial Age Weekly,* 4 (February 26, 1917): 693, and *Aircraft* 7 (May, 1917), 96–97.

39. Henry Woodhouse, *Textbook of Military Aeronautics,* 199. The Woodrow Wilson Airway was first displayed on a small-scale outline map compiled by American Geographical Society cartographer W. A. Briesemeister in November, 1917. See "Outline Map of the United States showing the route of the proposed Woodrow Wilson Aërial Highway and military camps and aviation stations," in Woodhouse, "Aëronautical Maps and Aërial Transportation," p. 334. This map also accompanied a notice in magazine indicating President Wilson's support of the plan. See "Woodrow Wilson Highway Plans Please President Wilson," *Flying* 6 (October 1917): 754.

40. As early as 1915, officials of the Lincoln Highway lobbied the Aero Club of America to have it airways "follow in a general way the Lincoln Highway. . . . It may

be of mutual interest and advantage to all." A. R. Pardington to Woodhouse, Feb. 5, 1915, box 57, Letters Received, Papers of Robert E. Peary, RG 401.

41. Woodhouse, *The Aero Blue Book and Directory of Aeronautic Organizations*, 9–25. For a variant reproduction of this map, see Henry Woodhouse, "U.S. is Ahead of World in Aerial Transportation," *Flying* 9 (August 1920): 436–43.

42. Woodhouse, "Aëronautical Maps and Aërial Transportation," 329–33. U.K. Ministry of Defence, *Technical History of the Admiralty Departments* (Taunton: British Hydrographic Department, n.d.), vol. 3, part 22, p. 43 and table 16. The author is grateful to Dr. Ian Mumford for this report.

43. In response to a request from Isaiah Bowman (director of the American Geographical Society) to Brig. General George O. Squier that an Air Service officer prepare an article on aeronautical maps for the *Geographical Review* ten days after America's entry into the war, a staff officer recommended to Squier that "Because of the feeble progress (next to nothing) achieved on this subject I would suggest diplomatically stalling this off." See file, Squier to National [*sic*] Geographical Society, April 19, 1917, E45, box 42, file 061—Aeronautical, Records of the Office of the Chief Signal Officer, Record Group 111, NARA (hereafter cited as RG 111).

44. *Report of the Chief Signal Officer to the Secretary of War 1914* (Washington, D.C.: GPO, 1914), 3.

45. Johnson, *U.S. Army Aviation through World War*, 20, 70, 135–36, 151–54.

46. Chief signal officer to commanding officer, Signal Corps Aviation School, San Diego, March 22, 1913, E44, box 416, file 32446, RG 111; Josephs E. Carberry to chief of engineers, U.S. Army, Aug. 3, 1913, E44, box 423, file 32944, RG 111.

47. Arthur S. Cowan to chief of engineers, August 5, 1913, E44, box 423, file 32944, RG 111.

48. U.S. Government, War Department, *Report of the Chief of Engineers, 1919* (Washington, D.C.: Government Printing Office, 1919), 89.

49. William Rossell to adjutant general, Sept. 16, 1913, E44, box 423, file 32944, RG 111; "Instructions for Military Mapping," adjutant general to department commander, March 24, 1917, E 146, box 77, file 170 WWI, RG 57; Benjamin Foulois to Glen S. Smith, "Aviation Field Notes," April 27, 1917, E 146, box 78, file 170.11 Signal Corps, RG 57.

50. Mitchell to adjutant general, "Aerial navigation map," Sept. 6, 1916, E103, box 2526, file 102721, Records of the Corps of Engineers, Record Group 77, NARA (hereafter cited as RG 77).

51. Alfred F. Hurley, *Billy Mitchell: Crusader for Air Power* (Bloomington: Indiana University Press, 1975), 19. Mitchell's interest in maps extended beyond aerial navigation. See, for example, William Mitchell, "Aviation and Geology," *U.S. Air Service* 8 (May 1923): 9–10.

52. Mitchell to adjutant general, "Aerial navigation map," Sept. 6, 1916; M. M. Macomb to chief of staff, Sept. 18, 1916, E 296, file 9759-1, Records of the War Department General and Special Staffs, Record Group 165, NARA (hereafter cited RG 165); and F. Murphy to chief of engineers, September 20, 1916, E103, box 2526, file 102721, RG 77.

53. Chief of Engineers to adjutant general, "Military Mapping Project, Fiscal year 1918," January 31, 1917, E146, box 77, file WWI, RG57, p. 3.

54. Alfred Goldberg, *A History of the United States Air Force* (Princeton, NJ: D. Van Nostrand Company, 1974), 19.

55. Chief signal officer to chief of engineers, November 15, 1917, E 103, box 2526, file 102721, RG 77, and E146, box 78, file 170.11, RG 57. To ensure that American army

pilots were familiar with the maps being used on the Western Front, the director of Military Aeronautics sent examples of French topographic maps to each training field. See, for example, director of Military Aeronautics to commanding officer, Taliaferro Field, Hicks, Texas, May 23, 1918, E 168, box 2640, file "Taliaferro Field 061," RG 18.

56. Lyle Brown to chief of staff, War Department, July 10, 1918, E 296, File 8236.13, RG 165.

57. Chief signal officer to chief of engineers, November 15, 1917, *idem.*

58. *U.S. War Department. Bulletin 64, December 19, 1918* (Washington, D.C.: Government Printing Office, 1918), paragraph 2, subsection 3, p. 2–3.

59. *United States Army. Maps. Conventional Signs. With Supplement Containing Changes No. 1 and 2, Dec. 28, 1918* (Washington, D.C.: Government Printing Office, 1918), p. 17–18.

60. See, for example, adjutant general to commanding general, Southern Department, "Aerial Navigation Maps," E 104, box 109, folder 061.1A (S. Dept) 1–100, RG 77; Rufus Putnam to department engineer, Northeastern Department, "Washington—Boston, Aerial Navigation Map," E103, box 2526, folder 102721, RG 77.

61. "List of Aerial Route Maps Available," January 19, 1918, E166, box 695, folder 393A Aerial Photography, RG 18; "Report of Operations Military Surveys and Maps, Northeastern Department," May 11, 1918, E104, box 105, Folder 061.1A (NE Dept.), RG 77.

62. R. C. Kuldell to chief of engineers, December 31, 1917, E 103, box 2526, file 102721, RG 77; Putnam to officer in charge, Central Map Reproduction Plant, Washington Barracks, January 17, 1918, 1st Ind., E103, box 2526, file 10271, RG 77.

63. Charles H. Ruth to chief of engineers, February 9, 1918, 2nd Ind., E 103, box 2526, file 102721, RG 77.

64. Paul J. Alexander, "Arms and the Map: A History of the Army Map Service," Typescript, folio 1 of 2, p. 12, G&M.

65. Putnam to chief signal officer, January 11, 1918, E 103, box 2526, file 102721, RG 18; Charles D. Walcott to George Otis Smith, March 10, 1917, E29, box 74, file 110.11.2 NACA, RG 57.

66. Erwin Raisz, "James Warren Bagley, 1881–1947," *Association of American Geographers* 37 (1947): 122; "List of Officers Who Have Been Commissioned," Nov. 1, 1917, E146, box 77, folder 170 WWI, RG 57; James W. Bagley, "Conditions to govern the taking of aerial photographs for use in making aviation maps," n.d., filed with chief signal officer to chief of engineers, November 15, 1917, E 146, box 78, file 170.11 (Signal Corps), RG 57; "Cameras for Photographic Surveying Designed by Major J. W. Bagley and Fred H. Moffet of the United States Geological Survey," filed with B. Marshall to E. H. Marks, February 11, 1919, E146, box 77, folder 170.1 Chief of Engineers, RG 57; "The Use of Aerial Photographs in Topographic Mapping, A Report of the Committee on Photographic Surveying of the Board of Surveys and Maps of the Federal Government, 1920, " *Air Service Information Circular* 2, no. 184 (March 10, 1921): p. 7; Mary Rabbitt, *Minerals, Lands, and Geology for the Common Defense and General Welfare, Volume 3, 1904–1939* (Washington, D.C.: Government Printing Office, 1986), 187.

67. Calvin E. Giffin to office of chief of engineers, Sept. 3, 1918, and Giffin to office of chief of engineers, "Report of Experience on Aerial Photography," February 17, 1919, E 104, box 109, file 061.1A (Southern Department), RG 77.

68. Wilfred L. Hinkle to commanding officer, Division of Military Mapping, Oct. 11, 1918, and Theodore A. Bingham to chief of engineers, June 27, 1918, E104, box 103, file 061.1 (Eastern Department), RG 77.

69. F. B. Wilby to Otto Praeger, Septembe• 17, 1920, File 061.1A (E. Dept), E104, box 103, RG 77; W. M. Black to Director of Military Aeronautics, March 6, 1919, E104, box 104, file 0A61.1A, RG 77. While no copies of the original nine-sheet aerial navigation map from Washington to Boston have been found, a blueprint of the index is filed among the records of the Corps of Engineers with the title "Polygonic Projection for Aerial Navigation Map Washington—New York—Boston/Scale 1:625,000," E 104, box 103, file 061.1A (Eastern Department), RG 77. It measures 46.5 × 14 inches, and is stamped with the date June 29, 1918.

70. C. L. Sturtevant to Chief of Engineers, August 5, 1918, E 104, box 103, 061.1 (Eastern Department), RG 77.

71. Commanding officer, Signal Corps Aviation School, Love Field, Texas, to Air Division, chief signal officer, Dec. 17, 1917, E 168, box 2043, RG 18.

72. Department Engineer, Fort Sam Houston, Texas to Chief of Engineers, July 2, 1917, E 103, box 2498, file 101950—1289, RG 77.

73. "[Aviation Map of] Ellington Field, 18 Miles S. E. of Houston," drawn by Joseph Palle, September 11, 1918, and "[Aviation Map of] Taylor Field 11 1/2 Miles S. E. of Montgomery, Alabama," drawn by Joseph Palle, September 17, 1918, E 143, box 18, folder: Houston, Texas, RG 18; "Cross Country Landing Fields Showing Their Relation to Carruthers Field," undated, E 143, box 22, folder: Plane Facilities, RG 18.

74. Walter H. Frank to Director, Military Aeronautics, Sept. 28, 1918, E 168, box 1590, folder: Ellington Field, RG 18. While none of these maps have been found, an example of a student pilot map based on a French map is, "Amiens, [France]," May 22, 1918, filed with E 168, box 2043, folder: Love Field, RG 18.

75. U.S. Post Office, *Annual Report of the Postmaster General for the Fiscal Year Ended June 30, 1919* (Washington, D.C.: Government Printing Office, 1919), 18; United States Postal Office, *Annual Report of Post Master General, 1920* (Washington, D.C.: GPO, 1920), 61; R. E. G. Davies, *Airlines of the United States Since 1914* (London: Putnam, 1972), 17–19.

76. *Annual Report Chief of Air Service 1925* (U-723) (Army Air Service, 1925), 71.

77. Quoted in Davies, *Airlines of the United States Since 1914*, p. 29.

78. The American Expeditionary Force (AEF) first Day Bombardment Group at Kelly Field, Texas, for example, requested 100 copies of the "Scarborough census map of Texas, suitable for use as a flying map." Ralph Cousins to Director of Air Service, Oct. 17, 1919, E 168, box 1782, filed "Kelly Field 061 Misc. A." RG 18.

79. Quoted in John Meyers and Ed Mack Miller, *Airways to Airlines: A 50 Year History of Commercial Aviation* (Seattle, WA: John T. Meyers, 1975), 12.

80. The author examined fifty-four procurement requests submitted by flight commanders and individual pilots to the Training & War Plans Division, Airways Section, Army Air Service, during the years 1922 and 1923. A breakdown of the map types requested reveals the following: Rand McNally commercial maps numbered 2646 (70.6 percent), Post Office Route Maps, 699 (18.6 percent), and U.S. Geological Survey quadrangles and maps, 81 (2.2 percent). The remaining maps were primarily aerial route maps produced for the model airway by the Air Service in 1923. They numbered 323 (8.6 percent). This study was based on the following files from E 168, 061.A, RG 18: box 1311, Bolling Field; box 1623, Fairfield Field; box 1871, Langley Field; box 1782, Kelly Field; box 2207, Mitchell Field; box 2428, Rockwell Field.

81. Helmuth Bay, "The Beginning of Modern Road Maps in the United States, *Surveying and Mapping* 12 (1952), 414; Walter W. Ristow, "A Half Century of Oil-Company Road Maps," *Surveying and Mapping* 24 (1964): 628. The Army Air Service also used Rand McNally road maps for locating sites of airplane crashes. See, for example, Augustine W. Robins to chief of Air Service, October 31, 1922, E 168, box 1623, file Fairfield Field, RG 18.

82. William McKiernan, Jr., "Pilot's Report, Cross Country Flight, [July 3–4, 1919]," E 128, box 61, flight 151, RG 18. See similar reports by Royal B. Woodelton, July 4, 1919 (box 60, flight 126); James W. Welch, July 4, 1919 (box 62, flight 179), and Jerome N. Machle, July 7, 1919 (box 64, flight 248), E128, RG 18.

83. George Van Deurs, "Aviators Are a Crazy Bunch of People," In E. T. Wooldridge, ed., *The Golden Age Remembered, U.S. Naval Aviation, 1919–1941* (Annapolis, MD: Naval Institute Press, 1998), p. 88.

84. Levi L. Beery, "Report on Cross Country Flight," Langley Field, Virginia, September 6, 1922, E 147, box 2, report #27, RG 18.

85. Frank to commanding officer, Langley Field, Sept. 20, 1922, filed with William Souza to chief of Air Service, Sept. 11, 1922, E 168, box 1871, Folder Langley Field Maps, RG 18; Harrison G. Crocker, "Report of Gulf to Border Flight," June 8, 1923, E 166, box 704, folder 373A Cross Country Flight, RG 18.

86. Samuel C. Skemp, "Report on air-route: Post Field–Oklahoma City–Dallas," June 6, 1922, E 143, box 16, folder "Cross Country Flights," RG 18; See also Captain Skemp's flight report dated June 20, 1922, box 14.

87. Crocker, "Report of Gulf to Border Flight."

88. James E. Fechet to chief, Information Division, E143, box 14, file Airway Markings, RG 18, NARA; Burdette Wright to Rand McNally Company, July 31, 1923, E 149, File Navigation Aids, RG 18.

89. Herbert H. Balkam, "Diary of Cross-Country Flight from Selfridge Field to Chanute Field," November 27, 1918, E 128, box 56, flight 11, RG 18, NARA.

90. Henry Abbey, Jr. to Director Air Service, Nov. 24, 1919, E 187, box 4, 319.1 TRT, Pilots Reports, RG 18, NARA.

91. Arthur Hecht, "Route Maps of the U.S. Postal Service of the 18th and 19th Centuries," *American Philatelist* (November 1979): 983–86; Harry S. New, "Engineering Work of the Post-Office," *Military Engineer* 17 (September–October 1925): 369; U.S. War Department, *Report of the Director of Air Service to the Secretary of War, 1920* (Washington, D.C.: Government Printing Office, 1920), 26.

92. "Had Only One Whole Day's Flying from New York to Omaha," *Air Service News Letter* 3 (October 25, 1919): 4.

93. H. C. Drayton, "Report Trans-Continental Airplane Reliability Test," Nov. 21, 1919, E 187, box 4, folder "319.1 TRT Pilots Reports," no. 47, RG 18, NARA.

94. The measurements were taken from the following series of belt maps: Chicago–Minneapolis via La Crosse; Chicago–St. Louis; Chicago–St. Louis via Rantoul; Cleveland–Chicago; Pittsburgh–Buffalo; New York–St. Louis; Reno–Salt Lake; Washington–Newark via Busleton Station; and Washington–Pittsburgh. They are filed in E163, box 88, folder 29.00 (Strip Maps), Records of the Post Office Department, Record Group 28, NARA (hereafter cited as RG 28).

95. Carl Spatz [*sic*] to director of Air Service, November 4, 1919, E 187, box 4, folder 319.1 TRT, Pilots Reports, RG 18.

96. In response to a request for Rand McNally maps, the army Air Service's Information Group responded that "this office has adopted the Post Office maps for . . . flying purpose." This letter concluded by conceding "that the Rand McNally Company markets a very fair map but one entirely unsuited for flying purposes." Ernest L. Jones to commanding officer, Mitchell Field, Long Island, New York, August 14, 1920, E 168, box 2207, folder 061.9 Mitchell Field Maps, RG 18.

97. Ray L. Bowers, "The Transcontinental Reliability Test," *Airpower Historian* 8 (January 1961): 45–54, (April 1961): 88–100; U.S. Air Service, *Report on First Transcontinental Reliability and Endurance Test Conducted by the Air Service, U.S.A. October 8 to October 31, 1919* (Air Service Information Circular, vol. 1, no. 2) (Washington, D.C.: Government Printing Office, 1920), 26, 33.

98. Three Post Office strip maps were provided to each pilot, covering the following route segments: Mineola, New York, to Cleveland, Ohio; Cleveland to Omaha, Nebraska; and Omaha to San Francisco. Thomas de W. Milling, "Memorandum for Information Group," December 3, 1919, E 187, box 1, file Transcontinental Reliability Test Flight, 061, RG 18. The flight reports are found in E 187, box 4, folder 319.1 TRT Pilots Reports, RG 18.

99. Robert Worthington to director of Air Service, November 4, 1919, E 187, box 4, folder "319.1 TRT Pilots Reports," RG 18.

100. War Department, office of the chief of Air Service, *Circular No. 62*, June 29, 1922 (Stencil V-4422, A. S.), E149, box 1, folder: "Airway Operation Model Airway," RG 18.

101. Claude H. Birdseye, *Topographic Instructions of the United States Geological Survey* (Bulletin 788) (Washington, D.C.: Government Printing Office, 1928), 161, 360–61.

102. Two maps were issued for California in 1929, and one for Texas in 1923. Peter L. Stark, *A Cartobibliography of Separately Published U.S. Geological Survey Special Maps and River Surveys* (Santa Cruz, CA: Western Association of Map Libraries, 1989).

103. Archie Miller to Director, U.S. Geological Survey, Dec. 2, 1919, E 146, box 78, file 170.11 (Signal Corps), RG 57.

104. "Four Fly 1300 Miles," *Air Service News Letter* 2 (May 10, 1919), 3–4; Souza to model airway control officer, Bolling Field, District of Columbia, February 6, 1923, E 143, box 19, file Model Airways 1922–23, RG 18.

105. Karl F. Smith to Director of Geological Survey, Jan. 23, 1922, E 146, box 57, file U.S. Navy Dept., RG 57; Theodore Macaulay to Horace Hickam, Feb. 20, 1919, E143, box 14, file Airway Markings, RG 18; chief signal officer to director, Geological Survey, December 15, 1917, E 168, box 1782, Filed Kelly Field 061 Misc. A, RG 18.

106. Fechet to executive, Air Service, August 9, 1922, E143, box 15, Field Airway Marking 2, RG 18.

107. Milton F. Davis to office of director of Military Aeronautics, Nov. 29, 1918, and Herbert A. Dargue to chief of training, Dec. 3, 1918, E 149, box 2, Navigation Aids, RG 18.

108. Fred Wieners to assistant to fourth assistant executive, U.S. Post Office Department, "Proposed Air Service Tactical Units," June 2, 1919, File 361.A2, RG 18; Praeger to director of Air Service, June 3, 1919, E 166, box 743, file 573, Flights All American Pathfinders, RG 18.

109. Mitchell to director, Air Service, June 27, 1919, and Oscar Westover to director, United States Railroad Administration, E 166, folder 361.A1, RG 18, NARA;

"Marking of Airways," *In Airways and Landing Facilities* (Air Service Information Circular, vol. 5, no. 404, March 1, 1923), 9–13; Lester J. Maitland, "Markings for American Airways," *Aeronautical Digest*, 2 (May 1923): 327–29, 381.

110. Maurer, *Aviation in the U.S. Army, 1919–1939* (Washington, D.C.: Office of Air Force History, 1987), 21–28.

111. Kenneth C. Leggett, Civil Operations Branch to Colonel Brereton, office of the director of Military Aeronautics, May 21, 1919, E 143, box 15, folder: "Aeronautical Policies," RG 18.

112. Davis to commanding officer, Post Field, Fort Sill, Oklahoma, January 14, 1919, E 168, box 2326, folder 061.9 Post Field—Maps, RG 18; War Department, *Circular No. 46*, (Air Service Stencil V—445), June 3, 1919.

113. A copy of this map is found in the Map Collection of the San Diego Historical Society Research Archives, filed as M1556. The author is indebted to Muriel Strickland for bringing this map to his attention.

114. J. E. Proctor to Tony Barone, May 9, 1919, E 166, box 661, file 061.9, 361A, Air Routes, RG 18, NARA.

115. "U.S. Army Air Service, Maj. Gen. C. T. Menoher, director, Aerial Map of the United States Showing Landing Fields. Compiled under the direction of the chief of Information Group, O. D. A. S., Washington, D.C. January 1920. Engineer Reproduction Plant 1920." Filed as RG 77, ERP 2546, Cartographic Section, Media Branch, NARA. The 1923 and 1924 copies examined are filed as "U.S. Airways 1923" and "U.S. Airways 1924," G&M.

116. Hickman to [blank], undated questionnaire, E143, box 14, Airway Markings, RG 18.

117. J. G. Rankin to Chief of Air Service, January 23, 1924, E 143, box 14, Airways, RG 18.

118. "Aviation Map of United States Featuring Landing Fields. Compiled for the National Aeronautic Association of U.S., Washington, D.C. by the United States touring Information Bureau, Inc. Waterloo, Iowa with the Cooperation of the Airway Section of United States Army Air Service and the Aeronautical Chamber of Commerce of America. Rand McNally, [1923]." This map was registered for copyright on April 18, 1923. See folder "U.S. Airways, 1923," G&M.

119. E. P. Gaines to Chief of Air Service, "Report on Airways Flight from Pope, N. C. to San Diego, California, and return," June 21, 1923, E 147, box 3, folder 74, RG 18.

120. B. K. Mount to Theodore Holcombe, May 17, 1923, E 143, box 14, RG 18, NARA.

121. *Airplane Landing Fields of the Pacific Coast* (Standard Oil Company of California, June 1, 1927), p. 1; Airplane *Landing Fields of the Pacific West*... (Standard Oil Company of California, Nov. 1, 1929).

122. John C. Mulford to Mason Patrick, July 21, 1923, Patrick to Charles Davis, July 27, 1923, and Davis to Donald Duke, March 3, 1926, E 146, box 3, folder "National Highway Association," RG 18.

123. James C. Edgerton to John Sullivan, May 6, 1919, E 1, box 29, file 1-38A Interdepartmental Committee on Aerial Surveying, Records of the National Aeronautics and Space Administration, Record Group 255, NARA (hereafter cited as RG 255).

124. Leary, *Pilot's Directions*, 28, 51.

125. Black to director of the Air Service, August 1, 1919, Sullivan to Advisory Board, May 24, 1919, Hickam to the executive, December 12, 1919, and Mitchell to the executive, March 16, 1920, all filed in E166, box 695, 373A Aerial Photography,

RG 18; Leggett to Ernest L. Jones, "Status of Aerial Mapping." January 7, 1920, E149, box 2, Navigation Aids, RG 18.

126. U.S. Government, War Department, *Report of the Chief of Engineers, 1917, Part I* (Washington, D.C.: Government Printing Office, 1917), 31; Alexander, "Arms and the Map," p. 9.

127. L. W. Miller to The Chief of Engineers, "Aerial Route Map—Freeport to Amityville, N.Y.," September 12, 1918, E104, box 103, file 061.1 (Eastern Department), RG 77. A copy of the one inch to one mile map is attached to this report.

128. F. B. Wilby to Praeger, September 17, 1920, E104, box 103, file 061.1A (Eastern Department), RG 77.

129. Marks to W. P. Stokey, May 15, 1919, E104, box 109, File 061.1A (Southern Dept.), RG 77.

130. Stokey to chief of engineers, May 20, 1919, E 104, file 319.12M (Southern Dept.), RG 77; Marks to Eugene Cassady, March 25, 1919, E104, box 103, 061.1A (Eastern Dept.), RG 77.

131. *Annual Report of Operations, Ending June 30th, 1920, Department Engineer, Southern Department*, E 104, box 110, file 319.12A (Southern Dept.), RG 77.

132. Sullivan to Col. Woods, "Development of Aerial Photography," January 24, 1919, and Fechet to W. C. Johnson, March 6, 1922, both filed as E166, box 695, 373A Aerial Photography, RG 18.

133. William J. Jacobi to chief of the Photographic Branch, April 3, 1919, E151, box 1, Memos to photo chief, RG 18; A. L. Fuller to Commanding Officer Post Field, Fort Sill, Oklahoma, March 6, 1919, E168, box 2326, field 061. (Post Field Maps), RG 18; Sullivan to Woods, "Development of Aerial Photography," January 24, 1919.

134. Committee on Photographic Surveying of the Board of Surveys and Maps of the Federal Government, *The Use of Aerial Photographs in Topographic Mapping* (Air Service Information Circular, Aviation), 2, no. 184 (March 10, 1920): 9, 25; and Morris M. Thompson, *Development of Photogrammetry in the U.S. Geological Survey* (Geological Survey Circular 218) (Washington, D.C.: 1952, Revised 1958), 1

135. Director of U.S. Coast and Geodetic Survey to chief, Material Division, Air Corps, July 31, 1934, E166, 061.9, RG 18.

136. Leggett to E. L. Jones, "Status of Aerial Mapping." January 7, 1920, E149, box 2, Navigation Aids, RG 18.

137. Edgerton to Sullivan, May 6, 1919, E1, box 29, file 1-38A Interdepartmental Committee on Aerial Surveying, RG 255.

138. Hickam to Praeger, Oct. 20, 1919, E 143, box 14, RG 18.

139. The following untitled blueline route maps were examined by the author: (1) Langley Field, Virginia-Mineola, Long Island; (2) Langley Field-Macon, Georgia; (3) Mineola, Long Island-Cleveland; (4) Mineola-Washington, D.C.; and (5) Richmond, Virginia-Washington. These maps are filed in E93, boxes 83–84, Classified Documents File, RG 255, and folder Y2001200, Misc. Navigation: Charts & Air Maps, Air and Space Museum Library, Smithsonian Institution.

140. J. Parker Van Zandt to Wright, March 24, 1922, and Wright to Van Zandt, March 30, E 143, box 14, Airway Markings, RG 18; Wright, "Notes on Model Airway, Feb. 8, 1922," E 149, box 1, folder Airway Operator, Model Airway, RG 18.

141. Hickam to commanding officer, Hazelhurst Field, Garden City, New York, Feb. 27, 1920, E 168, box 2207, File 061.09 Miscel. Maps, RG 18; Information Officer, Mitchel Field, Long Island, New York to Information Group, Air Service, March 22,

1921, E 168, box 2207, File Mitchel Field, RG 18; Praeger to chief draftsman, U.S. Geological Survey, July 13, 1920, E 146, box 80, file 172.1, RG 57; chief of Air Service to wing operations officer, Langley Field, Hampton, Va., Sept. 29, 1922, E168, box 1871, file Langley Field Maps, RG 18; chief of Air Service to commanding officer, Langley Field, Hampton, Va., Nov. 4, 1922, E 168, box 1871, file Langley Field, RG 18.

142. A. F. Hassan, "Memorandum for Chief Topographic Engineer," August 9, 1920, E 146, box 56, file 119, RG 57; Praeger to Hickam, Nov. 7, 1919, E 143, box 14, file Airways Markings, RG 18; *Report of the Director of Air Service to the Secretary of War* (Washington, D.C.: Government Printing Office, 1920), 4.

143. War Department, *Report of the Director of Air Service to the Secretary of War 1920* (Washington, D.C.: Government Printing Office, 1920), 26.

144. "The Aeronautical Bulletin," *Aviation* (May 7, 1923), copy in E143, box 1, RG 18.

145. Untitled set of nine maps of "Landing Fields," E 143, box 18, folder: "Houston, Texas," RG 18.

146. See for example, Grafton Higgins to director of Military Aeronautics, March 10, 1919, E 143, box 18, RG 18.

147. "Questionnaire for Landing Fields," February 10, 1922, E 166, box 661, folder 361 B, RG 18; Henry B. Clagett to chief of Air Service, "Sketch of Flying Field," July 7, 1922, E 143, box 21, File Montgomery, Alabama, RG 18.

148. Wright to Streett, April 12, 1922, E 143, box 23, RG18; Patrick to Deputy chief of staff, Jan. 10, 1923, p. 5, E 166, box 661, RG 18.

149. An example of an airport diagram pasted to a map is found on a composite Rand McNally Map-Tack Systems map of New York that was annotated in red to show the line-of-flight from Mineola to Ithaca. Near the town of Ithaca, the pilot pasted a photoprocessed Air Service airport diagram. This map is filed in E 128, box 58, folder 64, RG 18.

150. See, for example, Site Location Report for Haskins, Ohio, Oct. 4, 1920 (box 89, file Ohio #4) and supt. Air Mail Service to second asst. postmaster general, May 4, 1921 (box 87, file Diagrams of Landing Field), both E163, RG 28.

151. C. F. Egge to Wright, Airways Section, E143, box 14, RG18; "Emergency Landing Fields/Chicago to Cheyenne/U.S. Air Mail Service" and "Key to Conventional Signs Used in Emergency Landing Fields U.S. Air Mail Service," 2nd asst. P. M. Emergency Landing Fields, Chicago to Cheyenne, RG 28, Cartographic Section, Media Division, NARA.

152. Joseph Palle, "[Map of] Cross Country Landing Fields Showing Their Relations to Kelly Field," Sept. 10, 1918, E 168, box 1782, file Kelly Field, RG 18, NARA.

153. "Aerial Route Map/ Taliaferro Field, Texas to Oklahoma City, Oklahoma," filed with Macaulay to Director of Military Aeronautics, December, 1918, E 128, box 56, File Flight 14, RG 18; George W. Goddard with Dewitt S. Copp, *Overview: A Lifelong Adventure in Aerial Photography* (New York: Doubleday, 1969), 13.

154. Untitled strip map filed with Charles H. Danforth to Chief of Air Service, Sept. 12, 1922, filed with Major W. H. Frank to Commanding Officer, Langley Field, Sept 11, 1922, E 168, box 1871, Field Langley Field, RG 18.

155. Frank to commanding officer, Sept 11, 1922, E 168, box 1871, Field Langley Field, RG 18, NARA.

156. "Aerial Route Map/Washington to Aberdeen/District of Columbia–Maryland," E 104, box 103, folder 061.1A (Eastern Dept.), RG 77. A copy with red overprinting is found in E 143, box 14, folder Airway Markings, RG 18.

157. H. S. Hetrick to chief of engineers, Oct. 15, 1918, E 104, box 103, RG 77; Black to the director of Military Aeronautics, February 5, 1919, E 143, box 14, folder Airway Markings, RG 18.

158. Hickam to Marks, Dec. 19, 1919, E 149, box 2, folder Navigation Aids, RG 18, NARA; "Memo for Col. Watkins," c. March 11, 1920, E 104, box 110, folder 300.43 (Southern Dept.), RG 77.

159. "Aerial Route Map/Washington, D.C. to Xenia, Ohio/Sheet No 1 Washington, D.C. to Washington, PA/Engineer Reproduction Plant, 1920" and "Aerial Route Map Washington, D.C. to Xenia, Ohio/Sheet No 2 Washington, PA to Xenia, Ohio, Engineer Reproduction Plant [1920]," E143, box 20, folder Model Airways, 1921–22, and E149, box 2, folder Maps, RG 18.

160. Hickman to Lieutenant Wheeler, March 15, 1920, E 166, box 695, folder 373A Aerial Photography, RG 18.

161. Delos C. Emmons to chief of Air Service, July 31, 1922, E 143, box 15, RG 18.

162. Maurer, *Aviation in the U.S. Army, 1919–1939,* 150; St. Clair Streett, "The Air Service Model Airway," *Aeronautical Digest* 2 (May, 1923): 324–26, 381.

163. Hartney to Chief, "Establishment of Airways," Dec. 2, 1920, E 166, box 661, RG 18; "Rough Draft for Pamphlet on Airways and Landing Facilities for Aircraft," Feb 15, 1922, E143, box 22, Publications 1925, RG18, p. 2; Airways section, chief of Air Service, *Airways and Landing Facilities* (Air Service Information Circular, vol. 5, no. 404) (March 1, 1923): 1–13.

164. Paolo E. Coletta, "The Navy's Air Navigation Charts," *American Aviation Historical Society Journal* 5 (Spring 1960): 32.

165. "History of the Board of Surveys and Maps," *Military Engineer* 16 (March–April 1924): 158–59.

166. "Airways. Our Need in the United States/To be delivered by Captain B. S. Wright on January 22, 1923, by Radio Broadcast," typescript, p. 5, January 16, 1923, E 150, box 1, RG 18.

167. Van Zandt, "Aerial Cross Country Maps," Dec. 17, 1920, and Walter H. Frank, "Suggestions for Standard Map for Aerial Navigation," Feb. 21, 1922, E 143, box 14, RG 18; Thomas Robins to Director, U.S. Geological Survey, June 22, 1922, E 146, box 90, RG 57.

168. Wright to U. C. Thies, Feb. 24, 1922, and to Harold R. Harris, May 18, 1922, and Fechet, "Report as to Situation on Aerial Navigation Maps," March 14, 1923, E 143, box 14, RG 18; chief, Engineering Division to Chief of Air Service, June 5, 1922, and Emmons to chief of Air Service, July 31, 1922, E 143, box 15, RG 18.

169. Mitchell, "Report of Inspection Trip ... to Engineering Division, McCook Field, May 1st to 12th, 1922," papers of William Mitchell, box 44, folder McCook Field, Manuscript Division, Library of Congress; Wright to Harris, May 18, 1922, E 143, box 14, RG 18; Frank to chief, Engineering Division, June 3, 1922, and Thurman H. Bane to chief of Air Service, June 27, 1922, E 143, box 15, RG 18. Both companies responded to the government solicitation with sample maps (now lost) prepared in accordance with the Air Service specifications. See A. W. Klump, Rand McNally & Company to Thurman Bane, June 23, 1922, and George S. Clason to chief, Engineering Division, E 143, box 15, RG 18.

170. Wright to Harris, June 3, 1922, E 143, box 15, RG 18.

171. Patrick to director of Geological Survey, Oct. 18, 1922, E 29, folder 118.2 part VI, RG 57; Fechet to chief, Information Division, Dec. 4, 1922, E 143, box 15, RG 18;

Patrick to deputy chief of staff, Jan. 10, 1923, p. 6, E 166, box 661, RG 18. No copy of this map has been located.

172. Patrick to Glenn S. Smith, Sept. 27, 1922, E 29, box 28, folder 118.2 part IV, RG 57; Glenn S. Smith to Wright, Oct. 4, 1922, E 143, box 15, RG18.

173. Maitland to model airway control officer, Bolling Field, Mar. 10, 1923, E 149, box 2, folder Navigation Aids, RG 18.

174. Clarence E. Crumrine to chief, Flying section, Feb 19, 1923, E 149, box 2, RG 18.

175. Duke to chief, Training and War Plans Division, February 11, 1926, E143, box 16, Correspondence Requesting Maps of Air Routes, RG 18.

176. The switch to the Corps of Engineers was made for economic and political reasons. The cost of producing each map by the Engineer Reproduction Plant was $1,300 compared with $2,300 at the U.S. Geological Survey. At the same time, the War Department thought it efficacious to have the federal government's largest organization of aerial chart users working closely with its own mapmakers as it promoted air navigation. Office of chief of engineers to the adjutant general, "Map Project for the Fiscal Year 1927," E 146, box 61, RG 57.

177. Burdette S. Wright, "National Airways What the Army Air Service is Doing Toward Their Establishment," *Aeronautical Digest* (April 1923): 248–49; Fechet to executive, May 11, 1923, E 143, box 15, folder Aeronautical Bulletin before 1924, RG 18.

178. Colletta, "The Navy's Air Navigation Charts," 32; Wooldridge, *The Golden Age Remembered*, 3.

179. "Air Maps for Sea Flyers," *National Aeronautics* 4 (June, 1926), 88; Eugene F. Burkett, "The Need for Aviation Charts," *Military Engineer* 19 (Nov.-Dec. 1927): 501–3.

180. Isaiah Davies to chief of Air Corps, October 1, 1927, E166, O69.1, RG 18, NARA.

181. Lieut. Col. J. E. Fechet to the executive, February 9, 1923, E 143, box 15, file Aeronautical Bulletin Before 1924, RG 18; Burdette S. Wright to Col. A. J. Booth, July 10, 1923, E 149, box 1, file Aero Bulletin Domestic, RG 18.

182. *U.S. Coast and Geodetic Survey, Annual Report, 1930*, E33, box 776, Annual Reports, Records of the U.S. Coast and Geodetic Survey, Record Group 23, NARA, p. 35.

183. Harbold, *The Log of Air Navigation,* 28, 38, 49–50; Monte Duane Wright, *Most Probable Position. A History of Aerial Navigation to 1941* (Lawrence: University of Kansas Press, 1972), 120–27.

184. A. F. Hassan, "Report of the Committee on Aerial Navigation Maps, January 8, 1929. As amended Nov. 12, 1929," *U.S. Federal Board of Surveys and Maps of the Federal Government* (n.d.).

Chapter Seven

This paper, which was not delivered at the 12th Kenneth Nebenzahl, Jr. Lectures, draws heavily upon the author's earlier paper: "From Chinese Chariots to Smart Cars: 2,000 Years of Vehicular Navigation," which was presented at the Institute of Navigation (ION) 50th Anniversary Annual Meeting and published in the ION journal *Navigation* (Spring 1995, Vol. 42, No. 1, pp. 235–258, 1995). This paper also draws from "On-Board Navigation Systems," the author's unpublished 1997 lecture at the University of Texas at Arlington conference on "Mapping the Earth and Seas."

1. Lindley, J. A., "Urban Freeway Congestion Problems and Solutions: An Update," *ITS Journal* 59, 12 (1989): 21–23.

2. King, G. E., "Economic Assessment of Potential Solutions for Improving Motorist Route Following," *Federal Highway Administration*, Report FHWA/RD-86/029 (1986).

3. French, R. L., "Automobile Navigation Safety Issues," *Proceedings, VTI/TRB International Conference on Strategic Highway Research Program and Traffic Safety on Two Continents*, Gothenburg, Sweden, September 18–20, 1991: VTI Rapport 372A, part 1 (1991): 104–111.

4. French, R. L., "Land Vehicle Navigation and Tracking," in *Global Positioning System: Theory and Applications*, ed. B. W. Parkinson, J. J. Spilker, Jr., P. Axelrad, and P. Enge, vol 2, part 5, pp. 275–301 (Washington, D.C.: AIAA Press, 1996).

5. Lay, M. G., *Ways of the World* (New Brunswick: Rutgers University Press, 1992).

6. Ibid.

7. Ristow, W., "A Half Century of Oil-Company Road Maps," *Surveying and Mapping*, 24, 4 (1964): 617–37.

8. Needham, J. Science and Civilization in China, vol. 4, part 2, "Mechanical Engineering" (Cambridge: Cambridge University Press, 1965).

9. Anon., "Measuring Miles," *Oil-Power*, 3, 11 (1928): 164–68.

10. Needham, J., *Science and Civilization in China*, vol. 4, part 2, "Mechanical Engineering" (Cambridge: Cambridge University Press, 1965).

11. French, R. L., "Ancient Chinese South-Pointing Chariot: World's First Vehicular Navigation System," *Navigation News* 3, 1 (1988): 8. The author developed and patented (U.S. Patent No. 3,845,289, October 27, 1974) the first electronic version of the differential odometer before learning of the ancient mechanical version developed by the Chinese.

12. Needham, J., *Science and Civilization in China*, vol. 4, part 2.

13. Garner, H. D., "The Mechanism of China's South-Pointing Carriage," *Navigation, Journal of the Institute of Navigation*, vol. 40, no. 1 (1993).

14. French, R. L., "U.S. Automobile Navigation: Past, Present, and Future," *Proceedings, 16th International Symposium on Automobile Technology and Automation*, vol. 2, pp. 1–19, Florence, Italy, May 11–15, 1987.

15. Lindenthaler, F. J., and Protz, J., "Route-Indicator for Automobiles," U.S. Patent No. 915, 976 (1909).

16. Perry, H. W., "Some Remarkable Mechanical Road Guides," *Scientific American*104, 2(1911): 47–48.

17. Ellis, W. D., "Chadwick," *True's Automotive Yearbook*, 2 (1953): 88–89, 132–33.

18. Rhodes, J. B., "Route Indicator," U.S. Patent No. 1,005,474 (1911).

19. Faustman, J. D, "Automatic Map Tracer for Land Navigation," *Electronics*, Vol. 17, No. 11, pp. 94–99 (1944).

20. Rosen, D. A., Mammano, F. J., and Favout, R., "An Electronic Route Guidance System for Highway Vehicles," *IEEE Transactions on Vehicular Technology*, Vol. VT-19, pp. 143–52 (1970).

21. Command Systems Corporation was a fledgling minicomputer systems integration firm headed by the author.

22. French, R. L., and Lang, G. M., "Automatic Route Control System," *IEEE Transactions on Vehicular Technology*, Vol. VT-22, pp. 36–41 (1973).

23. French, R. L., "On-Board Vehicle Route Instructions via Plasma Display Panel," *Digest of Technical Papers,* 1974 Society for Information Display International Symposium, Vol. V, pp. 146–47, San Diego, California, May 21–23, 1974.

24. French, R. L., "Autonomous Route Guidance Using Electronic Differential Odometer Dead Reckoning with Vectorized Map Matching," *Proceedings, U.S. Army Artificial Intelligence and Robotics Symposium,* Part II, pp. O/1–30, Indianapolis, Indiana, September 12–13, 1984. Anon., "Computer-Controlled Delivery System Interests Circulators," *Editor & Publisher,* p. 33, August 12, 1972.

25. Silver, J., "The GBF/DIME System: Development, Design, and Use," paper presented at the 1977 Joint Annual Meeting of American Society of Photogrammetry and American Congress on Surveying and Mapping (1977).

26. French, R. L., "Map Matching Origins, Approaches, and Applications," *Proceedings, Second International Symposium on Land Vehicle Navigation,* pp. 91–116, Munster, Federal Republic of Germany, July 4–7, 1989.

27. French, R. L., "Land Vehicle Navigation and Tracking," *Global Positioning System: Theory and Applications,* ed. B. W. Parkinson, J. J. Spilker, Jr., P. Axelrad, and P. Enge, Volume II, Part 5, chapter 10, pp. 275–301, AIAA Press (1996).

28. The author demonstrated the role of GPS relative to dead reckoning and map matching in the Magellan PathMaster by disconnecting the GPS receiver antenna; the PathMaster continued to function normally for two days while driving around the Dallas–Fort Worth area.

29. French, R. L., Case, E. R., Noguchi, Y., Querée, C., Sakamoto, K., and Svidén, "The IVHS Race: How Does America Compare with Europe and Japan?," *IVHS Review,* pp. 99–111, Summer 1994.

30. Yumoto, N., Ihara, H., Tabe., T., and Naniwada, M., "Outline of the Comprehensive Automobile Traffic Control Pilot Test System," *Transportation Research Record 737,* National Academy of Sciences (1979).

31. Mitamura, K., and Chujo., S., "The Driver Guide System," SAE Paper 830660 (1983). Totani, S., Kato, T., and Muramoto, K., "Automotive Navigation System," SAE Paper 830910 (1983).

32. Tagami, K., Takahashi, T., and Takahashi, F., "'Electro Gyro-Cator,' New Inertial Navigation System for Use in Automobiles," SAE Paper 830659 (1983).

33. French, R. L., "The Evolution of Automobile Navigation Systems in Japan," *Proceedings, Institute of Navigation,* 49th Annual Meeting, pp. 69–74, Cambridge, MA, June 21–23, 1993.

34. Nojima, A., Kishi, H., Ito, T., Morita, M., and Komoda, N., "Development of the New Toyota Navigation System with Voice Route Guidance," *Proceedings, IVHS AMERICA 1993 Annual Meeting,* pp. 337–345, Washington, D.C., April 14–17, 1993.

35. Hirata, T., Hara, T., and Kishore, A., "The Development of a New Multi-AV System Incorporating an On-Board Navigation Function," SAE Paper 930455 (1993).

36. Braegas, P., "Function, Equipment, and Field Testing of a Route Guidance and Information System for Drivers," *IEEE Transactions on Vehicular Technology,* Vol. VT-9, No. 2, pp. 216–25 (1980).

37. Thoone, M. L. G., and Breukers, R. M. A. M., "Application of the Compact Disc in Car Information and Navigation Systems," SAE Paper 840156, 1984.

38. Pilsak, O., "EVA—An Electronic Traffic Pilot for Motorists," SAE Paper 860346 (1986).

39. Buxton, J. L., et al., "The Travel Pilot: A Second-Generation Automotive Navigation System," *IEEE Transactions on Vehicular Technology*, pp. 41, Vol. 40, No. 1 (1991).

40. Honey S. K., and Zavoli, W. B., "A Novel Approach to Automobile Navigation and Map Display," *Proceedings, NAV 85—Royal Institute of Navigation Conference on Land Navigation and Location for Mobile Applications*, Paper No. 27, York, England, September 9–11, 1985.

41. Buxton, J. L., et al., "The Travel Pilot: A Second-Generation Automotive Navigation System," *IEEE Transactions on Vehicular Technology*, pp. 41, Vol. 40, No. 1 (1991).

42. Collier, W. C., Kao, W. Kao and Hamahata, T., "In-Vehicle Route Guidance Systems: Prototype Results Toward a Marketable System," *Proceedings ION 46th Annual Meeting*, Atlantic City, NJ (June 1990).

43. French, R. L., and Krakiwsky, E. J., "Navigation Aids and Driver Information Systems," *Automotive Electronics Handbook*, 2nd Edition, R. K. Jurgen, editor, chapter 31, McGraw-Hill, Inc. (1999).

Contributors

James R. Akerman is director of the Hermon Dunlap Smith Center for the History of Cartography at the Newberry Library, Chicago.

Catherine Delano-Smith is a senior research fellow at the Institute of Historical Research, London, and editor of *Imago Mundi, The International Journal of the History of Cartography.*

Andrew S. Cook is Map Archivist in the India Office Records, The British Library.

Jerry Musich is the former director of the National Railroad Museum in Green Bay, Wisconsin and former director of the Indiana Donors Alliance (now Indiana Grantmakers Alliance)

Ralph E. Ehrenberg is the retired chief of the Geography and Map Division of the Library of Congress and former director of the Cartographic Records Division of the U.S. National Archives.

Robert L. French is a pioneer in the development of automobile navigation systems and principal of R&D French Associates, an intelligent transportation systems (ITS) consulting firm based in Nashville, Tennessee.

Index

Page numbers in italics refer to illustrations.